CHICAGO PUBLIC LIBRARY

R00978 52008

D1444861

REF

TD   Public reactions to
898.15   nuclear waste.
.P83
1993

$49.95

DATE			

CHICAGO PUBLIC LIBRARY
BUSINESS / SCIENCE / TECHNOLOGY
400 S. STATE ST.   60605

DISCARD

BAKER & TAYLOR BOOKS

Public Reactions to Nuclear Waste

Public Reactions to Nuclear Waste

Citizens' Views of Repository Siting

Edited by Riley E. Dunlap, Michael E. Kraft,

and Eugene A. Rosa

DUKE UNIVERSITY PRESS   *Durham and London* 1993

© 1993 Duke University Press
All rights reserved
Printed in the United States of America
on acid-free paper ∞
Typeset in Melior by Tseng Information Systems
Library of Congress Cataloging-in-Publication Data
appear on the last printed page of this book.

BUSINESS/SCIENCE/TECHNOLOGY DIVISION
Contents

R00978 52008

REF

List of Figures and Tables   vii

Preface   *Riley E. Dunlap, Michael E. Kraft, and Eugene A. Rosa*   xiii

Part I   The Context of Public Concern with Nuclear Waste

1. Public Opinion and Nuclear Waste Policymaking   *Michael E. Kraft, Eugene A. Rosa, and Riley E. Dunlap*   3

2. The Historical Development of Public Reactions to Nuclear Power: Implications for Nuclear Waste Policy   *Eugene A. Rosa and William R. Freudenburg*   32

3. Perceived Risk, Trust, and Nuclear Waste: Lessons from Yucca Mountain   *Paul Slovic, Mark Layman, and James H. Flynn*   64

Part II   Public Reactions to Preliminary Sites

4. Public Testimony in Nuclear Waste Repository Hearings: A Content Analysis   *Michael E. Kraft and Bruce B. Clary*   89

5. Sources of Public Concern About Nuclear Waste Disposal in Texas Agricultural Communities   *Julia G. Brody and Judy K. Fleishman*   115

6. Local Attitudes Toward Siting a High-Level Nuclear Waste Repository at Hanford, Washington   *Riley E. Dunlap, Eugene A. Rosa, Rodney K. Baxter, and Robert Cameron Mitchell*   136

Part III Public Reactions to the Yucca Mountain, Nevada Site

7. Perceived Risk and Attitudes Toward Nuclear Wastes: National and Nevada Perspectives   *William H. Desvousges, Howard Kunreuther, Paul Slovic, and Eugene A. Rosa*   175

8. The Vulnerability of the Convention Industry to the Siting of a High-Level Nuclear Waste Repository   *Douglas Easterling and Howard Kunreuther*   209

9. Nevada Urban Residents' Attitudes Toward a Nuclear Waste Repository   *Alvin H. Mushkatel, Joanne M. Nigg, and K. David Pijawka*   239

10. Rural Community Residents' Views of Nuclear Waste Repository Siting in Nevada   *Richard S. Krannich, Ronald L. Little, and Lori A. Cramer*   263

Part IV Summary and Policy Implications

11. Prospects for Public Acceptance of a High-Level Nuclear Waste Repository in the United States: Summary and Implications   *Eugene A. Rosa, Riley E. Dunlap, and Michael E. Kraft*   291

Index   325

Contributors   329

Tables and Figures

Figures

2–1 Annual number of regulations and amendments and annual number of nuclear reactor inspections   39

2–2 Public attitudes toward nuclear power in the United States (Louis Harris Associates)   48

2–3 Public attitudes toward nuclear power in the United States (Cambridge Reports)   49

2–4 Public attitudes toward construction of local nuclear power plants   51

2–5 Host community attitudes toward nuclear power plants before and after Three Mile Island (TMI)   53

3–1 Responses of Nevada residents when asked to rate their trust in federal, state, and local officials and federal agencies to do what is right with regard to a nuclear waste repository at Yucca Mountain   69

7–1 Conceptual framework of risk perceptions   177

7–2 Number of times read or heard about high-level nuclear wastes in past three months   182

8–1 Influence of a repository on convention planners' decision to hold a meeting in Las Vegas   212

8–2 Influence of a repository on convention attendees' decision to attend a meeting scheduled for Las Vegas   215

10–1 The study area   266

Tables

1–1   Major provisions of the Nuclear Waste Policy Act of 1982   10

3–1   Some viewpoints of experts regarding public perceptions of the risks from nuclear waste disposal   65

3–2   Survey details   67

3–3   Relationship between trust in DOE and perceived risk of transport accidents   69

3–4   Relationship between trust rating of DOE and perceived risk of transport accidents   70

3–5   Relationship between trust in DOE and response to the question "would you vote for a repository at Yucca Mountain?"   70

3–6   Images of a nuclear waste repository: totals for four surveys by superordinate and subordinate categories   73

3–7   Subordinate categories ordered by decreasing frequency   75

3–8   Relationship between affective rating of a person's first image of a nuclear waste repository and their response to the question "would you vote for a repository at Yucca Mountain?"   77

4–1   Public opposition to repository siting: reaction to the area recommendation report among those having read it   94

4–2   Public opposition to a nuclear waste repository: comprehensive measure   94

4–3   Public opposition to siting a nuclear waste repository within the state   95

4–4   Public opposition to DOE siting recommendations and in-state repository siting by state and group affiliation   95

4–5   Knowledge level of individuals testifying at public hearings on nuclear waste repository siting by group affiliation   97

4–6   Geographic focus of public comments   99

4–7   Emotive themes in public comments   99

4–8   Perceived credibility and competence of government and others   101

4–9   Technical criticisms of the Department of Energy's siting analysis made at public hearings   102

4–10   Political and social issues raised at public hearings   104

4–11   Perceived impacts of a nuclear waste repository by state   105

4–12   Acceptability of nuclear waste risk by group affiliation   108

4–13   Multiple regression of public opposition to siting proposals on independent variables   109

5–1   "Would you allow construction of a high-level nuclear waste repository . . . ?"   120

5–2   Percentages of finalist site residents who mentioned selected reasons for supporting or opposing the repository, 1986   121

5–3   Expected environmental risks, 1986   122

5–4   Expected socioeconomic impacts, 1986   123

5–5   Attitudes toward industrial development, preliminary sites, 1984   124

5–6   Regression of overall attitudes toward the repository on expected impacts, knowledge, uncertainty, and background characteristics   126

5–7   Regression of expected environmental risk on knowledge, uncertainty, and background characteristics   128

5–8   Regression of expected socioeconomic benefits on knowledge, uncertainty, and background characteristics   129

6–1   Summary of Washington State polls on nuclear waste repository siting at Hanford: statewide and tri-cities   141

6–2   Attitudes toward siting a high-level nuclear waste repository at Hanford   146

6–3   Likelihood of accidents associated with siting a repository at Hanford   149

6–4   Expected impacts of a HLNWR at Hanford   152

6–5   Regression of attitudes toward repository on demographic characteristics: bivariate correlation coefficients (r's), standardized regression coefficients (betas) and multiple correlation coefficient (R^2) 155

6–6   Regression of attitudes toward repository on cognitive characteristics: bivariate correlation coefficients (rs), standardized regression coefficients (betas) and multiple correlation coefficient (R^2) 156

6–7   Regression of attitudes toward repository on confidence in Department of Energy and knowledge about nuclear waste: bivariate correlation coefficients (rs), standardized regression coefficients (betas) and multiple correlation coefficient (R^2) 159

6–8 Regression of attitudes toward repository on perceived repository impacts and likelihood of accidents: bivariate correlation coefficients (rs), standardized regression coefficients (betas) and multiple correlation coefficient (R^2) 162

6–9 Regression of repository attitudes on sets of predictor variables: standardized regression coefficients (betas) and multiple correlation coefficients (R^2s) 165

7–1 Frequency distribution of seriousness of various sources of pollution: national sample vs. Nevada sample 180

7–2 Respondent's overall attitudes toward the repository: national sample vs. Nevada sample 184

7–3 Frequency distribution of seriousness of various health and safety risks: national sample vs. Nevada sample 186

7–4 Perceived risk characteristics: national sample vs. Nevada sample 188

7–5 Perceived likelihood of large accidental releases of radiation from repository: national sample vs. Nevada sample 190

7–6 Description of variables 192

7–7 Relationship between conceptual and actual risk perception models 196

7–8 Regression models on national data 197

7–9 Regression models on Nevada data 199

7–10 Voting behavior 202

7–11 Voting behavior model 1 203

7–12 Voting behavior model 2 204

8–1 Repository scenarios used in convention planner survey 220

8–2 Amenity factors in convention planner study scenarios 221

8–3 Changes in planners' ranking of Las Vegas under repository scenarios 222

8–4 Reported intent to choose a city other than Las Vegas among those planners who initially ranked Las Vegas first 224

8–5 Forecasts of percentage of planners who would choose a city other than Las Vegas 226

8–6 Associations included in convention attendees study 228

8–7   Relation between host city's imagery score and the decision to attend past meetings: results from logistic regression analysis   229

8–8   Relation between three risk factors and the decision to attend past meetings: results from logistic regression analysis   231

8–9   Willingness to attend a meeting after finding out that a noxious facility was located 100 miles away   232

9–1   Perceptions of repository risks   242

9–2   Opinion about building the repository   244

9–3   Attitudes about the Nevada Test Site (NTS)   246

9–4   Relationships between NTS attitudes and repository attitudes (Kendall's Tau)   247

9–5   Relationships between attitudes on government and repository attitudes (Kendall's Tau)   251

9–6   Relationships between repository attitudes and beliefs about governmental honesty in reporting radiological accidents involving radioactive materials (Kendall's Tau)   253

9–7   Percentage of respondents reporting extreme negative attitudes toward the repository by gender and race   254

9–8   Descriptions of multi-item indicators   255

9–9   Regression of repository risk index on explanatory variables   257

9–10   Regression of opinion to build the repository on explanatory variables   258

10–1   Sample size and response rates for rural Nevada community surveys   269

10–2   Distribution of perceived repository health and safety risks for six study communities   272

10–3   Distribution of support/opposition for construction of repository   273

10–4   Means, standard deviations, and ANOVA results comparing response patterns on measures of anticipated economic effects, trust in science, trust in government, and perceptions of NTS health effects   274

10–5   Zero-order correlations between dependent and independent variables for combined communities   276

10–6   Multiple classification analysis of health and safety concerns regarding the repository   277

10–7   Multiple classification analysis of support/opposition to the repository   279

10–8   Multiple classification analysis of support/opposition to the repository, with risk perception as a covariate   280

11–1   Summary of methods and samples for each chapter presenting original data   294

11–2   Summary of key findings in each chapter presenting original data   297

Preface

A full day's symposium at the 1989 annual meetings of the American Association for the Advancement of Science in San Francisco provided an in-depth examination of the social and political factors associated with the disposal of high-level nuclear waste. The symposium comprised two thematic panels consisting of nationally recognized experts in a variety of relevant fields. The first panel—consisting primarily of practicing policymakers—focused on the policymaking process, and the second—consisting of social scientists—examined the public's reactions to nuclear waste disposal and the repository siting process. Most of the chapters in this volume are revised and updated papers from the second panel, all of which report results of original empirical research on citizens' views of nuclear waste repository siting.

We have added an introductory chapter to provide background and context for those chapters and a concluding chapter to summarize and discuss the implications of the empirical chapters' findings. We have also included two chapters, not presented at the 1989 symposium, covering important features of high-level waste disposal: one on long-term trends in public attitudes toward nuclear energy and nuclear waste policy (by Eugene A. Rosa and William R. Freudenburg) and the other assessing the effects on Las Vegas convention business if a high-level nuclear waste repository (HLNWR) were sited in Nevada (by Douglas Easterling and Howard Kunreuther). The aim of the first of these two chapters is to locate public perceptions of nuclear waste within the evolution of public concerns with nuclear energy more generally. That of the second is to ensure coverage of an important economic consideration in siting a HLNWR. With the exception of the Rosa and Freudenburg chapter, which relies on available historical data, all other empirical chapters report original data, collected between 1986 and 1991, dealing with public perceptions, attitudes, beliefs, and opinions concerning nuclear waste and repository siting.

How Americans view nuclear waste, and especially how they feel about it, are critically important to waste disposal policy and management. Yet the character of public sentiment and its role in the policy process is not well understood. The importance of public input to nuclear waste policy, reflecting a commitment to democratic process and a response to the concerns of affected parties, is explicitly recognized by the Nuclear Waste Policy Act of 1982 (NWPA), with its provision for extensive participation by the public, affected states, and Indian tribes. Many members of Congress and other knowledgeable observers are convinced that successful implementation of a high-level radioactive waste repository program will be possible only if there is public support for the siting process and public confidence in the principal implementing agency, the U.S. Department of Energy (DOE). These convictions made their way into NWPA itself, as that act required DOE to conduct public hearings across the nation where sites were being considered and to engage in detailed discussion with states and Indian tribes as a part of a "consultation and cooperation" process.

In most respects these ambitious efforts have failed to generate confidence in the siting process or to convince the public that DOE is a technically competent and credible agency. The American public is deeply concerned about the risks associated with radioactive waste and deeply skeptical of DOE plans for its disposal. All signs indicate that these concerns will continue to be a major obstacle to the successful implementation of a waste repository program. This is apparent from the fact that nearly five years after Congress attempted to resolve the controversy over siting a high-level nuclear waste repository by limiting consideration to a single site at Yucca Mountain, the state and residents of Nevada (along with other critics) have managed to thwart the final siting. Furthermore, the disposal of *low-level* nuclear wastes has become ever more problematic. The failure to develop multi-state compacts for dealing with low-level waste on a regional basis has led the states of Washington and Idaho to threaten to prohibit importation of wastes from other states to existing low-level disposal sites within their borders.

Consequently, an improved understanding of public concerns, their source and resiliency, is crucial to the future of nuclear waste policy. It is also important for learning how to deal with the broader challenge of managing the wide range of technological risks that abound in modern society and the challenge of how to implement such management tasks through democratic procedures.

We hope that by assembling some of the best recent work by sociologists,

psychologists, political scientists, and economists on the topic we have provided a timely, original, and multidisciplinary assessment of the public's
views of nuclear waste and repository siting issues. Because the bulk of
the data and interpretations reported here focus specifically on repository
siting, the findings are especially relevant for policymakers and other professionals concerned with the public context of waste policy. The findings
should also interest social scientists concerned with nuclear energy issues,
scientists and engineers working on the technology of waste disposal, and
policymakers at all levels of government who need to understand public perceptions and opinions of risky technologies in order to discharge their policy
obligations in a publicly acceptable fashion. More generally, they should
prove to be of interest to anyone—scholars, policymakers, citizens—concerned with one of the major challenges to modern societies: how to manage
the growing number of risky technologies on which they have become so
dependent.

Among the most important themes to emerge from the various studies
reported in this volume is that radioactive waste is perceived to be a particularly dangerous hazard and that the public has little trust in DOE's ability
to protect it from this hazard. Furthermore, the high level of perceived risk
and the low level of trust are key factors underlying public opposition to
repository siting, more important in predicting opposition than is public
knowledge about waste disposal. The consistency of the findings, emerging
as they do from a diverse set of studies employing different sampling and
measurement techniques within different geographical locales and at various stages of the repository siting "saga," suggests that they have a high
degree of validity. Hence, we have considerable confidence in the policy
implications of these findings.

Many debts are incurred in the process of completing any book, and this
one is no exception. We especially thank the contributing authors for their
generosity, cooperation, and patience in responding to our continual and
sometimes fastidious editorial requests. They helped to transform what was
initially a collection of loosely related papers into an integrated book. We
also want to acknowledge the indirect contributions provided by John Petterson, of Impact Assessment, Inc., and C. P. Wolf, of the Social Impact Assessment Center. The research reported in chapter 6, conducted as part of their
innovative assessment of the social impacts of siting a HLNWR at Hanford,
was the original stimulus for this volume.

Financial and clerical assistance from institutions, typically important in
research and writing, was especially invaluable for this project. We grate-
fully acknowledge support from the University of Wisconsin's Urban Corridor Consortium, the Herbert Fisk Johnson Professorship in Environmental

Studies at the University of Wisconsin-Green Bay, the Everett McKinley Dirksen Congressional Leadership Research Center, and, at Washington State University, the departments of Sociology and Rural Sociology, and the Social and Economic Sciences Research Center. Thanks are due to Rebecca Spithill and Beth Hull, who provided valuable research assistance, and to Patti Waldo, Tammy Small, Donna Poire, and Darcie Young for able typing and proofreading. Special thanks are also due to Diana Wuertz for preparing the index. We also thank Richard Rowson, former director of Duke University Press and his successor, Larry Malley, for their understanding, encouragement, and patience in bringing this volume to fruition.

Riley E. Dunlap
Michael E. Kraft
Eugene A. Rosa
September 1992

Part I

The Context of Public Concern

with Nuclear Waste

1   Public Opinion and Nuclear Waste

Policymaking

Michael E. Kraft, Eugene A. Rosa, and Riley E. Dunlap

The safe disposal of radioactive wastes from nuclear power plants and national defense activity is a highly complex technical and managerial problem. It is also a profoundly difficult social and political issue. For the more than forty years during which high-level waste has accumulated in the United States and other nations, far more attention has been devoted to the technical and management problem than to the social and political one. Yet understanding the social and political dimensions of nuclear waste disposal may be far more crucial to the resolution of conflicts over repository siting. For that reason, this book brings together some of the most recent and extensive empirical research available on public concerns about radioactive waste and the U.S. Department of Energy's (DOE) repository program, both nationally and within selected states.

The role that public perceptions and opinions may play in nuclear waste policymaking is vividly illustrated by recent controversies over several different DOE programs. In the fall of 1988, a series of news reports documented widespread chemical and radioactive contamination at seventeen DOE nuclear weapons plants and laboratories in twelve states, attributable in large part to the department's previous neglect of health, safety, and environmental responsibilities. With site clean-up costs estimated at over $100 billion, the revelations provoked public outrage, lawsuits, and congressional investigations. That response, in turn, prompted Secretary of Energy James D. Watkins to call for a fundamental change in DOE culture, emphasizing a commitment to environmental protection and open decisionmaking that he hoped would improve the department's credibility with the American public (Watkins, 1989).

Also in 1988, DOE was forced to postpone the opening of its Waste Isolation Pilot Plant (WIPP) near Carlsbad, New Mexico (designed to store transuranic

wastes from atomic weapons production), and in 1989 DOE had to abandon
its initial two-year, $2 billion effort to assess the technical suitability of the
Yucca Mountain, Nevada, site being considered for the nation's only civilian
high-level nuclear waste repository (HLNWR). In the first case, experts both
within and outside of DOE, including the National Academy of Sciences, had
questioned the design and construction of the facility and the expected environmental impacts (Schneider, 1988), bringing further public scrutiny of the
project. In the second case, criticism of the quality of DOE's technical work by
the Nuclear Regulatory Commission (NRC) and others, and adamant opposition from the state of Nevada, led Secretary Watkins late in 1989 to overturn
the department's prior assessments of the site's suitability. He launched a
major reorganization of the DOE's Office of Civilian Radioactive Waste Management and promised to increase attention to "institutional" issues such
as public opinion and federal-state relations (Wald, 1989; Lippman, 1989;
DOE, 1989).

As these and similar developments in the late 1980s and early 1990s have
shown, public acceptability can be a vitally important and even a deter-
minative force in the nuclear waste policy process. For that reason alone,
the public's views of radioactive waste and repository siting merit careful
assessment by scholars and decisionmakers. We need to understand the public's perceptions, beliefs, and attitudes toward nuclear waste and toward a
repository to store it: the nature of these cognitions, how they are formed,
what forces shape them, how they change over time, and how they differ among various population subgroups. Such understanding is essential
to evaluating public acceptability of various waste disposal options; it is
also crucial to decisionmakers charged with formulating and implementing
nuclear waste policies.

The study of the public's views of nuclear waste is also an interesting
topic in its own right, particularly as a case of perception of risky technologies. The substantial literature on risk perception (e.g., Gould et al.,
1988; Slovic, 1987; Covello, 1983) raises intriguing questions about the public's view of technological risks in modern society, as does survey research
on public attitudes toward nuclear energy and radioactive waste (Nealey,
Melber, and Rankin, 1983; Walker, Gould, and Woodhouse, 1983; Freudenburg and Rosa, 1984; Nealey, 1990). The new research collected here follows
the path of previous work by attempting to describe and explain public perceptions of and attitudes toward nuclear waste, but it does so in a diversity
of geographic contexts, from a variety of perspectives, and through the use
of various methodologies. It also emphasizes questions related to repository
siting that we believe have significant implications for nuclear waste policy.

To underscore the relationship between public sentiment and the policy process, in this introduction we review the status of nuclear waste as a policy problem, trace the development of the Nuclear Waste Policy Act (NWPA) of 1982 and its amendment, the Nuclear Waste Policy Amendments Act of 1987, and assess the public's role in policy implementation to date. We also discuss conceptual and methodological issues in the study of public views of nuclear waste. We reserve questions concerning the policy implications of the research for the individual chapters that follow and for our concluding chapter.

Nuclear Waste as a Policy Problem

The need for a high-level nuclear waste repository derives chiefly from the federal government's decision forty years ago to promote and to subsidize heavily the commercial use of atomic energy. Since the first nuclear reactor was put on-line at Shippingsport, Pennsylvania, in late 1957, commercial wastes (unavoidable by-products of fission reaction) have been accumulating in water-filled basins at reactor sites around the nation (U.S. Office of Technology Assessment [OTA] 1985:21–36). The bulk of these wastes are in the form of spent fuel rods, removed from the reactor after they are no longer fissionable. The spent fuel rods, along with other material from civilian power plants, represent the majority of the waste (as measured by radioactivity) to be stored in the planned repository, although the quantity of military wastes is also substantial. Hence, the issue of nuclear waste disposal is inherently linked to the commercial use of nuclear power.

Over the past decade, most analysts assumed that commercial nuclear energy had a dismal future for a variety of reasons: economics principally, but also because of environmental, public health, and social factors (Freudenburg and Rosa, 1984). Nuclear power had ceased to be competitive due to inflation-driven cost overruns and the effect of restrictive regulatory requirements and delays caused by public opposition, among other factors (Morone and Woodhouse, 1989). After 1978, no additional nuclear plants were ordered by the nation's utilities, and all of those ordered after 1974 were eventually canceled. For these reasons, by 1988 DOE projected that growth in nuclear-generated electricity would decline by the year 2000 (DOE, 1988a, 1988b).

More recently, however, increasing concern over the environmental consequences of fossil fuel use (especially global climate change) and other energy policy considerations have led to re-evaluations of nuclear power and, from some analysts, to more optimistic projections of its future (Nealey, 1990). Consistent with these expectations, in mid-1989 the NRC approved the streamlined licensing procedures for power plants that, for years, the nuclear power industry had sought as a way of reducing the construction delays and startup costs of new plants. The main provisions of the new licensing procedures, which eliminated the need for an operating license separate from a construction license, were struck down in November 1990 by the United States Court of Appeals for the District of Columbia. The ruling remained under appeal in late 1992.

Notwithstanding the setback of the court ruling and the eventual outcome of that appeal, efforts toward revitalization of the nuclear industry received a strong boost in early 1991 from the Bush administration with the release of its National Energy Strategy (DOE, 1991). Perhaps most striking among its proposals was the call to expand nuclear capacity from the then current (as of December 31, 1990) 99.5 gigawatts (*Nuclear News*, 1991) to between 190 and 290 gigawatts by the year 2030 (a gigawatt is 1000 megawatts). Meeting these objectives would require the construction of between 100 and 200 new plants, at a current average plant size of 1000 megawatts, in that forty-year period—or between a doubling and tripling of the current 112 operating plants. The required number of plants would be considerably larger, by as much as two-thirds, if the industry adopts the midsize (600 megawatt) design recommended by DOE.

To fulfill these extraordinary projections, the National Energy Strategy, consonant with industry demands, emphasized regulatory reform: "An overriding theme behind these goals is to remove undue regulatory and institutional barriers to the use of nuclear power for generating electricity in the United States. These include some barriers to constructing new nuclear power plants, to extending the life of existing generating units, and to disposing of power plant radioactive waste" (DOE, 1991:108). Key among these efforts were proposals to reinstate the streamlined licensing procedures established by the NRC but struck down by the courts. In late 1992, Congress approved, and President Bush signed, the Energy Policy Act of 1992 (Pub. L. 102–486), which included these provisions.

Regardless of whether nuclear power enjoys a revival in the 1990s and the twenty-first century, the nation must deal with accumulated radioactive waste as well as future wastes to be generated over the life of the 112 commercial reactors in operation by late 1992. These highly dangerous wastes, now stored temporarily near reactor sites, must be isolated from the biosphere for up to 10,000 years or more. Assuming no new orders for reactors, DOE estimated in 1988 that the total amount of such high-level wastes by the year 2000 would be some 41,000 metric tons and, by the year 2020, nearly 87,000 metric tons (DOE, 1988b:2). Should there be a revival of nuclear power in the United States resulting in new reactor construction, the total amount of waste would greatly exceed these projections.

Our focus below is on nuclear waste within the United States only. Never-theless, other nations also are accumulating large quantities of nuclear waste, and some will look to the United States for policy guidance as well as to nations such as France, Sweden, and Canada, which have adopted different policies and procedures (Carter, 1987). What U.S. policymakers choose to do, and the way in which public reaction influences the process, may well hold important lessons for other countries, many of which are far more dependent on nuclear power than is the United States (Flavin, 1987; OECD, 1989).

Formulation of the Nuclear Waste Policy Act

Enactment of a nuclear waste policy in the United States is surprisingly re-cent, given civilian and military use of nuclear power for over forty years.[1] From the dawn of the nuclear age in the 1940s until well into the 1970s, the federal government and the nuclear industry considered waste disposal to be a manageable and noncontroversial technical problem. These perspec-tives very likely derived from the optimistic outlook on nuclear energy and its future common among both industry and government officials from the 1940s through the 1960s. A faith in the reliability and safety of nuclear power plants, a belief that reprocessing of spent fuel would greatly reduce its vol-ume, and a tendency to underestimate the importance of social and political aspects of nuclear power doubtless reinforced the assumption that disposal of nuclear waste would be easily handled when there was enough of it to worry about. Those beliefs, in combination with a highly favorable politi-cal climate, kept nuclear waste issues off the political agenda even though civilian use of nuclear energy, and thus waste generation, grew appreciably during this period. Contributing to the low salience of the issues was domi-nance over the atomic energy agenda by a closely knit nuclear subgovern-ment,[2] consisting of the nuclear industry, the Atomic Energy Commission (AEC), and the congressional Joint Committee on Atomic Energy (Temples, 1980; Jacob, 1990).

The consensual and relatively closed politics of nuclear energy began to change in the early 1970s, and eventually affected waste policy issues as well. Both changes may be traced to a shifting political climate, which began to give greater emphasis to environmental, health, and safety concerns in managing technological hazards. Three reasons for this new political cli-mate were especially important: (1) a growing quantity of nuclear waste and shortage of storage space at reactor sites; (2) increasing public fear of the risks associated with nuclear power and waste, due in part to problems of leakage at some military storage facilities (e.g., in Hanford, Washington), in part to the failure of an early effort to locate a waste repository at Lyons, Kansas, and in part to the Three Mile Island (TMI) nuclear accident; and (3) the emergence of an explicitly "antinuclear" movement opposed to nuclear power (Price, 1982) and its impact on public opinion and government policymakers (Freudenburg and Rosa, 1984; Jones and Baumgartner, 1989; OTA, 1984:chap. 8).

Two other developments hastened the need to formulate a nuclear waste policy. One was increasing apprehension in the nuclear power industry that failure to devise a national solution to the waste problem would jeopardize the future of commercial nuclear power. The industry's stance was understandable in light of actions by several states, such as a referendum in California, later upheld by the U.S. Supreme Court, that linked expansion of nuclear power to solving the waste problem (Carter, 1987). The second development was the Carter administration's decision to terminate reprocessing of spent fuel rods, which meant that a much larger quantity of waste than anticipated would have to be stored or disposed of (Metlay, 1985; Downey, 1985).

The combined effect of these changes shifted nuclear waste issues from the back to the front burner. Many more participants sought to shape waste policy, and these new policy actors represented a far more diversified set of interests than in the old days of the highly consensual nuclear subgovernment. Environmental risks and health concerns, in particular, were propelled to new heights as environmental and other public interest groups achieved significant political influence. Their success was aided by a series of governmental reforms from the late 1960s to the mid-1970s that democratized the policy process and made Congress as well as federal agencies far more receptive to the new political forces. The upshot of these political developments was that, by the late 1970s, achieving consensus on nuclear waste policy, and nuclear power in general, had become extraordinarily difficult (OTA, 1984:chaps. 8, 9). Congress was seemingly destined to settle for a flawed policy whose chance of success was slight.

The early formulation of nuclear waste policy in the Carter administration originally looked more promising. A Carter administration interagency review group (IRG) launched a detailed study of the problem and reported to the president after widely circulating its draft recommendations (IRG, 1979). Following an extended debate within the administration between the Council on Environmental Quality and nuclear power proponents in DOE, President Carter sent his plan to Congress in February 1980 (Colglazier, 1982; Carter, 1987). The president's proposals were in line with the previous federal emphasis on mined geologic repositories (OTA, 1985:39–80). They also endorsed the study of multiple sites in a diversity of geological formations to ensure selection of the technically most suitable location. Of special interest for our purposes is Carter's strong support for the concept of "consultation and concurrence" with the states. He referred to the "continuing role" of the states in federal repository siting, design, and construction and called for the "fullest possible disclosure to and participation by the public and the technical community" in the site selection process (Carter, 1980:220–24).

Coming less than one year after the Three Mile Island reactor accident, President Carter's plan could not help but rekindle concerns in Congress and the nation about the government's commitment to nuclear power and related issues, including reactor safety and nuclear waste disposal (Sills, Wolf, and Shelanski, 1982; Nealey, Melber, and Rankin, 1983; Freudenburg and Rosa, 1984). Congress had been struggling with nuclear waste policy since 1978, and it took another two years to resolve what were, by then, highly contentious issues. Disputes focused on the role of the states (especially whether a state should be granted a veto of sites within its borders), the extent of public participation, the number of sites to be characterized, the extent of environmental review to be required, the management of military waste, and whether to provide for interim federal storage of spent fuel (favored by the nuclear industry and opposed by environmentalists).

Under the threat of a filibuster as the Ninety-seventh Congress neared adjournment, agreement was reached on a bill in late December 1982. The political resolution in 1982 reflected the importance of new demands for public involvement in managing technological risks, but it also demonstrated the continuing power of what some have called the nuclear establishment (Jacob, 1990), which was determined to have an operational repository as rapidly as possible. Tensions between public concern over nuclear wastes and DOE and industry efforts to speed repository construction would continue throughout the 1980s (Carter, 1987, 1989).

As signed by President Reagan on January 3, 1983, the Nuclear Waste Policy Act (Pub. L. No. 97–425, 96 Stat. 2201–63) established a comprehensive national program for the permanent disposal of high-level commercial and military radioactive waste in mined geological repositories. The site selection process involved nationwide screening of potential locations using technically based evaluation guidelines and provided opportunities for state and public involvement in the siting decisions. The act set out a detailed schedule for site assessment and selection and called for recommendations to the president on a first site in the West by January 1985 and for a second site in the East by July 1987. The early dates would have important implications for the quality of site evaluations and the credibility of the department's studies. A full environmental impact statement, as mandated under the National Environmental Policy Act of 1969, was required only

Table 1-1 Major Provisions of the Nuclear Waste Policy Act of 1982

Established federal responsibility for managing high-level radioactive waste.

Created Office of Civilian Radioactive Waste Management in DOE to handle overall management of repository program and authorized NRC to regulate repository.

Established Nuclear Waste Fund to meet program costs through fees assessed against owners and generators of high-level wastes.

Required DOE to sign contracts with utilities to begin accepting spent fuel in 1998.

Specified means of disposal (geologic repositories using multiple natural and engineered barriers) and called for quick establishment of technical, demographic, and economic guidelines for site evaluation (within 180 days after passage of the act).

Directed secretary of DOE to develop and operate a nationwide transportation system for wastes.

Created a detailed and demanding timetable for initial screening and evaluation, nomination, characterization, and selection of the two repositories and authorized construction and operation of one.

Nomination of a site required preparation of an environmental assessment, which was to be a "detailed statement" of the basis for recommendation and of the probable impacts of site characterization, a comparative evaluation with other sites being considered, and an assessment of regional and local impacts of locating a proposed repository at the site.

States and Indian tribes were granted extensive rights to oversight and participation, and funds were authorized in support of those rights.

The secretary of energy was directed to study the need for and feasibility of a monitored retrievable storage (MRS) facility for civilian waste and to submit proposal for such a facility to Congress by January 1, 1985.

---

when a site was selected by DOE for recommendation to the president as a repository. At that time, the facility would also need to be fully licensed by the NRC. Prior to the final recommendation, all site evaluation activity was termed "preliminary decisionmaking" and only required less comprehensive and less rigorous environmental assessments. Table 1-1 summarizes the key provisions of the act.

Opportunities for public participation in the repository siting process were impressive, at least on paper. Public hearings—long the principal federal vehicle for public input—would be held in the vicinity of each candidate site, and DOE designed an elaborate program of federal-state consultation and cooperation (no longer "concurrence"). Some nuclear waste experts even suggested that the decisionmaking process set out in DOE's 1985 *Mission Plan for the Civilian Radioactive Waste Management Program* might serve as a model for citizen involvement in other technically complex policy decisions (Walker, 1986). As with any public policy, however, one must look to the details of, and controversies over, implementation to estimate the likely success of such ambitious technological undertakings.

Implementing the Nuclear Waste Policy Act

Putting into effect a program as complex as that mandated by the NWPA requires extraordinary managerial and technical capabilities and was bound to be both difficult and time consuming for DOE. Implementation efforts were also likely to be adversely affected by state and public opposition to DOE's decisions (Isaacs, 1989b). Not surprisingly, the department soon ran into formidable social and political obstacles that dwarfed the technological challenge of repository siting.

The Role of Public Acceptability

Missing Link? Pivotal to the success of national state policy is public acceptance, and this is certainly the case for any nuclear waste program. Public acceptance is a function of the correspondence between citizen preferences and policy outcomes. But seldom is the correspondence perfect, and almost never is public acceptance unanimous. Partly for those reasons, political scientists and other serious students of politics consider the relationship between public opinion and public policy to be crucially important for both theoretical and practical reasons. Theoretical interest stems from the fundamental tenet of democratic governance that governmental policies, if they are to be viewed as legitimate, must be responsive to the will of the citizens governed by those policies. Practical interest stems from the fact that wide disparities between citizen opinions and public policies can undermine the legitimacy of governmental leadership and obstruct the implementation of policy. Even the most totalitarian regimes must engender some level of public support to govern effectively.

Surveys of public opinion are one means for gauging the congruence between policy and public preferences. Their importance is aptly described by Bradburn and Sudman (1988:10): "Although public opinion is expressed in elections, they are not held frequently enough and are not refined enough to give ongoing guidance to popular feeling. . . . Public opinion polling provides another means of measuring vox populi; furthermore, it lays claim to being based on scientific principles rather than on political or commercial interests."

Owing to its central role in understanding democratic theory and policy, the question of the linkage between preferences of the mass public and policy decisions has historically attracted considerable scholarly interest.

Until recently, however, the link has been difficult to demonstrate—except indirectly and with generous interpretation (Alford and Friedland, 1975). In short, to the question, What is the link between public opinion and policy? the conventional answer has been a firm, It depends—it depends upon the issue, its salience, and how strongly one expects the linkage ought to be.

Recently, several empirical studies have clarified this picture considerably. Monroe (1979) demonstrated that public policies reflected the majority opinion in nearly two-thirds of the issues studied. Page and Shapiro (1983) found that when public opinion shifted by as much as 20 percent within a year, government policy changed in the same direction 90 percent of the time. Similarly, Goss and Kamieniecki (1983) found a high degree of congruence between public opinion and congressional actions—especially where public feeling was widespread and intense, but less so for issues on which elected representatives felt their constituents to be ill-informed. The Monroe and Goss and Kamieniecki studies are especially relevant because Monroe included energy among the issues he examined, and Goss and Kamieniecki focused exclusively on energy issues. Also of special relevance is a 1980 Harris survey (reported in Nealey, Melber, and Rankin, 1983) that compared the attitudes of five groups—the general public, Congress, federal regulators, investors and lenders, and corporate executives—on six questions pertaining to nuclear energy. The attitudes of Congress were most consistent with those of the general public; in comparison to the general public and Congress, the attitudes of the other three groups were markedly more pronuclear.

Taken together, the evidence makes it clear that government policies on a variety of energy issues tend to be at least roughly consistent with public preferences. But is rough consistency the same as direct influence? The short, simplistic answer is no: the attitudes of the general public are likely to have less direct influence than the efforts of intermediate actors who—whether as proponents, opponents, or members of other activist groups—devote their attention and efforts specifically to the issues involved. It is also possible to argue, in essence, that public attitudes may not shape public policies so much as being shaped by them (e.g., Jasper, 1988), although a strict version of this argument would be inconsistent with the history of nuclear power in the past decade or more in the United States. Since approximately the time of the Reagan inauguration, in January 1981, and despite the efforts of both the Reagan and Bush administrations to ease constraints on nuclear power development, public attitudes have become increasingly negative (see chapter 2 in this volume by Rosa and Freudenburg).

Normative Link? Whether public opinion "should" influence political leaders is also the topic of a time-worn, continuous debate—still vigorous and unresolved. Indeed, the origins of public opinion polling in the United

States are due, in no small part, to the efforts of early polling pioneers to improve democratic processes. For example, George Gallup, imbued with a midwest populism, thought that all poll results could be considered a mandate from the people and the key means for preventing democracies from degenerating into a government by elites. While the views of Gallup and other like-minded pioneers have found many supporters in the United States, they are not without their detractors. An early detractor was the political philosopher and distinguished journalist Walter Lippman, who— drawing on the philosophy of Edmund Burke—argued that, although leaders should be attentive to popular feelings, they should ultimately act according to their perceptions of the common good. Crespi (1989), who provides the historical account described above, also sides with Lippman in opposing the use of public opinion polls as mandates for policymakers.

The normative debate over the proper degree of public input to government policy is exacerbated in the case of technological policies. Complex technological issues, such as nuclear waste disposal and management, unavoidably require rigorous scientific, technical, and bureaucratic analysis: specialized knowledge beyond the interest, and often beyond the ken, of ordinary citizens. What then is the proper balance between the influence of technical expertise, which is essential to thoughtful policy, and the influence of citizens, which is almost always less informed and sometimes uninformed?

Answers to this question parallel those to the question of the proper role of public input to policy more generally. Some question the wisdom of providing a major role for the public (or even its elected representatives) in highly technical decisionmaking, where they prefer reliance on the expertise of scientists and bureaucratic officials (see discussions in Nelkin, 1977 and Kraft, 1988). Others express greater confidence in the public's capacity for assessing risks and participating in technological decisions (Fiorino, 1989; Shrader-Frechette, 1991) and place a high value on the democratic discourse that such participation fosters (Borgmann, 1984).[3] At least some empirical studies support the latter assessment of public participation in decisions affecting nuclear power and other technological risks (Hill, 1992; Kraft and Clary, 1991; Hager, 1989).

These broader debates about technological democracy also find expression in expectations about the impact of public opinion on successful implementation of policies like the NWPA. Many factors affect implementation, including (1) the tractability of the problem addressed (e.g., the adequacy of technical knowledge for solving it); (2) the clarity and consistency of statutory objectives (e.g., whether policies contain incompatible elements or unusually ambiguous statements); (3) the resources available to the implementing agency (especially a knowledgeable and experienced staff and sufficient financial resources); (4) the commitment and leadership skill of key officials (dedication to program goals and ability to manage the agency effectively); and (5) changing socioeconomic conditions and technology, such as an economic downturn or insufficient technological innovation (Mazmanian and Sabatier, 1983).

Given our knowledge of public attitudes toward technological risks and barriers to effective risk communication with the public (Slovic, 1987; National Research Council, 1989), it seems reasonable to hypothesize that the public is especially likely to influence implementation under the following conditions: (1) the policy as enacted was controversial and key elements were opposed by major political interests; (2) the statute provides for formal access by the public or organized interest groups (e.g., through hearings); (3) there is a low level of trust and confidence in the implementing agency's competence and/or credibility; (4) experts both within and outside of government disagree on the extent of risk to the public and/or the environment posed by the agency's activities; and (5) public attitudes on the policy issues are intensely held and highly salient (Mazmanian and Sabatier, 1983).

All of these conditions apply in varying degree to DOE's implementation of the NWPA, which suggests that public acceptance is critically important to the success of the repository siting program. The congressional authors of the NWPA seemed to grasp that fact in specifying that "[s]tate and public participation in the planning and development of repositories is essential in order to promote public confidence in the safety of disposal of . . . waste and spent fuel" (96 Stat. 2207). Similarly, DOE itself has acknowledged that "[p]ublic understanding and participation are essential to the success of the radioactive waste management program" (DOE, 1988b:5). Yet the department discovered, to its dismay, that building public confidence in its repository siting program would be exceptionally difficult, if not impossible, and that, in the absence of such confidence, public opposition threatened to undermine its implementation efforts.

DOE Implements Nuclear Waste Policy: 1982–1987
The DOE began its implementation of the NWPA under what were, by any standard, adverse conditions. Created as a cabinet department only in 1977, DOE was established to consolidate energy-related activities from across the federal government and thus promote coordinated and rational decisionmaking during a time of repeated energy crises. The NRC was assigned regulatory responsibilities for the safety of civilian nuclear power originally handled by the AEC, and DOE assumed regulatory roles for military nuclear facilities

as well as managerial authority for radioactive waste from civilian nuclear plants. DOE also inherited the AEC's role of promoting civilian nuclear power activities and sponsoring research and development to that end. The reorganization plan did not achieve the noble objectives of the planners. Critics contended that DOE had too many (and inconsistent) policy mandates and too little authority to implement them. Others suggested it lacked influential constituencies and congressional support essential to effective policy implementation and that it suffered from low staff morale, inexperience with many of its new programs, and poor leadership (Katz, 1984; Rosenbaum, 1987).

One year before passage of the NWPA in 1982, the Reagan administration first tried to abolish DOE, claiming it was unnecessary (an action opposed by Congress), and then targeted its environmental and energy conservation programs for severe cutbacks (Axelrod, 1984). One result of these programmatic changes was that DOE continued to be dominated by a pronuclear staff, which it had inherited from its predecessor agencies (the AEC and the Energy Research and Development Administration). That organizational heritage and bias helps to explain why DOE has been criticized so often for neglecting health, safety, and environmental concerns; for ignoring public opinion and social and political issues in nuclear power; and for poor working relations with the states (Metlay, 1985; Shrader-Frechette, 1980). It was hardly unexpected that the same charges would be leveled at DOE's management of the nuclear waste programs.

Despite what many outside the department saw as substantial institutional handicaps of this kind, DOE tried to move quickly to implement the NWPA, and cited the congressionally imposed deadlines as justification for emphasizing a "fast-track" timetable for site evaluation (Isaacs, 1989a:4). The department issued draft siting guidelines and its draft mission plan (DOE, 1985a, 1985b) amid complaints that it was moving too quickly for sound scientific judgments to be made. By December 1984, DOE faced mounting and often intense public opposition as the department directed its attention to specific repository sites.

On December 19, 1984, DOE ranked nine candidate sites in six states for the first of the two repositories authorized by NWPA, and it named the top three for site characterization: Deaf Smith County, Texas; Yucca Mountain, Nevada; and Hanford, Washington. Within days, Nevada, Texas, and several environmental groups filed suit against DOE challenging the legality of the designation process, and they were joined by Washington within six months. Much of the dispute centered on the environmental assessments. Criticisms, including that from the National Academy of Sciences Board on Radioactive Waste Management and the NRC, were directed especially at omissions and other deficiencies in the data used, the methodology for site evaluation and comparison, and DOE bias in the evaluation process (Clary and Kraft, 1989; Colglazier and Langum, 1988).

Similar controversies surrounded the second round siting process in the East. On January 16, 1986, DOE announced that twenty locations in seven states in the upper Midwest or East had been selected from its original list of 235 potential sites in seventeen states. Once again DOE defended its choices in terms of the official siting guidelines. Public hearings in the potential host states in the spring of 1986, attended by thousands of individuals, group representatives, and state and local government officials, revealed massive public opposition to DOE's siting proposals (see chapter 4 in this volume by Kraft and Clary). One study of a second round state, Wisconsin, documents an even earlier shift to entrenched state opposition and a policy of "noncooperation" with the federal government following DOE's "thwarting" of "good faith efforts" and the department's "bungling" of requests for technical information (Schaefer, 1988). The Chernobyl nuclear accident in April 1986 did little to improve DOE relations with a public that already feared possible contamination by radioactive waste and that was becoming increasingly skeptical about nuclear power itself (see Nealey, 1990 and chapter 2 by Rosa and Freudenburg in this volume).

Intense opposition of this kind in the East, and fears of political repercussions in an election year, led the Reagan White House in late May 1986 to put the siting process for the eastern states on hold (Hershey, 1986). Many members of Congress up for reelection and eager to avoid the political heat generated by the repository program supported that action. Chances for successful implementation of the NWPA diminished further when the western states objected strenuously to what they viewed as inequitable treatment that left them vulnerable to imposition of a repository.

As a result of these and other criticisms, Congress was forced to revise the act in 1987. Facing sharply conflicting proposals to declare a siting moratorium and study the issues further or to press ahead with a "sequential" siting process that would examine only a single site (at Yucca Mountain, Nevada), Congress chose the latter (Carter, 1989; Kraft, 1992). The Nuclear Waste Policy Amendments Act of December 1987 canceled authorization for the second repository, directed DOE to evaluate the Yucca Mountain site, and stipulated that the department cease consideration of the other western sites in Texas and Washington. DOE was to report to Congress between 2007 and 2010 on the need for a second repository. The amendments also authorized a monitored retrievable storage (MRS) facility (for temporary storage of waste), linking its construction to completion of the repository itself, and provided for compensation to the state of Nevada, including assistance for the state to participate in site evaluation (Pub. L. 100–203; Carter, 1987, 1989; Cooper, 1989).

Opposition in Nevada: 1987–1992

The amendments act of 1987 appeared promising at first, and it was defended by its congressional sponsors as both less expensive and as a potentially more rapid route to repository siting—since only a single site would have to be studied in depth. It was also attractive politically because the other forty-nine states had little reason to criticize DOE or the program and thus risk reopening the issue of repository location (Davis, 1987; Cooper, 1989). This optimistic prognosis depended, of course, on the willingness of Nevada to cooperate with DOE. As it turned out, more than five years of bitter relations between the state and DOE followed, which many attributed to Nevada's "wounded psyche" at being singled out for the repository. Even now, few state officials believe that cooperating with DOE is in their political self-interest, suggesting that the logic behind Congress's grant of modest financial compensation—as a supposed inducement for cooperation by the host state—is flawed.

Much of the controversy in Nevada focused on conflicting estimates of the risk of locating the repository at Yucca Mountain and the adequacy of DOE's massive characterization plan, which weighed in at eight volumes and 6300 pages (Malone, 1990; Loux, 1989). Originally estimated at some $200 million in 1982, the cost of characterization for this one site rose to $6.3 billion by 1992 (U.S. General Accounting Office, 1992). The scope and cost of site characterization reflected the status of the project as the first of its kind as well as the demanding standards the repository would have to meet under Environmental Protection Agency and NRC rules. Given those conditions, such a scientific investigation could not help but generate a multitude of technical questions about long-term environmental risks. Critics would inevitably have abundant opportunities to voice their dissent over the DOE investigation and its conclusions.

The most important issues in these disputes, however, were political rather than scientific. Nevada had few reasons to cooperate with DOE on program implementation. Relations between the state and DOE deteriorated steadily until the Nevada legislature, in mid-1989, passed AB222 which, in effect, vetoed placement of the repository in Nevada. As a consequence, state agencies refused to issue environmental permits essential for DOE excavation and testing at the site, and all site activity was brought to a halt. The issue was taken to federal court, where Nevada lost its case, and, as a consequence, in June 1991 the state issued the first of a series of permits to allow DOE to study the Yucca Mountain site.

Policy Failure and Prospects for the Future

Few policies can be expected to follow what Mazmanian and Sabatier (1983) have called an *effective implementation scenario,* that is, situations where policies are implemented efficiently, smoothly, and generally without conflict. As noted above, such an outcome is especially unlikely for a nuclear waste disposal policy, which is certain to arouse public concern and stimulate intense opposition. Public opposition is considerably more likely when the implementing agency is viewed as lacking credibility. Hence, it is not surprising that DOE's record from 1982 on has been widely characterized as a policy failure (Clary and Kraft, 1989; Power, 1989; Lippman, 1989). The "redirection" of policy in 1987 did not solve the seemingly intractable problem of securing state cooperation, as even DOE admitted by late 1989 (DOE, 1989; Wald, 1989).

Some observers believe the program faces so many obstacles at this time that little hope remains for its success in the near term (see chapter 3 by Slovic, Layman, and Flynn in this volume). Others believe that significant changes in DOE organization and behavior might save the repository siting effort even at this late date (Cook, Emel, and Kasperson, 1990). Our focus here is on the role that public reaction—including the public's perceptions, attitudes, beliefs, and opinions—will likely play in the various scenarios one can imagine for the next decade or two.

There are several different directions the program might follow. A likely one would be a continuation of the program's current trajectory with implementation resembling what Mazmanian and Sabatier (1983:279–80) term the *gradual erosion scenario.* In this model, a statute mandates measure that would create significant change for the affected constituency, and the responsible agency must work through other governmental bodies to achieve its objectives while remaining dependent on external parties for the development of new technologies needed for implementation success. These requirements create a large number of veto points in the decisionmaking process. In addition, the agency typically is staffed at the beginning with enthusiastic supporters of the program, but they move so aggressively to develop rules and regulations that they undermine the agency's credibility, which in turn diminishes the staff's enthusiasm and commitment to the program. The agency may also lose its supportive constituency over time while opposition may become "vociferous, well organized, and persistent." Even the intervention of new managers or a policy "fixer" does not alter the long-term erosion of the program. Eventually the policy fails: goals and objectives are not met.

A more optimistic path for the future would resemble a *rejuvenation sce-*

nario. Here the initial implementation process parallels the history of the nuclear waste program thus far, a gradual erosion scenario, where ambiguities or contradictions in the statute in combination with a lack of political support lead to a long period of ineffectual activity. At some point, however, changing socioeconomic and political conditions create new program proponents who are able to strengthen the legislative mandate and provide a clearer guide to further implementation.[4] The agency staff also learns to be more effective in working with the affected parties and thus is better able to achieve program objectives. In short, reorganization, provision of new resources or leadership, and legislative changes may rebuild a crumbling policy structure (Mazmanian and Sabatier, 1983:281–82). Interestingly, a National Academy of Sciences panel in mid-1990 urged a host of such basic reforms in the waste disposal program, including far more flexible work schedules, technical specifications, and regulatory standards; the use of outside experts to review and critique the program; greatly expanded risk communication to the public; and meaningful dialogue with states and Indian tribes. The academy panel warned that the program was "bound to fail" if pushed ahead without such reforms (National Research Council, 1990).

Under this scenario, Congress might once again reformulate nuclear waste policy, which could mean establishment of a different basis for federal-state relations in Nevada, consideration of sites other than Nevada, or creation of alternatives to geologic disposal such as MRS facilities (recommended by the MRS Review Commission in 1989) or expansion of storage capacity at reactor sites, where most commercial wastes are now kept, through the use of dry cask storage.

What will determine which of these several paths is followed in the future? Initially the choice will be shaped by the way in which DOE manages the waste program. Much depends on the department's ability to address fundamental institutional issues, such as public fear of nuclear waste and state opposition to siting, and to learn from its experience of the past decade, particularly how relationships with the states and the public might be improved and how its credibility can be enhanced (Cook, Emel, and Kasperson, 1990; Whipple, 1989).

There were signs by 1989 that DOE had indeed become more attentive to public concerns with nuclear waste. In a January 1989 report on implementation, a top official in the program acknowledged that "widespread misperception among the general public" created a "monumental challenge" for the department to build public understanding of the nuclear waste problem and the repository program, especially due to its complexity and the inevitability of scientific uncertainty. He admitted that meeting that challenge was as vital to the program's success as building the repository itself (Isaacs,

1989a:21). Whether the department's proposal for new public information and public involvement strategies will prove sufficient remains to be seen (DOE, 1990).

An effective nuclear waste policy is clearly essential for the nation, irrespective of whether or not nuclear power enjoys a commercial revival in the future. A critical ingredient in making that policy work is improvement in our understanding of the public's views of nuclear waste and repository siting issues. Contributing to this understanding is our chief objective in preparing this book. We want to know more about the perceptions, beliefs, attitudes, and opinions people hold concerning nuclear waste issues; how they are formed, how they change, and how they affect support for public policy.

Assessment of Public Viewpoints:
The Empirical Evidence

The first part of this volume's title, "Public Reactions to Nuclear Waste," is intentionally both ambiguous and broad: the ambiguity reflects the complexity and multifaceted nature of the volume's topic, while the breadth demonstrates our explicit recognition of the expansive continuum of cognitive and behavioral traits embedded in public sentiment toward controversial technologies such as nuclear waste repositories. Perceptions, beliefs, opinions, attitudes, values, and behavioral responses are the various mental domains underpinning public reactions to nuclear wastes. The chapters that follow address the repertoire of issues subsumed under our broad title from a multiplicity of vantage points, focusing on the variety of cognitive domains through the empirical examination of different variables. The richness of those vantage points are, in large part, a consequence of the wide variety of disciplines—and their respective perspectives—represented in this volume. Some chapters focus on perceptions, seeking to understand some of the basic building blocks of people's cognitive processing of nuclear waste issues. Other chapters emphasize opinions, examining public opinion with hypothetical ballot items designed to measure support or opposition to a repository. Others focus on attitudes and beliefs about nuclear wastes and repository siting, including beliefs about the likely impacts (both positive and negative) of a repository, and attitudes toward the institutions entrusted with siting it. Still others combine various of these foci into summary measures to reflect in one variable the multifaceted and multilayered features of public sentiment toward nuclear waste and repository siting.

The chapters are also richly varied in other important ways, including the range of geographical locales considered and by the diversity of populations sampled. As a result, the chapters document, extensively, citizen views on repository siting in all three of the finalist site areas, as well as in some site areas that attracted preliminary consideration. Equally well-documented are the range of views, from the residents of small rural communities, living in the proximity of a proposed repository, to the nation as a whole. Similarly, the locales examined vary in the expected receptivity to a local nuclear facility; they range from Hanford, Washington, where strong support was expected, to the Texas Panhandle, where staunch opposition was found.

In addition, sharing a common quest to uncover the antecedents of views toward repository siting, the chapters examine an assortment of variables presumed to shape those views. The shaping variables include trust in the federal government and its lead agencies (such as DOE), knowledge about nuclear wastes, faith in science and technology, and the expected impacts (positive and negative) from a repository. The effects of social background variables—especially those demographics shown to influence attitudes toward nuclear power, such as sex, age, and parenthood—are also examined.

Each of the chapters is a self-contained report of a well-executed empirical study that can adequately speak for itself. Recognizing that, the following section maps the remainder of the volume by presenting a brief overview of each chapter. Summarizing and organizing the findings emerging from these studies is the purpose of the concluding chapter.

The Larger Context of Repository Siting
The goal of chapter 2, by Eugene A. Rosa and William R. Freudenburg, is to set the historical and social context of nuclear power in the United States, a context preceding—and partly shaping—the nuclear waste problem, the principal focus of this volume. Their analysis is based upon an examination of key historical documents and events as well as on a comprehensive review of the best available longitudinal data on attitudes toward nuclear power. Of the several conclusions to emerge from their analysis, three are especially relevant to the question of the social acceptability of a high-level nuclear waste repository: following an initial stage of enthusiastic support, nuclear power fell into disfavor with Americans, consistently attracting a majority of opposition for nearly all of the past decade; the siting of nuclear power plants and nuclear facilities of *any kind* anywhere in the United States is problematic due to the NIMBY syndrome ("not in my back yard") and other factors; and the problem of nuclear wastes was the last of a series of nuclear issues to reach the public consciousness. This last conclusion is especially revealing, for it shows that the nuclear waste problem emerged in a social climate that had already become distinctly inhospitable to nuclear power.

That a majority of Americans were already concerned that nuclear plants were not safe enough—that is, were too risky—warned that the siting of a HLNWR would not likely be greeted with open arms. Indeed, the evidence pointed to especially stormy times for the repository program.

The most significant change resulting from the 1987 amendments to the NWPA was a reduction in finalists, from three to one. The single remaining candidate site to receive a comprehensive characterization and evaluation—including geological and other physical characteristics, engineering features, and social and economic impacts—was at Yucca Mountain, Nevada. Located approximately ninety miles northwest of Las Vegas, the site has been undergoing continuous evaluation since 1987. Chapter 3, by Paul Slovic, Mark Layman, and James H. Flynn, based on research conducted as part of that evaluation, examines the mental imageries associated with underground nuclear wastes. The goal of their research was to get beneath perceptions of risk at the surface of conscious thought to reveal the mental content of those perceptions. Using a word association technique, the authors collected a statewide sample of Nevada residents' images; for comparative purposes, they also administered the word associations to a national sample, and to two samples in areas that are a major source of tourism to Nevada: Phoenix, Arizona, and southern California.

The results reported by Slovic, Layman, and Flynn, though not necessarily counterintuitive, are nevertheless profound and disturbing: the images of underground nuclear wastes for *all* samples (not just Nevada) are overwhelmingly negative, suggesting that perceptions are deeply rooted in fear and dread. The findings have broad implications for the siting of a nuclear waste repository anywhere in the United States, for, as the authors' conclude, "siting and development of a national repository for disposal of radioactive wastes is not politically feasible in the foreseeable future."

Public Reactions to Preliminary Sites
In its original form, as noted above, the NWPA called for the siting of two permanent HLNWRs, one in the East and the other in the West. Implementing the citizen participation provisions of the NWPA required DOE to conduct preliminary assessments of candidate sites to assess physical, environmental, and social impacts of all phases of the siting process. By 1984, having completed the first round of these assessments, DOE issued draft reports for public comments in six states in the West and South—covering the area considered for the western site. Two years later, the second-round preliminary assessment process—for a site in the East—was completed, resulting in the identification of twenty candidate sites in seven eastern states. Draft reports of preliminary site assessments were prepared for public review.

Chapter 4, by Michael F. Kraft and Bruce B. Clary, is a detailed examination of testimonies at DOE-held public hearings in four of the seven states: Wisconsin, Maine, North Carolina, and Georgia. Though the data is not representative of the general public, owing to the self-selection of the testimonies, it is representative of one critical public: the *attentive public*, or that segment of the populace with particular interest in repository siting. An examination of the attentive public is important because it is the segment of the public that is typically knowledgeable about the issues and often the one most likely to pressure the government. Content analysis of the public testimonies, the method employed by the authors, provides an in-depth sounding of public views and concerns—often producing insights not revealed in general surveys. Quite expectedly, those testifying were overwhelmingly opposed to the DOE recommendations contained in the draft reports. The authors investigate the basis of this finding and ask whether it was due to the NIMBY syndrome or other factors.

Until amended in 1987, the NWPA mandated the exhaustive characterization and assessment of three locations as the site for the nation's first HLNWR. The first site, to be located in the West, was to be selected from three finalists: Yucca Mountain, Nevada; Hanford Nuclear Reservation in Washington State; and Deaf Smith County, located in the Texas Panhandle. Assessing the public acceptability of a high-level repository began in earnest first in Texas. Chapter 5, by Julia G. Brody and Judy K. Fleishman, reports an analysis of two surveys, conducted in 1984 and 1986, comparing residents of Deaf Smith County to residents of similar proximate rural counties and to a nearby urban area. As the first systematic assessment of public acceptability of a nuclear waste repository, the results of the Texas research were something of a harbinger of things to come—an early warning signal of difficult social obstacles to the siting of a repository. Brody and Fleishman's principal discovery was a deep and widespread opposition to the repository, not only among Deaf Smith residents, as expected, but also, unexpectedly, among the residents of all the comparison communities.

The third finalist site in the West was the Hanford Nuclear Reservation located on the Columbia River in the south central area of Washington State. Originally established in 1943 as part of the Manhattan Project, Hanford has since been the site for a wide variety of federal research and production activities associated with nuclear energy. Thus, for over forty years, Hanford has been a major economic presence in the area, providing steady employment—especially scientific, engineering, and other highly skilled employment—and other economic benefits. Historically, owing to a supportive "nuclear culture" (Loeb, 1986) and because of the deep economic dependency of local communities, activities at the reservation enjoyed strong political and community support in the Hanford area. Indeed, the announcement of the three finalists in the West sparked a controversy over the inclusion of Hanford because its geological characteristics were not as suitable as some of the disqualified sites (Keeney, 1986). The reason given for ranking Hanford ahead of these geologically superior sites was "social acceptability": since the activities at Hanford (including the production of plutonium for making bombs) had always received uncritical local support, it was thought that the same would be true for the proposed waste repository.

Chapter 6, by Riley E. Dunlap, Eugene A. Rosa, Rodney K. Baxter, and Robert Cameron Mitchell, begins by examining the pervasive belief that local citizens, as they always had in the past, would support the siting of a re-pository at Hanford. Using public opinion polls sponsored by newspapers, Dunlap et al. were able to chart trends in attitudes of Hanford area residents toward the repository over time and to compare these local attitudes to attitudes of residents in the entire state. While the findings show that locals were, as expected, more supportive of the repository than statewide residents, the level of support was surprisingly lower than predicted by many observers: government officials, other politicians, the media, activists, and attentive citizens. The main section of the chapter presents results from a December 1987 Hanford-area survey, part of the site evaluation, devoted to a detailed examination of attitudes toward the proposed repository and the factors shaping them, including perceptions of the likely impacts of a repository.

Public Reactions to the Yucca Mountain, Nevada, Site
With the passage of the Nuclear Waste Policy Amendments Act in 1987, Yucca Mountain, Nevada, became the only site to be considered for the nation's first HLNWR. As a consequence it became an exemplary case study of the nation's most difficult—literally—siting decision to date. The focal point of chapter 7, by William H. Desvousges, Howard Kunreuther, Paul Slovic, and Eugene A. Rosa, is public perceptions of risks associated with the proposed Nevada HLNWR. How do members of the public perceive the uncertainties associated with the HLNWR? How likely, do they believe, is it that the repository will generate adverse effects? Should adverse effects be generated, how serious and widespread do they think they will be? Also, how much confidence does the public have in the institutions responsible for the repository? Risk perception, incorporating the foregoing questions and others, has been the topic of a rich tradition of research which has created an orderly picture of how laypersons cognitively organize a wide variety of risks (Slovic, 1987). Building upon this tradition, chapter 7 exam-

ines perceptions of risks associated with the proposed repository revealed in a statewide survey of Nevada residents, and compares risk perceptions to a variety of other factors presumed to influence people's views of this type of facility. Because there is no absolute standard against which to gauge levels of risk perception, a second survey, sampling the entire nation, was obtained for comparative purposes. The featured analyses of the chapter, therefore, are comparisons between the views of citizens of the host state, Nevada, and those of citizens of the United States more generally. Key among the detailed findings are that Nevadans, similar to Americans generally, perceive the repository vis-à-vis other technological hazards to be especially risky, believe that accidents are highly likely in the near future, and have little confidence that the federal government can construct and manage the repository safely.

Lying but ninety miles southeast of Yucca Mountain is the city of Las Vegas, Nevada. This gambling and entertainment mecca is not only the state's major source of economic income, but also the main population concentration in the state; the over 600,000 residents in the Las Vegas Standard Metropolitan Statistical Area constitute approximately 60 percent of Nevada's total population. Thus, building and maintaining a repository at Yucca Mountain could affect directly both the tourism that drives the economy of Las Vegas and the large, rapidly growing population that resides there. Chapter 3 in this volume, by Slovic, Layman, and Flynn, examines risk images and concerns among samples taken from two geographical areas that supply a sizable number of tourists to Las Vegas: Phoenix, Arizona, and southern California.

The effect on tourism is examined more directly and thoroughly in chapter 8, by Douglas Easterling and Howard Kunreuther. Using two carefully crafted samples, one of convention planners and one of convention attendees, the authors attempt to assess the impact on the Las Vegas convention industry of a HLNWR sited at Yucca Mountain. Such an assessment is crucial since conventions and trade shows are a major source of revenue for Las Vegas and the state of Nevada. The findings are consistent for both samples: each gives a clear indication that the siting of a HLNWR at Yucca Mountain could "trigger major dislocations of Las Vegas's convention industry."

Chapter 9, by Alvin H. Mushkatel, Joanne M. Nigg, and K. David Pijawka, addresses the effect on Las Vegas residents themselves by examining survey data collected via intensive face-to-face interviews with a sample of randomly selected residents in the Las Vegas metropolitan area. Consistent with the conclusions of the other chapters, residents of Las Vegas—like residents elsewhere in the state and in the nation—were found to be seriously concerned by the health and other risks associated with a HLNWR and are deeply opposed to the construction of the facility.

While Las Vegas is the largest concentration of people in proximity to Yucca Mountain, other populations—much smaller and rural—are closer and potentially more vulnerable to the risks posed by the repository. Chapter 10, by Richard S. Krannich, Ronald L. Little, and Lori A. Cramer, the last empirical chapter in this volume, examines the receptivity to the repository among six communities in southern Nevada located either near Yucca Mountain or on a probable transportation route of wastes going to the repository. Rural communities like these are often faced with a difficult dilemma. On the one hand, they welcome the employment and economic benefits that come with growth (even boomtown growth), but, on the other hand, they are reluctant to accept hazardous and other noxious facilities (the NIMBY syndrome). Thus, how such communities resolve this dilemma, and what factors underlie the resolution, is especially interesting to stakeholders and policymakers—and particularly challenging to researchers attempting to predict the outcome. Examining perceptions of health and safety risks and viewpoints toward the repositories, the authors found considerable variation among the six communities in both perceptions of risks and support for the repository.

Summary and Policy Implications

Finally, in the eleventh and concluding chapter of this volume, we three editors summarize and organize the findings from the previous chapters with the aim of establishing empirical generalizations that clearly emerge from the wide range of studies reported here. In addition, we review the key social and political issues associated with the siting of a high-level nuclear repository vis-à-vis the empirical evidence amassed from the studies. We match those issues against the themes and patterns of findings from the chapters to draw out the important substantive and policy implications and, especially, to draw attention to those aspects of conventional policy wisdom that are at odds with the evidence accumulated from the various studies. Finally, we go on to make recommendations for a more deeply informed repository siting policy.

Notes

1   The long history of neglect of nuclear waste issues by nuclear promoters is a recurrent theme in Jacob's (1990) thoughtful, penetrating political history of nuclear waste management. For example, he notes: "Forty years after the first nuclear chain reaction was produced at the University of Chicago, Congress passed the first piece of legislation to explicitly address the management of high-level nuclear waste" (1990:45).

2   The subgovernment is sometimes referred to as the "nuclear establishment." We believe

this practice inappropriate since the latter term is typically more comprehensive, comprising, for example—as in Jacob's scheme (1990)—the national state, the economy, and civil society.

3   Even where the public is permitted direct access to technological decisionmaking, it may still be in a disadvantaged position vis-à-vis experts and bureaucrats. Public policy issues, when they are narrowly defined as scientific or technical problems, favor experts, on the one hand, and deflect attention away from the broader political and social issues, on the other hand. The NWPA, according to some observers, defined the siting of a HLNWR principally as a scientific and technical problem in order to reduce the influence of political factors on the siting process. Jacob, for example, observes: "Thus, the emphasis on technology and technocratic decisionmaking was used to undermine state and local opposition. Technical substance was placed in opposition to 'parochial' interests and politics, implying that opposition was motivated by self-interest whereas those in control of technologies served the public interest" (1990:136).

4   Such a possibility, as noted above, is made more credible in view of a growing concern over global warming—largely due to the burning of fossil fuels—and nuclear energy's potential for ameliorating the trend.

References

Alford, Robert R., and Roger Friedland. 1975. "Political Participation and Public Policy." *Annual Review of Sociology,* 1:429–79.

Axelrod, Regina, S. 1984. "Energy Policy: Changing the Rules of the Game." In Norman J. Vig and Michael E. Kraft, eds., *Environmental Policy in the 1980s.* Washington, D.C.: CQ Press. 203–28.

Borgmann, Albert. 1984. "Technology and Democracy." *Philosophy and Technology* 4:211–28.

Bradburn, Norman, and Seymour Sudman. 1988. *Polls and Surveys: Understanding What They Tell Us.* San Francisco: Jossey-Bass.

Carter, Jimmy. 1980. "Appendix A: Presidential Message and Fact Sheet of February 12, 1980." In E. W. Colglazier, ed., *The Politics of Nuclear Waste.* Elmsford, N.Y.: Pergamon. 220–41.

Carter, Luther J. 1987. *Nuclear Imperatives and Public Trust: Dealing with Radioactive Waste.* Washington, D.C.: Resources for the Future.

———. 1989. "Nuclear Waste Policy and Politics." *Forum for Applied Research and Public Policy* 4 (Fall): 5–18.

Clary, Bruce B., and Michael E. Kraft. 1988. "Impact Assessment and Policy Failure: The Nuclear Waste Policy Act." *Policy Studies Review* 8:105–15.

———. 1989. "Environmental Assessment, Science, and Policy Failure: The Politics of Nuclear Waste Disposal." In Robert V. Bartlett, ed., *Policy Through Impact Assessment.* Westport, CT: Greenwood. 37–50.

Colglazier, E. W., ed. 1982. *The Politics of Nuclear Waste.* Elmsford, NY: Pergamon.

Colglazier, E. W., and R. B. Langum. 1988. "Policy Conflicts in the Process of Siting Nuclear Waste Repositories." *Annual Review of Energy* 13:317–57.

Cook, Brian J., Jacque L. Emel, and Roger E. Kasperson. 1990. "Organizing and Managing Radioactive Waste Disposal as an Experiment." *Journal of Policy Analysis and Management* 9 (Summer): 339–66.

Cooper, Benjamin S. 1989. "The Nuclear Waste Policy Amendments Act of 1987." Paper presented at the annual meeting of the American Association for the Advancement of Science, San Francisco, January 17.

Covello, Vincent T. 1983. "The Perception of Technological Risks: A Literature Review." *Technological Forecasting and Social Change* 23:285–97.

Crespi, Irving. 1989. *Public Opinion, Polls, and Democracy.* Boulder: Westview.

Davis, Joseph A. 1987. "Nevada to Get Nuclear Waste; Everyone Else 'Off the Hook.' " *Congressional Quarterly Weekly Report,* December 19, 3136–38.

Downey, Gary L. 1985. "Federalism and Nuclear Waste Disposal: The Struggle Over Shared Decision Making." *Journal of Policy Analysis and Management* 5 (Fall): 73–99.

Fiorino, Daniel J. 1989. "Technical and Democratic Values in Risk Analysis." *Risk Analysis* 9 (September): 293–99.

Flavin, Christopher. 1987. "Reassessing Nuclear Power." In Lester Brown et al., *State of the World 1987.* Washington, D.C.: World Watch Institute. 57–80.

Freudenburg, William R., and Eugene A. Rosa, eds. 1984. *Public Reactions to Nuclear Power: Are There Critical Masses?* Boulder: Westview/American Association for the Advancement of Science.

Goss, Carol, and Sheldon Kamieniecki. 1983. "Congruence Between Public Opinion and Congressional Actions on Energy Issues, 1973–74." *Energy Systems and Policy* 7:149–70.

Gould, Leroy C., Gerald T. Gardner, Donald R. DeLuca, Adrian R. Tiemann, Leonard W. Doob, and Jan A. J. Stolwijk. 1988. *Perceptions of Technological Risks and Benefits.* New York: Russell Sage Foundation.

Hager, Carol J. 1989. "Technological Democracy? Grassroots Participation and Technological Innovation in West German Energy Policy." Paper presented at the annual meeting of the Western Political Science Association, Salt Lake City, March 30–April 1.

Hershey, Robert D. 1986. "U.S. Suspends Plan for Nuclear Waste Dump in East or Midwest." *New York Times,* May 29, 1, 10.

Hill, Stuart. 1992. *Democratic Values and Technological Choices.* Palo Alto: Stanford University Press.

Interagency Review Group on Nuclear Waste Management. 1979. *Report to the President.* TI 29442. Washington, D.C.: GPO, March.

Isaacs, Thomas H. 1989a. "Implementing the Nuclear Waste Policy Act and Its Amendments." Paper presented at the annual meeting of the American Association for the Advancement of Science, San Francisco, January.

———. 1989b. "DOE Goals: Excellence, Openness." *Forum for Applied Research and Public Policy* 4 (Fall): 36–38.

Jacob, Gerald. 1990. *Site Unseen: The Politics of Siting a Nuclear Waste Repository.* Pittsburgh: University of Pittsburgh Press.

Jasper, James M. 1988. "The Political Life Cycle of Technological Controversies." *Social Forces,* 67:357–77.

Jones, Bryan D., and Frank R. Baumgartner. 1989. "Changing Images and Venues of Nuclear Power in the United States." Paper presented at the annual meeting of the Midwest Political Science Association, Chicago, April 14–16.

Katz, James Everett. 1984. *Congress and National Energy Policy.* New Brunswick, NJ: Transaction.

Keeney, Ralph L. 1986. "An Analysis of the Portfolio of Sites to Characterize for Selecting a Nuclear Repository." Report in the Decision Analysis Series. Los Angeles: Systems Science Department, University of Southern California.

Kraft, Michael E. 1988. "Evaluating Technology Through Public Participation: The Nuclear Waste Disposal Controversy." In Michael E. Kraft and Norman J. Vig, eds., *Technology and Politics*. Durham: Duke University Press. 251–77.

——. 1992. "Technology, Analysis, and Policy Leadership: Congress and Radioactive Waste." In Gary C. Bryner, ed., *Science, Technology and Politics: Policy Analysis in Congress*. Boulder: Westview. 65–85.

Kraft, Michael E. and Bruce B. Clary. 1991. "Citizen Participation and the NIMBY Syndrome: Public Response to Radioactive Waste Disposal." *Western Political Quarterly* 44 (June): 3 299–328.

Lippman, Thomas W. 1989. "Between a Rock and a Hard Place." *Washington Post National Weekly Edition*, October 16–22, 9.

Loeb, Paul. 1986. *Nuclear Culture: Living and Working in the World's Largest Atomic Complex*. Philadelphia: New Society.

Loux, Robert R. 1989. "A View on Siting Issues from Nevada." *Forum for Applied Research and Public Policy* 4 (Fall): 29–31.

Malone, Charles R. 1990. "Implications of Environmental Program Planning for Siting a Nuclear Waste Repository at Yucca Mountain, Nevada, USA." *Environmental Management* 14:25–32.

Mazmanian, Daniel A., and Paul A. Sabatier. 1983. *Implementation and Public Policy*. Glenview, IL: Scott, Foresman.

Metlay, Daniel S. 1985. "Radioactive Waste Management Policymaking." In U.S. Office of Technology Assessment, *Managing the Nation's Commercial High-Level Radioactive Waste*. Washington, D.C.: GPO. 199–232.

Monroe, Alan D. 1979. "Consistency Between Public Preferences and National Policy Decisions." *American Politics Quarterly*, 7:3–19.

Morone, Joseph G., and Edward J. Woodhouse. 1989. *The Demise of Nuclear Energy? Lessons for Democratic Control of Technology*. New Haven: Yale University Press.

National Research Council. 1989. *Improving Risk Communication*. Washington, D.C.: National Academy Press.

——. 1990. *Rethinking High-Level Radioactive Waste Disposal: A Position Statement of the Board on Radioactive Waste Management*. Washington, D.C.: National Academy Press.

Nealey, Stanley M. 1990. *Nuclear Power Development: Prospects in the 1990s*. Columbus, OH: Battelle.

Nealey, Stanley M., Barbara D. Melber, and William L. Rankin. 1983. *Public Opinion and Nuclear Energy*. Lexington, MA: D. C. Heath and Co.

Nelkin, Dorothy. 1977. *Technological Decisions and Democracy*. Beverly Hills: Sage.

Nuclear News. 1991. "World List of Nuclear Power Plants" (February).

OECD. 1989. *Nuclear Energy in Perspective*. Paris: Nuclear Energy Agency of the Office of Economic Cooperation and Development.

Page, Benjamin I., and Robert Y. Shapiro. 1983. "Effects of Public Opinion on Policy." *American Political Science Review* 77:175–90.

Power, Max S. 1989. "Politics and Science in Siting Battle." *Forum for Applied Research and Public Policy* 4 (Fall): 19–25.

Price, Jerome. 1982. *The Antinuclear Movement*. Boston: Twayne.

Rosenbaum, Walter A. 1987. *Energy, Politics, and Public Policy.* Washington, D.C.: CQ Press.

Schaefer, Jame. 1988. *State Opposition to Federal Nuclear Waste Repository Siting: A Case Study of Wisconsin, 1976–1988.* Green Bay: Center for Public Affairs, University of Wisconsin-Green Bay, July.

Schneider, Keith. 1988. "U.S. Delays Start of Plant to Store Nuclear Wastes." *New York Times,* September 14, 1.

Shrader-Frechette, Kristin. 1980. *Nuclear Power and Public Policy: The Social and Ethical Problems of Fission Technology.* Boston: D. Reidel.

———. 1991. *Risk and Rationality: Philosophical Foundations for Populist Reforms.* Berkeley: University of California Press.

Sills, David L., C. P. Wolf, and Vivien B. Shelanski, eds. 1982. *Accident at Three Mile Island: The Human Dimensions.* Boulder: Westview.

Slovic, Paul. 1987. "Perception of Risk." *Science* 236:280–85.

Temples, James R. 1980. "The Politics of Nuclear Power: A Subgovernment in Transition." *Political Science Quarterly* 95 (Summer):239–60.

U.S. Department of Energy. 1985a. Office of Civilian Radioactive Waste Management, *Record of Responses to Public Comments on the Draft Mission Plan for the Civilian Radioactive Waste Management Program.* Vol. 2. Washington, D.C.: GPO, June.

———. 1985b. Office of Civilian Radioactive Waste Management, *Mission Plan for the Civilian Radioactive Waste Management Program.* 3 vols. Washington, D.C.: GPO.

———. 1987. Office of Civilian Radioactive Waste Management, *Annual Report to Congress.* Washington, D.C.: GPO, April.

———. 1988a. Energy Information Administration, *Energy Facts 1987.* Washington, D.C.: GPO, June.

———. 1988b. Energy Information Administration, *Commercial Nuclear Power 1988: Prospects for the United States and the World.* Washington, D.C.: GPO, September 21.

———. 1989. Office of Civilian Radioactive Waste Management, *Report to Congress on Reassessment of the Civilian Radioactive Waste Management Program.* Washington, D.C.: GPO, November 29.

———. 1990. Office of Civilian Radioactive Waste Management, "Institutional Strategy Working Group Meeting, Summary Meeting Minutes." Washington, D.C.: GPO, January 4.

———. 1991. *National Energy Strategy: Powerful Ideas for America.* Washington, D.C.: GPO.

U.S. General Accounting Office. 1992. *Nuclear Waste: DOE's Repository Site Investigations, a Long and Difficult Task.* GAO/RCED-92-73. Washington, D.C.: GPO, May.

U.S. Office of Technology Assessment. 1984. *Nuclear Power in an Age of Uncertainty.* Washington, D.C.: GPO.

———. 1985. *Managing the Nation's Commercial High-Level Radioactive Waste.* Washington, D.C.: GPO.

Wald, Matthew L. 1989. "U.S. Will Start Over on Planning for Nevada Nuclear Waste Dump." *New York Times,* November 29, 1, 14.

Walker, Charles A. 1986. "Review of DOE Mission Plan for the Civilian Radioactive Waste Management Program." *American Scientist* 74 (January–February): 79–80.

Walker, Charles A., Leroy C. Gould, and Edward J. Woodhouse, eds. 1983. *Too Hot to Handle? Social and Policy Issues in the Management of Radioactive Wastes.* New Haven: Yale University Press.

Watkins, James D. 1989. "Remarks by James D. Watkins, Secretary of Energy, June 27." Washington, D.C.: GPO.

Whipple, Chris. 1989. "Reinventing Nuclear Waste Management: Why 'Getting It Right the First Time' Won't Work." Paper presented at Waste Management '89, Tucson, Arizona.

2 The Historical Development of Public Reactions to Nuclear Power: Implications for Nuclear Waste Policy

Eugene A. Rosa and William R. Freudenburg

Historical Background

Introduction

High-level nuclear wastes appear at the back end of the nuclear fuel cycle—after fissionable materials have been mined, enriched, processed, fabricated, and converted into heat to generate electricity. As a public policy issue as well, nuclear wastes appeared at the back end of a cycle: a cycle of shifting public concerns over nuclear technology (Rosa and Freudenburg, 1984). Other key issues—the local siting of nuclear facilities, reactor safety and risk of catastrophic accidents, and weapons proliferation—preceded public concern over radioactive wastes and, consequently, established the broader climate for radioactive waste policy.

To ask about how the public views nuclear waste, a central theme of this volume, is thus to imply a set of antecedent questions: What is the public climate for nuclear power in the United States more generally? What has been the role of public opinion toward nuclear power historically? What are the overall sentiments of the American public toward nuclear technology? The principal aim of this chapter is to address these questions with historical analyses, drawing upon two main sources of evidence: government and other historical documents, on the one hand, and trends in public opinion about key aspects of nuclear acceptability, on the other hand. As our analyses will show, the climate of public acceptance for nuclear power is far short of propitious—indeed, it has become distinctly foreboding. As a consequence, implementation of a publicly acceptable radioactive waste management program, especially where siting decisions are involved, is in for stormy times.

Secrecy and Techno-optimism

That waste management and disposal would be the last controversial issue to appear on the public's nuclear agenda can be traced to two principal historical sources: secrecy (Hertsgaard, 1983; Zinberg, 1984) and a steadfast preoccupation with the technology's commercial development. Though the two emerged in different epochs, the first set the stage for and deeply influenced the second.

Nuclear wastes began to accumulate with the production of bomb-grade material in the 1940s. The application of nuclear science to the development of military weapons was shrouded in secrecy—total blackout secrecy.[1] Given the wartime conditions and the urgency to develop the means to end the war, the importance of secrecy seemed reasonable—and acceptable— to Americans.[2] Secrecy persisted after Japan surrendered, however, for the Cold War was on, and the United States and Russia were engaged in a break-neck arms race. A major provision of the Atomic Energy Act of 1946 justified atomic secrets in the name of national security. With its reauthorization in 1954, the act empowered the Atomic Energy Commission (AEC) and its successors, the Energy Research and Development Administration (ERDA) and the U.S. Department of Energy (DOE), to continue operating under a canopy of secrecy. This canopy continued to shield the nuclear *subgovernment* (a closely knit group consisting of the nuclear industry, federal agencies overseeing the technology, and key congressional committees) until decades later, when the activities of public interests groups and changed legislation forced greater public disclosure. Ironically, it was the issue of secrecy itself that attracted the first public complaints about nuclear power and the establishment promoting it (Weart, 1988).[3] Thus, nuclear wastes, perhaps even more than other nuclear management issues, were long hidden from the public view.[4]

The problem of nuclear wastes raised little concern within the nuclear subgovernment itself. During the war, of course, the preoccupation was with creating weapons for bringing about its end—not with a bit of atomic garbage left over. Furthermore, the volume of wastes was small and remained so even after the war—at least for a while.

Even if someone had asked, What's to be done with the wastes? the entire nuclear subgovernment had the ready answer, No Problem! After all, hadn't a group of American scientists and engineers just accomplished the most remarkable scientific and engineering feat in the history of the human race? Could anyone doubt their ability to get rid of some piles of nuclear garbage? Apparently not, for no one did. Pervading this atmosphere of technological euphoria was the absolute certainty that the technologies to dispose of wastes would be developed when needed. Not until 1954, with the re-authorization of the Atomic Energy Act, was there any effort whatsoever to examine waste disposal. It took nearly two more decades to attract serious government attention to waste management. This unmistakably deep "techno-optimism," inherited from the successful building of bombs, became the dominant ethos of the nuclear establishment. It also became the guiding perspective in the commercial development of nuclear power that followed.

Commercialization as Idée Fixe

In the eyes of many nuclear power proponents, the majority of current problems of the nuclear industry can be attributed to "overregulation" and to an excessive sensitivity of government to public concerns and fears (Szalay, 1984; Koch, 1985). At least in the past, however, the tendency appears to have been for the public to exert relatively little influence over the industry or its regulators. During the early years of the effort to develop nuclear power, for example, the major public-vs.-private debates over nuclear power focused, not on the role of the broader public in the enterprise, but on the question of whether commercial nuclear power ought to be run strictly by the government (i.e., "publicly") or by private industry. The backers of private enterprise won that debate. In other respects as well, public health and safety concerns received less attention at the time than did the difficulties of establishing a viable commercial enterprise. Public interest, insofar as it was considered at all, was singularly defined as providing consumers with limitless supplies of cheap electricity—which nuclear power, once a commercial success, would gladly provide.

The Atomic Energy Act of 1954, while dealing at length with commercial matters such as patent rights, ownership of fissionable materials, and "free competition in private enterprise" (see Pub. L. 83–703, 42 U.S.C. 2011 et seq.), said little more about regulating nuclear power plants than to note broadly that the "health and safety of the public" should be protected. The provisions for licensing plants were copied "almost word for word from the Federal Communications Act of 1934, which had established procedures for the federal licensing of radio stations" (Ford, 1982:44).

An instructive contrast is provided by the issue of liability for accidents—the focus of the Price-Anderson Act of 1957. In March of that year, the AEC completed a "worst-case" analysis of conceivable nuclear power plant accidents: the analysis concluded 3400 people could die, another 43,000 could be injured, and $7 billion in property damage could result, from such a worst-case accident (AEC, 1957). While this estimate was not in any sense a prediction of likely outcomes, it showed that certain chains of events, while unlikely, could prove to be prohibitively costly to the fledgling industry. It would have been beyond the financial capability of the industry to run the risk of accidents, or even to obtain insurance against them, while keeping costs within a competitive range. Later that year, Congress responded by establishing a $500 million fund to supplement the $60 million in private insurance then available. Up to this $560 million maximum would be apportioned among nuclear accident victims. The only provision for damages beyond that amount was that victims were prevented from recovering damages from parties responsible for building and operating the plant—even though the $560 million limitation represented less than 10 percent of the property damage estimate (alone) of a worst-case accident.

Supporters of this legislation argue that it has proved justified in retrospect: for almost thirty years after the Price-Anderson Act was passed, not even the accident at Three Mile Island (TMI) in Pennsylvania—which cost nearly $50 million, all from private insurance—exceeded the act's limitations for damages experienced outside the plant (U.S. Senate, 1988). Critics nevertheless found the act objectionable in principle, arguing that it "effectively repealed every citizen's common-law right to sue for damages caused by someone else's negligence" (Ford, 1982:45). A major amendment to the law in 1988 (see Pub. L. 100–408, 42 U.S.C. 2210 et seq.), while still setting limits on the maximum liability levels, now provides for substantially more compensation, a maximum of $7.2 billion in damages to victims of nuclear incidents. Whatever one's position over the appropriateness or the adequacy of even the revised Price-Anderson limitations, the early legislation scarcely can be seen as "overregulating" the industry. Instead, it was written to "regulate" the public's ability to intervene in the industry's development and to provide the industry with a great deal of protection and support.

The Atomic Energy Act of 1946 established the AEC and charged it with the dual tasks of promoting and regulating the new industry. In the eyes of critics, the agency long showed a tendency to take the first task more seriously than the second, devoting as much attention to regulating public opinion as to regulating nuclear power. Partly in response to the utopian visions for the technology being promoted by the AEC, press coverage of the industry was uniformly positive, if not enthusiastic, for the next decade and a half. An oft-noted example is a 1954 speech by the then-chairman of the AEC, Lewis Strauss, who spoke of "unlimited power," of electricity "too cheap to meter," and of an era in which famines would be remembered only as matters of history. Thanks to the atom, Strauss argued, people would "travel effortlessly over the seas and under them and through the air with a minimum of danger and at great speeds and [would] experience a life span far longer than ours." Atomic power, in his view, promised "an age of peace" (Strauss, 1954:9).[5]

The public relations efforts of the AEC were not restricted to enthusiastic speeches: a later press release noted that more than 40 million people had attended screenings of AEC films during the 1960s, and another 158 million watched the films on television (AEC, 1970:5). Nor were the public relations efforts confined to the United States. The AEC developed a set of exhibits demonstrating the "wondrous" benefits of atomic energy that played "to huge crowds in Karachi, Tokyo, Cairo, Sao Paolo, Teheran, and many other capitals" (Weart, 1988:163).

Intense efforts to promote nuclear power might have been justified in the early years, at least in view of the subgovernment's desire to convince the citizenry that the most destructive technology known to history could indeed be converted to peaceful uses. But it continued at a fervent pace for decades. A General Accounting Office report even noted that, in the keen commitment to nuclear commercialization,

> Federal agencies have gone to great lengths to deny the uncertainty inherent in nuclear technology. Comic books and glossy publications have extolled the virtues of nuclear power, and the Energy Research and Development Administration went so far as to distribute 78,000 copies of a promotional pamphlet apparently attempting to turn voters against a 1976 California referendum proposal on nuclear power. According to the General Accounting Office, a nonpartisan Congressional agency, the ERDA pamphlet was inaccurate "propaganda" that "misleads the reader into believing that technical, environmental, and social problems of storing high and low level wastes have been solved" (quoted in Woodhouse, 1983:166).

Unmistakable from the historical record was the nuclear subgovernment's unfaltering efforts to make nuclear power a commercial success.

In contrast, safety regulations for the nuclear power plants were left almost entirely to the discretion of private industry. In releasing its early "three basic regulations" for the industry, the AEC emphasized its intent was to "open the way to all who are interested" in nuclear power (AEC, 1955:1). In the words of the critical, but scrupulously documented, history written by Daniel Ford: "The AEC simply asked the companies . . . to work, with good sense and good faith, toward a common goal, which it vaguely specified as 'reasonable assurance that the health and safety of the public will not be endangered' . . . [The AEC] staff made it clear to the industry that safety was the industry's responsibility" (Ford, 1982:52–54).

Even after the mid-1960s, when utilities finally started placing commercial orders for nuclear reactors, this approach did not change. Under AEC rules, the AEC staff and the utility worked out their differences beforehand; the public, by contrast, was not even permitted access to the agency's data about potential safety problems of the plants. Despite these conditions, the hearings on proposed reactors highlighted a number of safety issues that the AEC found itself unable to resolve, and again, the AEC's response scarcely indicated "excessive sensitivity" to public concern. Rather than holding up the issuance of permits until the questions could be answered, the AEC decided that if a question covered several plants, it no longer needed to be decided in an individual licensing case. Instead, it would be treated as a "generic" safety issue, the resolution for which would be sought through the ongoing research of the AEC and the industry. In the meantime, the plant could be built and operated.

The AEC effectively treated the safety issue as irrelevant to licensing hearings. The net result was that, even in cases where members of the general public managed to overcome the lack of access to AEC records and to identify issues the AEC staff found legitimate, attention to these issues was anything but assured: the larger the number of plants affected, the lower the likelihood that the issue would actually be dealt with in licensing cases. (Many "generic" safety issues remain unresolved to this day.) A 1973 evaluation of the AEC licensing process funded by the National Science Foundation concluded, "The whole process as it now stands is nothing more than a charade, the outcome of which is, for all intents and purposes, predetermined" (Ebbin and Kasper, 1974:264).

Regulation

By the early 1970s, however, it was becoming clear that the AEC policy of industry self-regulation was not working. The AEC may have been correct in its assumption that the industry itself often stood to lose as much as the general public from serious accidents—a point that was later driven home by the accident at TMI, which, though it released only minor amounts of radiation to the surrounding environment, left the utility with an expensive problem of radioactive waste disposal where a billion-dollar capital investment had stood only a few days earlier. The utilities, however, simply were not accustomed to the intensity of management required, and they sometimes showed indifferent compliance even with the relatively minimal AEC procedures of the time. A special management review in 1975 found that approximately half of all plants violated explicit safety conditions of their federal operating licenses; none had comprehensive written instructions, none had reactor safety committees that were doing the work expected of them, and none had fully implemented the quality-assurance program seen by the AEC as the key to safe operation (Ford, 1982:205).

These problems, in combination with a number of other developments that took place during the early-to-middle 1970s, led to a substantial change in the federal approach to the regulation of nuclear power during that decade. The National Environmental Policy Act of 1969 gave the public greater ability to influence regulatory decisions, an interpretation explicitly supported by federal courts. The nation's increasing environmental awareness and post-Watergate distrust of the government made it more difficult for the AEC to avoid public scrutiny (Dunlap, 1992; Lipset and Schneider, 1983). The AEC's hearings on a key safety system in nuclear power plants, the emergency core cooling system, revealed what critics regarded to be evidence of agency bias and efforts to cover up unfavorable evidence.

In early 1975, the AEC was abolished and its regulatory functions were transferred to the newly created Nuclear Regulatory Commission (NRC). Its development and promotional functions were taken over by the newly formed ERDA, which was subsequently absorbed by the DOE. Administratively separating the AEC's regulatory functions from its promotional functions was an attempt at resolving the image problem the AEC had developed. The AEC's staff and existing regulations were transferred intact to the NRC. Within two months, however, a serious fire owing to a safety failure at the Brown's Ferry nuclear power plant in Alabama put increased pressure on the NRC to toughen its regulations, and the widely publicized resignations of several key officials further increased the new agency's difficulties.

The net result of these and other events was to accelerate further a trend that had emerged in the mid to late 1960s, a trend toward greater public regulation of what had once been a more private undertaking. As early as 1967, while regulatory power was still under the old AEC, there were clear signs that a regulatory transformation was already underway. Figure 2-1 presents two indicators of change in the regulatory climate: the number of regulations and amendments enacted each year and the annual number of reactor inspections conducted by the AEC or the NRC.

Between 1957, the year of the first available data, and 1967, a meager number of regulations or amendments, between 4 and 16, were enacted each year: an average of 11 per year. Then, in 1968, began a steady, though erratic, increase in regulatory activity, which reached a peak of 61 in 1978 and a second peak of 51 in 1982. Over this entire period the number of regulations had increased to an average of 18 per year. In the years following the first peak in 1978, the annual average was even higher: 37 per year. Two of the years of available data experienced a significant decline in the number of regulations, 1975 and 1979. The first decline in 1975 coincides with the

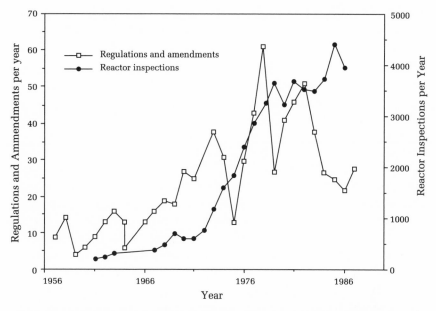

Figure 2-1   Annual number of regulations and amendments and annual number of nuclear reactor inspections. Source: Baumgartner and Jones (1991) and other data supplied by them.

transference of regulatory responsibilities from the AEC to the NRC. As for the second decline, in 1979, we do not fully understand the reasons producing it, but it may have been associated with the TMI accident in March of that year. The accident prompted a careful reassessment of regulatory activities that may have postponed the passage of further regulation until after the completion of the reassessment.

As for the other indicator of the regulatory climate, the annual number of reactor inspections conducted by AEC or NRC staff, the picture is much the same. In the early 1970s, shortly after the rise in annual regulations began, the volume of reactor inspections skyrocketed—as shown in Figure 2-1. Inspections for available years up to 1973 averaged fewer than 500 per year, but soon afterward shot upward to a peak of over 4000 in 1985. As pointed out by Baumgartner and Jones: "By almost any measure available, the regulatory environment had gone through a major transformation. . . . (1991:1059).

According to proponents of the industry, the growth in regulations was generated mainly by "public concern and the political pressure it generates" (Cohen, 1983:225). It is, of course, possible to debate the degree to which other factors were also involved, but it is clear that by the mid 1970s—even before the time of the 1979 accident at TMI—the situation had changed to the point where public concerns were beginning to play a much more important role in the fate of the nuclear power industry. What was the genesis of those concerns and how might they affect nuclear waste policy?

The Role of Public Opinion

Public Acceptability
The acceptability of nuclear power to Americans will affect the volume and management of nuclear wastes both directly and indirectly. The direct links include the fact that the nuclear industry in the United States is currently at a standstill: not a single reactor has been ordered since 1978. This is due, at least in part, to public disenchantment with the technology (Freudenburg and Rosa, 1984; U.S. Office of Technology Assessment [OTA], 1984:chap. 8). Were there to be a sharp turnaround in public sentiment—say, because of concern over global climate change due to the burning of fossil fuels or concern over the geopolitics associated with imported oil—waste management would be directly influenced. A sizable proportion of the volume of radioactive wastes is produced by the nuclear generation of electricity, and this accounts for much of the substantial current inventory. With a renewed commitment to nuclear-generated electricity, the volume of wastes would inevitably increase, perhaps dramatically.

As for the indirect effects, the connections are less obvious. It is clear that the congruence between public attitudes and public policies will often be far less than perfect. Nevertheless, the evidence reviewed in chapter 1 by Kraft, Rosa, and Dunlap indicates that public attitudes can have a greater influence on public outcomes than is sometimes recognized and that policies appreciably out-of-step with the public mood are difficult, sometimes impossible, to implement. Furthermore, because of a growing recognition of the need for public support, those organizations and institutions most directly involved with nuclear waste have become increasingly accessible to public influence. Moreover, the 1982 Nuclear Waste Policy Act itself requires greater public involvement. More broadly, the impacts of that involvement on nuclear waste management policies will be felt by the two agencies directly affected by the act: the DOE and NRC. Public opinion will also exert an influence in several other institutional arenas—Congress, state utility commissions, the investment community, and public interest groups.

U.S. Department of Energy. The DOE, created in 1977, has the principal governmental responsibility for the development of all energy sources, including nuclear power, and for managing nuclear wastes. DOE is charged with developing the technology for disposing of high-level nuclear waste, for locating suitable repository sites, and for devising environmentally safe repository management programs. It is required by law to issue environmental impact statements that lay out expected effects of its major actions, whether in siting waste repositories or in developing other nuclear facilities. These assessments require the DOE to hold public hearings that allow for the airing of views by concerned citizens or groups. Furthermore, as pointed out by Kraft, Rosa, and Dunlap in chapter 1 of this volume, the 1982 Nuclear Waste Policy Act mandated that public hearings be held in the vicinity of each candidate repository site and provided for an elaborate program of oversight and cooperation with state governments. It appears unlikely that any future federal nuclear waste legislation would depart significantly from the 1982 act, at least in providing an opportunity for public participation. In order to assess and implement facility sitings, the DOE will need to establish its presence in local communities. But it is not clear whether the DOE will be able to develop sufficient openness and public involvement to win public confidence. Even well-intentioned DOE public involvement efforts could in fact backfire, reinforcing already high levels of government mistrust, heightening existing concerns over nuclear technology, and seriously obstructing siting and other waste management decisions. It is clear also that the time horizon for establishing a waste repository is sufficiently long, and the list of uncertainties to be encountered is sufficiently lengthy, to present many opportunities for public objections to be raised.

   The Nuclear Regulatory Commission. Whereas DOE is responsible for the building and operation of nuclear facilities, including nuclear waste facilities, the NRC is responsible for licensing them. Through its licensing powers the NRC is the main oversight body of civilian nuclear power, entrusted with ensuring public health and safety more generally than DOE. As previously noted, nuclear power proponents have argued that the NRC tends to be "excessively" sensitive to public opinion in discharging its responsibility. While the five NRC commissioners are appointed rather than elected, they are obligated by law at least to consider public viewpoints.

   For thirty years, regulations required the NRC (and the AEC before it) to follow a two-step procedure in licensing nuclear plants: a construction permit first had to be approved before a plant could be built, and a second permit had to be issued before the plant could begin operations. Since each step in the procedure mandated public hearings, the NRC was required to obtain public input on at least two occasions. In 1989, after over a decade of effort, the NRC was successful in streamlining this licensing procedure, combining power plant permits for construction and operation. This reduced to one the number of public hearings required. This NRC decision was viewed by opponents of nuclear power as the last in a growing number of examples illustrating NRC's inattention to the public welfare. One response was near-immediate. A public interest group, the Nuclear Information and Resource Service, brought suit in federal court, arguing that the NRC did not have the authority under the Atomic Energy Act to change the licensing process. In 1990, the streamlined, one-step procedure was struck down by the United States Court of Appeals for the District of Columbia Circuit. The three judge panel ruled unanimously that, under the Atomic Energy Act, the NRC must hold public hearings before reactor operation "to consider significant new information that comes to light after initial licensing." Thus, while upholding much of the one-step rule, the court threw out the major contested portion—the portion eliminating public involvement in operations licensing (Wald, 1990:7).

Despite the recent court ruling, the final word on the streamlining of licensing has yet to be uttered. The ruling is under appeal. Beyond this immediate response, it is important to recognize that streamlining, along with other regulatory reform, has long been the mainstay of industry demands to the NRC and other key federal agencies. And recent presidential administrations, sympathetic to industry claims, have demonstrated a steady commitment to regulatory reform. The National Energy Strategy (DOE, 1991) issued by the Bush administration reaffirmed this commitment with its call for a reinstatement of the 1989 NRC one-step licensing procedure, along with a call for a twenty-year extension of operating licenses.

The key appeal of a streamlined licensing procedure to the nuclear industry and its investors is obvious: there would no longer be opportunity for lengthy and costly delays, due to public hearings, after a plant is built. This would, of course, reduce the chances of investing in a nuclear plant that, once built, was not approved for operations, as was the case with the Shoreham plant on Long Island in New York State. Far less obvious is the claim by the industry and other nuclear proponents that a streamlined licensing procedure enhances, rather than limits, public input. To them, the two-step process was "very fragmented" and therefore an impediment to organized and focused public input. In contrast, the one-step procedure is in the public interest, according to the general counsel for Nuclear Management and Resources Council, an industry group, because "[i]t is better for everybody for a decision to be made up front, rather than after a significant amount of time, money, and resources have been spent" (Robert Bishop, as quoted in Drumheller, 1990:B3).

Whether the DOE and NRC are inattentive to public concerns, as opponents of nuclear power argue, or whether they are overly protective of the public, as proponents argue, will doubtless be the basis of continuing debate. One major focus of that debate will be the licensing process. And the general issue of debate will be, not so much one of whether public concerns need to be taken into account by the DOE and NRC, but the depth of public influence. The depth of public influence is less of an issue when we move from the two federal agencies to the other institutional arenas where public influence is significant.

Congress. Congress is the third institution. While the past two presidential administrations proposed the streamlining of many regulations for the licensing and operation of nuclear power plants and while these proposals are quite similar to the recommendations made by the commercial nuclear power industry over the past several years, many changes could not have been implemented without direct action by Congress (Szalay, 1984). Congress, meanwhile, consists of elected representatives whose views and actions tend to be reasonably responsive to the public will. Recent evidence, for example, shows that attitudes of the professional staffs of Congress and the executive branch toward technological risk are remarkably similar to public attitudes and quite different from industry attitudes (Dietz and Rycroft, 1989). If further legislation is a prerequisite for the renewed health of the industry—or if further safety or managerial problems prompt a demand for frameworks that are more restrictive, rather than less so— then congressional sensitivity to public sentiment will help determine what outcomes emerge.

State Utility Commissions. While federal legislation gives the NRC ultimate authority over nuclear safety issues, a 1983 U.S. Supreme Court decision upheld the authority of California (and presumably other states) to regulate the economic aspects of utility investments in nuclear power plants. In practice, nuclear power has proven to be far more expensive than earlier estimates, and economic concerns may have been heightened by the accident at TMI Unit 2 (DOE, 1983; Cook, 1985). (The accident at TMI, which forced that reactor to shut down prematurely, effectively meant that electricity produced by this reactor during its short operating lifetime was some of the most expensive in history.) In addition, state utility commissioners tend to be elected to their offices, and thus negative public opinion is likely to provide an incentive for utility commissioners to give particularly tough scrutiny to plans for the expansion of nuclear-generating capacity. In fact, it may have been concern over the sensitivity of utility commissioners to public opinion that led the U.S. Justice Department to argue in its unsuccessful brief against the Supreme Court decision that "the result [of the decision] would be the virtual elimination of nuclear power as a potential energy source" (Wermiel, 1982:13).

Investors. Investors are a less obvious but nonetheless significant group that bears the influence of public opinion. Nuclear power plants are major investments that require massive commitments of capital. The $2.25 billion default of the Washington Public Power Supply System on a pair of unfinished nuclear power plants was the largest default in municipal bond history, and many utilities constructing nuclear power facilities have come perilously close to default. These considerations have attracted understandable attention in the investment community. Nuclear power is now widely seen as a risky investment; to compensate for the risks involved, a Merrill Lynch proposal to buy a portion of the financially troubled Seabrook Unit 1 asked for a 40-percent annual return on investment, far in excess of the normal 10 to 17 percent range (Wessel, 1985). Particularly given the long lead time (more than a decade) that is required to build nuclear power plants, either local or national opposition to the construction of a facility can add markedly to the financial risk involved. Negative attitudes may increase the likelihood that facilities will experience costly regulatory delays, be held up in court, or perhaps even be prevented from operating (Faltmayer, 1979). Such increases in risks could scarcely fail to be noted by investors who have the option of finding safer places to invest their funds.

Public Interest Groups, Public Protest Groups, and Other Organized Constituency Groups. While not an institution in the same clearly defined way as the other five sets of social actors examined, partisan groups are nevertheless crucially important to the direction of public policy. Through lobbying efforts, intervention in licensing and regulatory procedures, legal actions, and, where such efforts fail, civil protest, these groups exercise considerable influence over policy outcomes (OTA, 1984:214–15). At the very least, such groups look to public opinion as a source of support and legitimacy. Their legitimacy is strengthened by the extent to which they can demonstrate that the views they propound have widespread public support. The wider the support, the greater their ability to claim they are speaking for the general public, and not for special interests with narrow, self-serving objectives (Dunlap, 1992:89–90).

Public concerns with nuclear power are dominated by questions of safety and its inverse, risk (OTA, 1984:chap. 8). Many observers have argued that these concerns rest on a deeper foundation of fear and dread. Insofar as this argument is valid, it could affect, directly and markedly, the success of organized protest groups to attract supporters. For example, addressing himself to the question of technological risks more generally, Winner (1986:140–41) points out that

> [t]he fundamental issue here is fear—fear of injury, disease, death, and the prospect of having to live in deteriorating surroundings. For that reason arguments about impending dangers are often useful in attempts to unite people who have little in common other than shared fears. Con-

temporary environmentalism and consumerism have taken advantage of this opportunity to marshal support for their causes . . . The attempt to parlay the principle *salus populi suprema lex* (Latin, meaning "Let the welfare of the people be the supreme law") into a full-blown political movement has become a familiar approach among contemporary activists. By calling attention to a possible danger, one hopes to attract support for a broader program of social criticism and reform.

Thus, when partisan positions are in step with the public mood, as revealed by public opinion surveys, the potential for attracting broad-based support is enhanced appreciably.

The Evolution of Public Opinion

Public Opinion on Nuclear Waste
As noted above, nuclear waste disposal emerged as one of the last public concerns associated with fission technology. The public, it is clear, remained insulated from the problem of radioactive waste management for well over a decade after the first commercial reactor was made operational. Precisely when the problem reached the public's consciousness, however, is open to some interpretation. Rankin and Nealey (1978), Nealey and Hebert (1983), and the National Research Council (1984) emphasize an early public awareness—as early as November 1974. Their conclusion is based chiefly upon a national survey by Opinion Research Corporation that asked: "Which of these, if any, do you consider to be serious problems associated with nuclear power plants—thermal pollution, radiation discharge, nuclear accident, or disposal of radioactive wastes?" Over half (52 percent) named radioactive wastes, while the seriousness of none of the other options exceeded 20 percent. In an identical follow-up survey three months later, Opinion Research Corporation found virtually identical results. Both surveys were conducted after national media attention had focused on 115,000 gallons of radioactive wastes that had leaked from the Hanford reservation in 1973.

As compelling as this evidence appears, other survey data preceding and following these surveys paint a somewhat different picture. In particular, both of the major polling firms that have most consistently tracked public opinion on nuclear energy, Louis Harris and Associates and Cambridge Reports, have used open-ended questions that ask the opinionated public why it favors or opposes the technology. Throughout the 1970s, of those expressing opposition, only between 3 and 16 (an average of less than 8 percent over a dozen surveys) volunteered any mention of nuclear waste as the reason for their opposition to the technology. Most frequently mentioned were concerns that the technology was too dangerous or, alternatively, not safe enough, especially with respect to reactors. Thus, this evidence suggests that at least throughout the 1970s nuclear wastes were far from salient in the public mind.

Since it is well known (Bradburn and Sudman, 1982) that a question form (open or closed) can greatly affect the responses people make, the above discrepancy is not entirely unexpected. A recent set of experiments (Schuman and Scott, 1987) have raised considerable doubts about the validity of closed-ended survey responses compared to open-ended answers. The recent findings support Bradburn and Sudman's (1982) earlier observation that closed-form questions on complex issues, because they often obviate the need for respondents to search their memories and organize their thoughts, frequently produce superficial responses.

Thus, the key validity issue pertinent to the question of *when* nuclear wastes became a public concern rests on what one means by *concern*. Public concerns over nuclear waste may reflect any one of three principal cognitions: awareness, salience, or name recognition. If by *concern* we mean the public's ability to select wastes from a list of important problems associated with nuclear power, then the public was concerned as early as 1974. However, it is possible that these were transitory concerns, the residual effects of the media coverage of the Hanford leak. If by *concern* we mean that the issue was salient—at the forefront of the public mind—then concerns over wastes appeared much later—not until the late 1970s, perhaps early 1980s. Finally, if by *concern* we mean the public had become aware of nuclear waste problems and this awareness was a durable feature of their concern, partially determining their overall attitudes, then, in our view, the weight of evidence favors the later date, too. The history of the period bears this out.

The early 1970s—the period of contention—were overwhelmingly dominated by concerns over reactor safety. In 1971, the Union of Concerned Scientists, an expert watchdog group of M.I.T. scientists, began raising serious doubts about the emergency core cooling system. Congressional hearings were held in 1973–74 on reactor safety. Also in 1974, Ralph Nader convened a national conference devoted to nuclear safety, following it up in 1975 with his Critical Mass, a coalition of hundreds of local and national groups. Growing concerns over reactor safety during the period prompted the AEC to commission a comprehensive study of the risks of a catastrophic nuclear accident (NRC, 1975). Directed by M.I.T. professor Norman Rasmussen, the study led to a final report—the *Reactor Safety Study*, WASH-1400—which was released in 1975 amidst massive media coverage. The dominant issue of the time, this evidence makes clear, was reactor safety, not nuclear wastes. The public was apparently in tune with the dominating events for, as already noted, when asked to name its most serious concerns those most often mentioned had to do with reactor safety. Insofar as the public was concerned about nuclear wastes, the concern was episodic.

Trends in Public Opinion

Attitudes

For nearly two decades, nuclear power has been embroiled in controversy. For almost as long, the controversy has been the focus of attention by social scientists and, as a result, our understanding of the topic, while far from perfect, is considerable. Numerous surveys, including time-series studies, have assessed a wide variety of public attitudes (Nealey, Melber, and Rankin, 1983; Mazur, 1981; Freudenburg and Rosa, 1984; Nealey 1990). Trends in attitudes are, therefore, an effective means for examining the crystallization of public mood and for assessing the broad public climate for nuclear power.[6]

Three types of attitudinal trend data have particular relevance to nuclear energy policy. The first category, the broadest, includes attitudes of the national public toward the general idea of nuclear power—the public's overall assessment of the technology as an energy source. The second category includes surveys that ask the national public about "nearby" facilities. The third category also has to do with nearby facilities but is limited to surveys of persons in actual "host communities"—the localities in or near which nuclear power facilities are under construction or have recently been completed.

National Public Attitudes Toward Nuclear Power Plants "In General"

Data in this first category are the type most frequently cited in policy documents (e.g., OTA, 1984:chap. 8), and this is also the category for which the longest and most consistent time-series data are available. Two national polling organizations have played a particularly important role in documenting national attitudes toward the general idea of nuclear power plant construction—ABC News-Louis Harris and Associates and Cambridge Reports. The time-series data from the two organizations are presented as Figures 2-2 and 2-3. Both figures show quite similar patterns over the period for which comparable data are available: supporters outnumbered opponents by margins averaging 20 to 30 percentage points in all twenty-five separate surveys taken before the accident at Three Mile Island. It is clear that nuclear power's positive position was disrupted by TMI, with the number of opponents increasing significantly, as indicated by the spike in the trend data shortly after the accident. At the same time there was a corresponding decline in support and a smaller decline in "undecided" respondents.

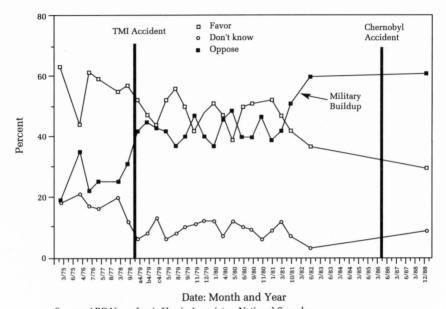

Date: Month and Year

Source: ABC News-Louis Harris Associates; National Samples

Figure 2-2   Public attitudes toward nuclear power in the United States. Item 1: In general, do you favor or oppose the building of more nuclear power plants in the United States?

The net effect of TMI on public opinion was a shift from an almost two-to-one margin of support before the accident to a roughly even split between supporters and opponents of the technology immediately thereafter, although supporters still outnumbered opponents in most polls. Many close observers of these trends (including the two of us) thought it was quite possible for public support to rebound somewhat, for nuclear power to recover at least part of the lost ground as TMI receded from public consciousness. Others argued that because the accident was such a dramatic signal of the dangers of nuclear power, opposition would not dissipate in the near future.

In the years immediately following TMI, the data appeared to favor the first expectation. Figures 2-2 and 2-3 show that support and opposition see-sawed up until late 1981, with support frequently exceeding opposition and inching back toward pre-TMI levels. Then, toward the end of 1981, opposition again surged upward, increasing by 9 percentage points between 1981 and 1982 in the Harris polls, as shown in Figure 2-2. Figure 2-3 illustrates the abrupt change even more clearly: according to the more fine-grained Cambridge polls, opposition increased by 12 percentage points between the fourth quarter of 1981 and the first quarter of 1982. Mazur (1990) has presented convincing evidence that this rise in opposition was due to public concerns over the buildup of American military strength, including nuclear arms, during this time. Ronald Reagan had just assumed office, and he began his foreign policy with a bellicose posture toward the Soviet Union. This posture, accompanied by the arms buildup, evoked immediate protests, first throughout Europe and then in the United States. During 1981 and 1982, for example, two groups of activists dedicated to halting the international nuclear arms race emerged in the United States: Ground Zero and the Nuclear Weapons Freeze Campaign.

The longer-term effects on opinions due to the military buildup can best be seen with the Cambridge polls in Figure 2-3, since they were repeated regularly throughout the 1980s. Rather than showing any tendency to rebound toward earlier levels of favorability, public attitudes toward nuclear

Figure 2-3 Public attitudes toward nuclear power in the United States. Item 2: Do you favor or oppose the construction of more nuclear power plants? *

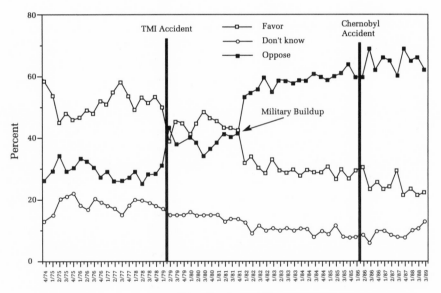

Date: Quarter and Year

*Sometime after 1983, the wording was changed to: "Please tell me if you generally favor or oppose a proposal to build more nulear power plants as a way of dealing with the energy crisis." Our considerable efforts failed to determine why or precisely when this wording change was introduced. Nor do we have a foolproof way of determining the effects of the wording change on responses. Nevertheless, we can make some reasonable assumptions about the direction of possible response effects. Since with the revised wording the idea of "crisis" is introduced, it seems highly unlikely that this would increase opposition to nuclear power. Rather, since crisis implies the need for immediate action, any wording effect is more likely to be toward inflating support. Source: Cambridge Reports; National Samples

power appear to have become durably negative. Since early 1982, opposition has consistently exceeded support by margins of as much as three to one—almost the complete inverse of pre-TMI attitudes.

A number of observers (e.g., Schneider, 1986; Morone and Woodhouse, 1989) have argued that the Chernobyl accident in the Soviet Union in April 1986 had a marked influence on nuclear power attitudes in the United States. Looking at Figure 2-3, the more reliable of our two time-series because of its consistency and frequency, we are left with a different conclusion. It appears that, while Chernobyl may have increased opposition slightly, opposition already exceeded support by a margin of two to one. Moreover, nearly that level of opposition preceded the Chernobyl accident for nearly four years, so that by the time of the accident it scarcely would have been possible for public attitudes to have become markedly more negative.

Another approach to measuring public sentiments toward nuclear technology is to compare nuclear attitudes with attitudes toward other energy technologies. This approach also fails to show significant support for nuclear power. In fact, nuclear power tends to be the least popular energy supply option, save for importing more oil (Farhar-Pilgrim and Freudenburg, 1984; Rosa, Olsen, and Dillman, 1984). Furthermore, nuclear power is viewed as being by far the most dangerous energy source and the least acceptable for widespread use. A Roper survey in 1985 compared nuclear power to three other energy supply technologies (coal, natural gas, and oil) with a series of nine questions. When asked to name the energy source *most dangerous* to human life, 82 percent of the respondents named nuclear power, while only 4 percent named coal, 5 percent named natural gas, and 1 percent named oil. Similarly, when asked to identify the source *least acceptable* for widespread use, 59 percent chose nuclear power, compared to 20 percent choosing coal, 4 percent natural gas, and 9 percent oil (Roper, 1985).

The evidence reviewed above paints a clear and consistent picture: a majority of Americans are opposed to the further growth of nuclear power. Furthermore, the amount of current opposition so exceeds support that it is doubtful that the levels of support the technology once enjoyed can be regained—at least not for the foreseeable future. It appears as if attitudes have passed through some threshold, beyond which levels of opposition so overwhelm support that the chances for a reversal are very much in doubt.

National Public Attitudes Toward Nuclear Power Plants "Nearby"
Given that nuclear facilities have become almost a textbook illustration of locally undesirable land uses, or LULUs (Popper, 1981), many people are surprised to learn that even the idea of a nuclear power plant in a respondent's own "back yard" remained relatively uncontroversial until the mid-1970s.

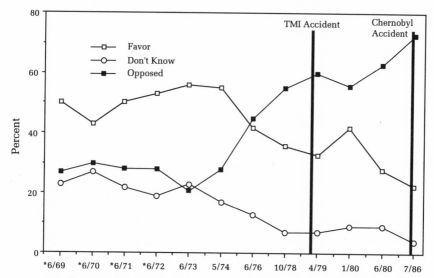

*Month survey conducted not reported in original sources. Mid-year dates were assumed.
Source: Variety of surveys summarized in Melber et al. (1977), Rankin et al. (1984) and
Nealey (1990); National Samples

Figure 2-4   Public attitudes toward construction of local nuclear power plants.

As can be seen from the data in Figure 2-4, polls in the early 1970s showed considerable support when national samples were asked about having "the local electric company . . . build a nuclear power plant in this area."[7] Indeed, in five of the polls, 55 percent responded favorably, nearly twice the level of opposition. Opposition to local nuclear facilities, however, began to increase as early as 1974. By perhaps 1976, and certainly by 1978, before the TMI accident, opposition to the idea of nearby nuclear facilities actually exceeded support for the first time. Figure 2-4 shows that TMI nevertheless had a marked effect: polls found considerably greater opposition to the idea of a nuclear power plant "in this area" after the accident.

A second approach to measuring the acceptability of a "local" power plant, a comparative one, shows the same pattern of local opposition. A Resources for the Future "national environmental survey" performed by Roper and Cantril for the President's Council on Environmental Quality in 1980 attempted to measure the "tolerance distance" for five types of facilities: an office building, a large factory, a nuclear power plant, a coal-fired power plant, and a disposal site for hazardous waste chemicals. Respondents were asked to say how close such facilities could be built to their homes "before you would want to move to another place or to actively protest." Even the disposal site for hazardous waste chemicals was slightly more popular (at a mean distance of 81.4 miles) than was the nuclear power plant, which had the greatest mean tolerance distance, 91 miles (Mitchell, 1980). In short, the same general pattern appears: from two different approaches representative national surveys of the U.S. public show considerable opposition to the idea of having a nuclear power plant constructed nearby.

Attitudes of Nuclear Host Community Residents
While national attitudes toward nuclear power—both general attitudes and those toward local plants—establish the broad context for the future of the technology, a more specific context may often prove to have greater relevance. If nuclear power plants or nuclear waste facilities are to be built anywhere, they will be built in specific ("host") communities. Even if a majority of Americans have strong objections to nuclear power, and even if a majority objects to having nuclear power facilities in their own back yards, such objections can be overlooked if the facilities are planned for someone else's back yard and if that someone else has little objection. Historically, moreover, nuclear host communities have typically been far more favorable toward nuclear power development than have cross-sections of the American public.

Recent evidence strongly suggests, however, that host-community support can no longer be taken for granted. While Melber et al. (1977) found high levels of support in almost all nuclear host communities surveyed before the TMI accident, surveys since the accident have shown a very different pattern. Freudenburg and Baxter (1984) summarized the results from thirty-six host community surveys (at fifteen individual reactor sites) that were conducted after local residents became aware of the facilities but before the facilities had completed their first six months of operation. While methodological variations across surveys are substantial, the overall pattern is quite clear, as can be seen from Figure 2-5. In none of the surveys conducted before the TMI accident was opposition found among more than 33 percent of the population; in no known surveys since TMI has there been *less than 50 percent opposition.* A dummy variable, indicating whether the survey was pre- or post-TMI, was entered into a regression analysis to determine the effects of TMI on host-community attitudes. Over 70 percent of the variation in attitudes across surveys was due to TMI—despite the substantial variations across studies in question wording, survey locations, and other factors (Freudenburg and Baxter, 1984).

Whatever hopeful signs proponents could find in the historical evidence from host communities, where support had typically been solid, has been dashed with the post-TMI data. Host communities, once supportive of nuclear facilities sited in their area, have completely flip-flopped; the attitudes

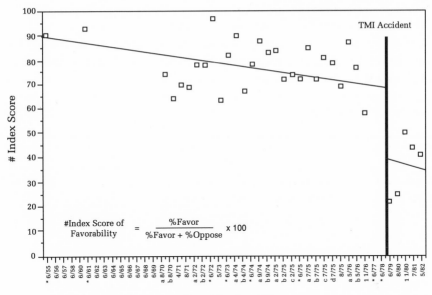

Month survey conducted not reported in original sources. Mid-year dates were assumed.
Source: Freudenberg and Baxter, 1984

Figure 2-5   Host community attitudes toward nuclear power plants before and after Three Mile Island (TMI).

of their residents, similar to the attitudes of the general public toward nuclear power, have become clearly unsupportive.

As the evidence reviewed shows, the American public shows deep reluctance to have a nuclear plant constructed locally (also see Erikson, 1990). While nuclear waste has not been the focus of as many surveys as has nuclear power, the available evidence supports the conclusion that it may prove to be even more difficult to find acceptance for a nuclear repository. An early Battelle Human Affairs Research Centers study (Lindell et al., 1978), surveying political activists rather than the general public, asked respondents to assess tolerance distances for eight industrial facilities—including a nuclear power plant and nuclear waste repository. The Battelle respondents were apparently somewhat more tolerant of nuclear plants than the post-TMI national sample described above; while the mean tolerance level for the national sample was 91 miles, just over 40 percent of the Battelle's respondents were willing to live within 10 miles of a nuclear plant. But the activists showed considerably less tolerance toward nuclear waste facilities. A nuclear waste repository was the least tolerable of the eight facilities assessed, with only 29 percent of the respondents saying they would be willing to live within 10 miles of a repository, and with 32 percent saying they would not be willing to live within 100 miles.

Another survey that included a direct comparison between the siting of a nuclear power plant and waste repository was conducted in the Texas Panhandle, once a potential site for the nation's first high-level nuclear waste repository (Texas Department of Agriculture, 1984; also see chapter 5 of this volume by Brody and Fleishman). Of a dozen potential industrial developments that were specifically mentioned in the survey, all but three were favored by a majority of residents. The three exceptions, all attracting overwhelming opposition, were nuclear facilities: the least favored of the three was a high-level waste repository, followed by a low-level waste repository and then by a nuclear power plant.

Discussion

Overview

From a vantage point of a half-century of hindsight on government nuclear policies and nearly two decades of data on nuclear attitudes, we are well positioned to outline the social and political climate for one of the most controversial technologies of our time—perhaps of all time. In the early days of nuclear power, owing to a pervasive secrecy inherited from a wartime mindset, public concerns were all but nonexistent. Secrecy meant nuclear activities were out of sight for Americans and, therefore, out of mind—or at least out of serious concern. Then, once commercialization began, Americans showed strong support for the technology. Well into the 1970s, sizable majorities supported the growth of nuclear power, often by margins of two to one. Since about the time of the TMI accident in 1979, however, clouds of uncertainty have welled up around nuclear power. Since at least early 1982, there have been consistently high levels of opposition to further growth in fission technology, suggesting that public attitudes have crystallized around a deeply negative evaluation. At present, nuclear power is at its lowest ebb to date. Whether discussing facilities planned for the nation generally, or for one's own or a nearby community, Americans everywhere are expressing increasingly grave misgivings about the technology.

While this conclusion seems both clear and strong, other observers have reached contradictory conclusions. Perhaps the strongest statements have continued to come from the utility industry. For example, Jerry Farrington, chairman of the board and chief executive of Texas Utilities Company, has gone so far as to reach the following conclusion:

A myth of a public opposed to nuclear energy . . . contrary to popular belief, acceptance [of nuclear power] is widespread. Americans have indicated in poll after poll that they believe that nuclear energy will be— and *should be*—important, and that nuclear power plants will have to be built in the years ahead (Farrington, 1989; emphasis in the original).

The principal basis for Farrington's claim is the results of a question asked repeatedly in national surveys from 1985 to 1989: "How important do you think nuclear energy plants will be in meeting this nation's electricity needs in the years ahead?" Sizable majorities (between 69 percent and 80 percent) have answered either "very important" or "important (U.S. Council on Energy Awareness, 1989). These results, together with other selective evidence, have been interpreted by nuclear proponents (Bisconti, 1989) to mean that Americans have not only renewed their faith in nuclear technology but also have high expectations about its future prospects.

Whether the recent results favorable to nuclear power are principally due to wording effects or other methodological artifacts, or are the harbinger of an authentic shift in public mood, can only be determined by the test of time. It may be that at some future date Americans will once again embrace nuclear power with great enthusiasm. Nuclear supporters, wishing to pin their hopes on this expectation, have read this conclusion into the recent data. In the meantime, however, *all other available data* converge on the conclusion that that optimistic future is not now in evidence; the explicit findings summarized above show overwhelming opposition to the technology. Thus, a conclusion that the public mood has taken a complete turnaround and is now supportive of nuclear power is more reflective of hopes for the future than the evidence of the past.

The Source of Attitudes. In the studies of public attitudes toward technology it is customary to ask, once the overall levels of support and opposition have been identified, about the cognitive underpinnings and social variation in relevant attitudes. Four sets of factors, presumed to be the source of variation in nuclear attitudes, have typically been examined: sociological, social-psychological, psychological, and institutional.

The most common sociological approach to the explanation of survey data is to discuss variations across sociodemographic categories such as age, income, race, education, geographic place of residence, the presence of children, and gender. In discussions of attitudes toward nuclear power, the variations across categories are too few to be worthy of mention. Indeed, the principal finding of sociological approaches to explaining nuclear attitudes is a *nonfinding*. Demographically, to the question, Who's opposed to nuclear facilities? the answer is, Nearly everyone: opposition to nuclear power cuts across nearly every major sociodemographic category. The only finding of note is that women are nearly always more opposed than men, though even here most studies show only modest effects (Nealey, Melber, and Rankin, 1983).

The social-psychological factors most frequently investigated fall into two broad categories: efforts to explain concerns about nuclear technologies in terms of anxiety, emotionality, or irrationality, on the one hand, or in terms of personal values and ideologies, on the other. As was the case for the socio-demographic variables, the principal finding also tends to be a nonfinding: little evidence appears that links nuclear attitudes to social-psychological factors. Even in the case of irrationality, the most consistently researched of these factors, studies have consistently reported little to no correlation (Mitchell, 1984; Dunlap and Olsen, 1984; Freudenburg and Jones, 1991). Measures of ideology and of personal values have fared slightly better, show-ing somewhat more consistent, but still modest effects. For example, slightly higher opposition to nuclear power is found among political liberals, among those having Democratic rather than Republican political identifications, and among citizens committed to egalitarian forms of political governance (Freudenburg and Pastor, 1992).

As for the psychological foundation of nuclear attitudes, its cornerstone is the risks associated with nuclear technologies. The extensive work in the psychometric tradition by Slovic and his colleagues (e.g., 1984, 1987) has shown that, compared to a wide variety of risky technologies, activities, and substances, nuclear risks are perceived to be the riskiest—and are the most dreaded. The most recent empirical findings support this general conclusion and, furthermore, show that the dread of nuclear power is shared by citizens of other industrial societies as well (Kleinhesselink and Rosa, 1991).[7] Thus, as noted briefly above, nuclear power has generated not only electricity, but also widespread and enduring fear.

The cumulative evidence points to a conspicuous unevenness: we have a considerable understanding of the psychological underpinnings of nuclear attitudes but a surprisingly meager understanding of the effects of social variables on those attitudes. An alternative approach has emerged with the goal of closing the gap in our understanding of the social context of nuclear attitudes. A key factor in this approach is the level of trustworthiness—or, alternatively, recreancy[8]—possessed by major institutional actors, such as business or industry, and, in particular, scientists and government regula-tors. Early results from this approach are encouraging. Whereas sociodemo-graphic and ideological independent variables tend to show little relation-ship with variation in attitudes, typically accounting for less than 10 percent in samples of the general public, perceived recreancy and trustworthiness have been shown to explain two to three times as much variance (Freudenburg, 1990). If these preliminary findings are borne out by future analyses, then issues such as secrecy, mismanagement, and even misinterpretation of the truth could prove to make the future course of nuclear waste management even more problematic than has been the case to date.[9]

Implications

The difficulty of removing the hazy pall over nuclear power is deepened by two fundamental asymmetries in how public perceptions change. The first of these is the apparent asymmetry in how feelings of trust are formed or lost. It is much easier to lose people's trust than to regain it, once lost (Rothbart and Park, 1986). Because public opposition is apparently due, in no small part, to the lack of trust in the federal government and its lead agencies, especially the DOE (Freudenburg, 1990; Lipset and Schneider, 1983; Hohenemser, Kasperson, and Kates, 1977; and chapter 3 of this volume by Slovic, Layman, and Flynn), it is highly unlikely that the support the technology once enjoyed can be easily regained. It is even doubtful whether, over the short term, support levels can regain any appreciable part of their lost ground. Thus, despite recent legislation to facilitate greater public involvement and despite an increased openness by lead government agencies, public mistrust—the predictable result of objectionable past practices—will likely not disappear soon.

The second asymmetry occurs between perceptions of fear and calm. While ostensibly semantic opposites, the perceptions themselves are not equally enduring. Calm can be turned to fear with a single, dramatic mishap. The regaining of calm, once lost, can take markedly more time and patience. Thus, a single nuclear event can have a much deeper and lasting impact on public perceptions than years of safe reactor operation. Indeed, many close observers believe that, if there is another accident having the serious implications of TMI or Chernobyl, nuclear power may be permanently doomed.

In part, the deep impact of untoward events stems from the fact that nuclear power is not a salient concern for laypersons: it is not at the forefront of most people's daily awareness. Rather, thoughts about nuclear power are submerged below the exigencies of regular routines. Safe operations are typically out of mind. But with the occurrence of a mishap or the emergence of a controversy, awareness is reawakened with a "signal" that all is not well with nuclear power, that instead, "the technology is dangerous." For example, experiments by Slovic et al. (1984) have asked subjects to rate thirty types of accidents on the degree to which they "serve as a warning signal for society, providing new information about the probability that similar or even more destructive mishaps may occur within this type of activity." The three nuclear accidents included in the thirty were judged to have the highest signal potential. Furthermore, because the technology is now perceived to be unacceptably risky by large segments of the American public, the template of citizen predisposition is to view nearly all newsworthy nuclear events in a negative light. Finally, the media, shown to play a central role in shaping nuclear attitudes (Mazur, 1981, 1990), are inclined to cover dramatic events and those elements of news stories that are easily conveyed to a public audience. The undramatic is not newsworthy; there is not much to report when nuclear power works smoothly and safely.

There is little doubt that general attitudes toward nuclear power will condition public perceptions of nuclear waste. Furthermore, public perceptions will likely influence waste policies: the only question will have to do with the depth of influence. The overall climate for nuclear power is decidedly unfavorable, and opposition to the technology is not confined to some narrow segment of the general public but spans the entire social landscape. To the extent that waste management programs require public cooperation or acceptance, the evidence suggests that troubled times lie ahead. Where programs call for the siting of waste facilities, the prognosis is even more dismal.

As one of the unavoidable management tasks of nuclear power, waste disposal is doubtless one of the most difficult technological challenges facing the nation for the remainder of the century. The picture emerging from the evidence analyzed here strongly suggests that it may also be one of the nation's most difficult public policy challenges.

Notes

We wish to thank Kathy Charlee, Tom Dietz, Riley Dunlap, Scott Frey, Greg Hooks, Mike Kraft, Ragnar Lofstedt, Allan Mazur, Lisa McIntyre, Max Power, and Spencer Weart for their help or comments on earlier versions of this chapter and Frank Baumgartner and Bryan Jones for kindly sharing their data.

1   Secrecy was so tight that not even Vice-President Harry Truman, though knowledgeable about the existence of something called the Manhattan Project (actual code name "The Manhattan Engineering District"), knew anything about the making of atomic bombs. It wasn't until April 13, 1945, within twenty-four hours after Roosevelt's death, that the newly sworn president was informed (Rhodes, 1986).

2   The American public became equally committed to secrecy. Posters, reflecting this commitment, abounded: "A slip of the lip can sink a ship," "Careless talk costs lives," and from Britain came "Tittle-tattle lost the battle" (Zinberg, 1979).

3   Hertsgaard (1983), too, notes the importance of secrecy but goes further by arguing

that the root cause of the eventual public disenchantment with commercial nuclear technology was the repeated abuse of secrecy.

4   While the secrecy complaint has long been submerged below other public issues, it nevertheless continues to shadow the technology in two significant ways. First, it was the issue of secrecy that first attracted public mistrust to the government agencies responsible for the management of nuclear power—a mistrust that not only continues to this day but has deepened considerably. Second, there has been a steady release of previously classified information—due to the Freedom of Information Act, passed by Congress in 1966, the activities of public interest groups, and the efforts of lead agencies, such as the DOE to regain public confidence through greater openness. The availability of this information has reinforced public mistrust of those government agencies and increased the awareness of hazards (previously secret) associated with the technology. With release of the information has come the steady reminder that secrecy not only concealed knowledge of nuclear power from enemy agents, but also from the nation's own citizens—often those citizens who bore the greatest nuclear risks in the past.

5   Identical promotional rhetoric, such as the declaration by C. P. Steinmetz of the General Electric Company that electricity would become "so cheap that it is not going to pay to meter it" (Weart, 1988:12), leaves little doubt that the call and response between government and industry were in tune to the same gospel.

6   Since the 1960s, numerous surveys of a wide variety of nuclear and related attitudes have been conducted, many of which have been summarized in Nealey, Melber, and Rankin (1983) and Nealey (1990). Because small changes in question wording can sometimes produce large changes in responses, a problem afflicting nuclear questions in particular (Nealey, Melber, and Rankin, 1983), we have deliberately tried to select time-series where questions have been repeated verbatim or nearly so.

7   The results of general surveys in Connecticut and Arizona by Gould et al. (1988) are an exception to the conclusions of the psychometric tradition. While nuclear weapons were perceived to be the riskiest technology, nuclear power ranked fourth—behind handguns and industrial chemicals.

8   In this context, *recreancy* refers to the actions of individuals or organizations that have been entrusted by the broader society with specialized responsibility but where their actions taken to fulfill that responsibility have not met expectations; consequently, they fail to merit public trust. The failure may be unintentional or indirect, as well as intentional or direct.

9   As a concrete example, recent events surrounding the Hanford Nuclear Reservation in Washington State clearly illustrate how the secrecy legacy continues to shape public sentiment toward nuclear technology. Hanford was initially established in 1943 to produce plutonium for the making of atomic bombs. It continued to be a major plutonium producer after the war, and subsequently expanded its operations markedly to include a wide variety of other activities associated with nuclear power. In 1985, under the Freedom of Information Act, a coalition of public interest groups requested the records documenting radiation releases in the 1940s and 1950s in the Hanford area. Their request, honored in 1986 with the release of 19,000 pages of documents, showed that dangerously high levels of radiation had been released during those earlier decades. Furthermore, at least two of the releases were deliberate, the result of experiments exposing nearby residents to massive amounts of iodine 131. The evidence was sufficiently compelling to prompt the secretary of energy, Admiral James Watkins,

to admit for the first time that Hanford releases in the late 1940s were large enough to create a health risk to people living near Hanford (Steele, 1990). Once this information was released, citizens who had lived downwind of Hanford during the time of releases were understandably alarmed—and naturally critical and skeptical of government trustworthiness, especially the DOE. "There was denial all the way," said one resident. "Now they are beginning to tell the truth." Another lamented: "We trusted the government to protect us, and they put our lives at risk. We were worth more" (Steele, 1990:A5). For a recent historical analysis of early Hanford releases and their long-term impacts—medical, social, and political—see Stenehjem (1990).

References

Baumgartner, Frank R., and Bryan D. Jones. 1991. "Agenda Dynamics and Policy Subsystems." *Journal of Politics* 53:1044–74.

Bisconti, Ann S. 1989. "The Public Opinion Environment for Advanced-Design Nuclear Energy Plants." Remarks for the U.S. Global Strategy Council Symposium on Advanced Nuclear Reactors and U.S. Energy Security in the 1990s, Washington, D.C., June 22.

Bradburn, Norman M., and Seymour Sudman. 1988. *Polls and Surveys: Understanding What They Tell Us.* San Francisco: Jossey-Bass.

———. 1982. *Asking Questions: A Practical Guide to Questionnaire Design.* San Francisco: Jossey-Bass.

Cohen, Bernard L. 1983. *Before It's Too Late: A Scientific Case for Nuclear Energy.* New York: Plenum.

Cook, J. 1985. "Nuclear Follies." *Forbes,* February 11, 82–100.

Dietz, Thomas M., and Robert W. Rycroft. 1989. "Who's Representing Whom?: A Comparison of Risk Attitudes Among Corporate Executives, the General Public, and Risk Professionals." Presented at the Annual Meeting of the Society for Risk Analysis, San Francisco, October.

Drumheller, Susan. 1990. "Citizens Challenge N-Plant Licensing." *Spokane Spokesman-Review,* September 27.

Dunlap, Riley E. 1992. "Trends in Public Opinion Toward Environmental Issues: 1965–1990." In Riley E. Dunlap and Angela G. Mertig, eds., *American Environmentalism—The U.S. Environmental Movement.* Washington, D.C.: Taylor and Francis. 89–116.

Dunlap, Riley E., and Marvin E. Olsen. 1984. "Hard-path Versus Soft-path Advocates: A Study of Energy Activists." *Policy Studies Journal* 13:413–28.

Ebbin, Stephen, and Raphael Kasper. 1974. *Citizens Groups and the Nuclear Power Controversy.* Cambridge, MA: M.I.T. Press.

Erikson, Kai T. 1990. "Toxic Reckoning: Business Faces a New Kind of Fear." *Harvard Business Review* (January–February): 118–26.

Farhar-Pilgrim, Barbara, and William R. Freudenburg. 1984. "Nuclear Energy in Perspective: A Comparative Assessment of the Public View." In William R. Freudenburg and Eugene A. Rosa, eds., *Public Reactions to Nuclear Power: Are There Critical Masses?* Boulder: Westview/American Association for the Advancement of Science. 183–203.

Faltmayer, Edmund. 1979. "Nuclear Power After Three Mile Island." *Fortune,* May 7, 114–22.

Farrington, J. S. 1989. "Public Acceptance of Nuclear Energy." Draft of speech. Washington, D.C.: U.S. Council for Energy Awareness, August 31.

Ford, Daniel F. 1982. *The Cult of the Atom: The Secret Papers of the Atomic Energy Commission.* New York: Simon and Schuster.

Freudenburg, William R. 1990. "Risk and Recreancy: Weber, the Division of Labor, and the Rationality of Risk Perceptions." Presented at the 12th annual Department of Energy Conference on Low-Level Radioactive Waste, Chicago, August.

Freudenburg, William R., and Rodney K. Baxter. 1984. "Host Community Attitudes Toward Nuclear Power Plants: A Reassessment." *Social Science Quarterly* 65:1129–34.

Freudenburg, William R., and Eugene A. Rosa, eds. 1984. *Public Reactions to Nuclear Power: Are There Critical Masses?* Boulder: Westview/American Association for the Advancement of Science.

Freudenburg, William R., and Timothy R. Jones. 1991. "Attitudes and Stress in the Presence of Technological Risk: A Test of the Supreme Court Hypothesis." *Social Forces* 69:1143–68.

Freudenburg, William R., and Susan K. Pastor. 1992. "Public Responses to Technological Risks: Toward a Sociological Perspective." *Sociological Quarterly* 33.

Gould, Leroy C., Gerald T. Gardner, Donald R. DeLucca, Adrian R. Tiemann, Leonard W. Doob, and Jan A. J. Stolwijk. 1988. *Perceptions of Technological Risks and Benefits.* New York: Russell Sage Foundation.

Hertsgaard, Mark. 1983. *Nuclear, Inc.: The Men and Money Behind Nuclear Power.* New York: Pantheon.

Hohenemser, Christopher, Roger E. Kasperson, and Roger W. Kates. 1977. "The Distrust of Nuclear Power." *Science* 196:25–35.

Kleinhesselink, Randall R., and Eugene A. Rosa. 1991. "Cognitive Representation of Risk Perceptions." *Journal of Cross-Cultural Psychology* 22:11–28.

Koch, L. J. 1985. "A Grim Fairy Tale of Government Regulation." *Public Utilities Fortnightly.* (January 24): 18–23.

Lindell, Michael K., Timothy C. Earle, John A. Hebert, and R. W. Perry. 1978. *Radioactive Wastes: Public Attitudes Toward Disposal Facilities.* Seattle: Battelle Human Affairs Research Centers.

Lipset, Seymour Martin, and William Schneider. 1983. *The Confidence Gap: Business, Labor, and Government in the Public Mind.* New York: The Free Press.

Mazur, Allan. 1981. *The Dynamics of Technical Controversy.* Washington, D.C.: Communications Press.

———. 1990. "Nuclear Power, Chemical Hazards, and the Quantity-of-Reporting Theory of Media Effects." *Minerva* 28:294–323.

Melber, Barbara D., Stanley M. Nealey, Joy Hammersla, and William L. Rankin. 1977. *Nuclear Power and the Public: Analysis of Collected Survey Research.* Seattle: Battelle Human Affairs Research Centers.

Mitchell, Robert C. 1980. *Final Results of the Resources for the Future National Environmental Survey for the President's Council on Environmental Quality.* Washington, D.C.: Resources for the Future.

———. 1984. "Rationality and Irrationality in the Public's Perception of Nuclear Power." In William R. Freudenburg and Eugene A. Rosa, eds., *Public Reactions to Nuclear Power: Are There Critical Masses?* Boulder: Westview/American Association for the Advancement of Science. 137–79.

Morone, Joseph G., and Edward J. Woodhouse. 1989. *The Demise of American Nuclear*

Power: Learning From the Failure of a Politically Unsafe Technology. New Haven: Yale University Press.

National Research Council. 1984. *Social and Economic Aspects of Radioactive Waste Disposal.* Washington, D.C.: National Academy Press.

Nealey, Stanley M. 1990. *Nuclear Power Development: Prospects in the 1990s.* Columbus, OH: Battelle.

Nealey, Stanley M., and John A. Hebert. 1983. "Public Attitudes Toward Radioactive Waste." In Charles A. Walker, Leroy C. Gould, and Edward J. Woodhouse, eds., *Too Hot to Handle? Social and Policy Issues in the Management of Radioactive Wastes.* New Haven: Yale University Press. 94–111.

Nealey, Stanley M., Barbara D. Melber, and William L. Rankin. 1983. *Public Opinion and Nuclear Energy.* Lexington, MA: D.C. Heath and Co.

Popper, Frank J. 1981. "Siting LULUs." *Planning* (April): 12–15.

Rankin, William L., and Stanley M. Nealey. 1978. "Attitudes of the Public about Nuclear Wastes." *Nuclear News* 21:112–17.

Rhodes, Richard. 1986. *The Making of the Atomic Bomb.* New York: Simon and Schuster.

Rosa, Eugene A., and William R. Freudenburg. 1984. "Nuclear Power at the Crossroads." In William R. Freudenburg and Eugene A. Rosa, eds., *Public Reactions to Nuclear Power: Are There Critical Masses?* Boulder: Westview/American Association for the Advancement of Science. 3–37.

Rosa, Eugene A., Marvin E. Olsen, and Don A. Dillman. 1984. "Public Views Toward National Energy Policy Strategies: Polarization or Compromise?" In William R. Freudenburg and Eugene A. Rosa, eds., *Public Reactions to Nuclear Power: Are There Critical Masses?* Boulder: Westview/American Association for the Advancement of Science. 69–93.

Rothbart, Myron, and Bernadette Park. 1986. "On the Confirmability and Disconfirmability of Trait Concepts." *Journal of Personality and Social Psychology* 50:131–42.

Schneider, William. 1986. "Public Ambivalent on Nuclear Power." *National Journal* 18: 1562–63.

Schuman, Howard, and Jacqueline Scott. 1987. "Problems in the Use of Survey Questions to Measure Public Opinion." *Science* 236:957–59.

Slovic, Paul, Sarah Lichtenstein, and Baruch Fischhoff. 1984. "Modelling the Societal Impact of Fatal Accidents." *Management Science* 30:464–74.

Slovic, Paul. 1987. "Perception of Risk." *Science* 236:280–85.

Steele, Karen Dorn. 1990. "Perseverance Finally Revealed Hanford Secret." *Spokane Spokesman-Review,* July 25.

Stenehjem, Michele. 1990. "Indecent Exposure." *Natural History* (September):6–22.

Strauss, Lewis. 1954. Remarks prepared for delivery at the Founders Day Dinner, National Association of Science Writers, New York, September 16.

Szalay, Robert A. 1984. "A Nuclear Industry View of the Regulatory Climate." In William R. Freudenburg and Eugene A. Rosa, eds., *Public Reactions to Nuclear Power: Are There Critical Masses?* Boulder: Westview/American Association for the Advancement of Science. 295–306.

Texas Department of Agriculture. 1984. *Panhandle Residents' View of High-level Nuclear Waste Storage Part II: Survey Questions and Responses.* Austin: Texas Department of Agriculture.

U.S. Atomic Energy Commission. 1955. "AEC Announces Three Basic Regulations for

Civilian Atomic Industry." Press Release 622. Washington, D.C.: Atomic Energy Commission.

————. 1957. *Theoretical Possibilities and Consequences of Major Accidents in Large Nuclear Power Plants.* Washington, D.C.: GPO.

————. 1970. Press Release S-18-170. Washington, D.C.: Atomic Energy Commission.

U.S. Council on Energy Awareness. 1989. *U.S. Public Opinion on Nuclear Energy.* Washington, D.C.: Council on Energy Awareness.

U.S. Department of Energy. 1983. *1983 Survey of Nuclear Power Plant Costs.* Washington, D.C.: GPO.

————. 1991. *National Energy Strategy: Powerful Ideas for America.* Washington, D.C.: GPO.

U.S. Nuclear Regulatory Commission. 1975. *Reactor Safety Study: An Assessment of Accident Risks in U.S. Commercial Nuclear Power Plants.* WASH-1400. Washington, D.C.: GPO.

U.S. Office of Technology Assessment. 1984. *Nuclear Power in an Age of Uncertainty.* OTA-E-216. Washington, D.C.: GPO.

U.S. Senate. 1988. "Price-Anderson Act Amendments of 1988." U.S. Senate Report 100–70.

Wald, Matthew L. 1990. "Pre-licensing of Nuclear Plant is Barred." *New York Times,* November 3.

Weart, Spencer R. 1988. *Nuclear Fear: A History of Images.* Cambridge, MA: Harvard University Press.

Wermiel, S. 1982. "High Court Faces Big Nuclear-Power Cases, Including Ruling on California Restrictions." *Wall Street Journal,* November 26, 13.

Wessel, D. 1985. "Investor Group Seeks to Buy 9.7% of Seabrook Plant." *Wall Street Journal,* February 13, 25.

Winner, Langdon. 1986. *The Whale and the Reactor.* Chicago: University of Chicago Press.

Woodhouse, Edward J. 1983. "The Politics of Nuclear Waste Management." In Charles A. Walker, Leroy C. Gould, and Edward J. Woodhouse, eds., *Too Hot to Handle? Social and Policy Issues in Management of Radioactive Wastes.* New Haven: Yale University Press. 151–83.

Zinberg, Dorothy. 1979. "The Public and Nuclear Waste Management." *Bulletin of the Atomic Scientists* 34–39.

————. 1984. "The Public and Nuclear Waste Management Policy: A Struggle for Participation." In William R. Freudenburg and Eugene A. Rosa, eds., *Public Reactions to Nuclear Power: Are There Critical Masses?* Boulder: Westview/American Association for the Advancement of Science. 233–53.

3   Perceived Risk, Trust, and Nuclear Waste:

Lessons from Yucca Mountain

Paul Slovic, Mark Layman, and James H. Flynn

By the year 2000, the United States will have a projected 40,000 metric tons of spent nuclear fuel stored and awaiting disposal at some seventy sites. By 2035, after all existing nuclear plants have completed forty years of operation, there will be approximately 85,000 metric tons (Technical Review Board, 1991). The amount of spent fuel needing disposal will continue to grow with the relicensing of existing nuclear plants and the possible construction of new facilities. The U.S. Department of Energy (DOE) has been under intense pressure from Congress and the nuclear industry to dispose of this accumulating volume of high-level waste since the passage of the Nuclear Waste Policy Act in 1982 and its amendment in 1987, by which Yucca Mountain, Nevada, was selected as the only candidate site for the nation's first nuclear waste repository. The lack of a suitable solution to the waste problem is widely viewed as an obstacle to further development of nuclear power and a threat to the continued operation of existing reactors, besides being a safety hazard in its own right.

Yet, to this time, the DOE program has been stymied by overwhelming political opposition, fueled by perceptions of the public that the risks are immense (Flynn et al., 1990; Kasperson, 1990; Kunreuther, Desvousges, and Slovic, 1988; Nealey and Hebert, 1983; and this volume). These perceptions stand in stark contrast to the prevailing view of the technical community, which believes that nuclear wastes can be disposed of safely, in deep underground isolation. Officials from DOE, the nuclear industry, and technical experts are profoundly puzzled, frustrated, and disturbed by the public and political opposition, which many of them believe is based upon irrationality and ignorance (see Table 3-1).

A number of important events during the past several years underscore the seriousness of this problem.

Table 3-1   Some Viewpoints of Experts Regarding Public Perceptions of the Risks from
Nuclear Waste Disposal

"Several years ago . . . I talked with Sir John Hill, . . . chairman of the United Kingdom's
Atomic Energy Authority. 'I've never come across any industry where the public perception of the problems is so totally different from the problems as seen by those of us in the
industry . . . ,' Hill told me. In Hill's view, the problem of radioactive waste disposal was,
in a technical sense, comparatively easy" (Carter, 1987:9).

"Nuclear wastes can be sequestered with essentially no chance of any member of the public receiving a non-stochastic dose of radiation. . . . Why is the public's perception of the
nuclear waste issue at such odds with the experts' perception?" (Weinberg, 1989:1–2).

"The fourth major reason for public misunderstanding of nuclear power is a grossly unjus-
tified fear of the hazards from radioactive waste. . . . Often called an 'unsolved problem,'
many consider it to be the Achilles' heel of nuclear power. Seven states now have laws
prohibiting construction of nuclear power plants until the waste disposal issue is settled.
On the other hand there is general agreement among those scientists involved with waste
management that radioactive waste disposal is a rather trivial technical problem" (Cohen,
1983:119).

"It is possible to estimate the risk [of a high-level nuclear waste repository] if the material
is buried as planned. It turns out it is ridiculously low. . . . The risk is as negligible as it is
possible to imagine, yet the clamor about the subject has paralyzed the decision-making
authorities, and there is still no consensus solution. It is embarrassingly easy to solve the
technical problems, yet impossible to solve the political ones" (Lewis, 1990:245–46).

---

1. Official opposition by the state of Nevada has increased substantially.
In June 1989, the Nevada legislature passed Assembly Bill 222, making it
unlawful for any person or governmental entity to store high-level radioactive waste in the state. The Nevada attorney general subsequently issued
an opinion that the Yucca Mountain site had been effectively vetoed under
a provision of the Nuclear Waste Policy Act. The governor instructed state
agencies to disregard DOE's applications for environmental permits necessary to investigate the site. The state and DOE initiated federal lawsuits over
continuation of the program and issuance of the permits needed for on-site
studies. In September 1990, the Ninth U.S. Circuit Court of Appeals ruled
that the state had acted improperly and ordered Nevada officials to issue the
permits.

2. In November 1989, the DOE, admitting dissatisfaction with its earlier assessments of the Yucca Mountain site, announced that it would essentially
start over with, "for the first time," an integrated, responsible plan. This plan
would subject technical studies to close outside scrutiny to ensure that de-

cisions about Yucca Mountain would be made "solely on the basis of solid scientific evidence" (Moore, 1989).

3. In July 1990, the National Research Council's Board on Radioactive Waste Management issued a strong criticism of the DOE program, charging that DOE's insistence on doing everything right the first time has misled the public by promising unattainable levels of safety under a rigid schedule that is "unrealistic, given the inherent uncertainties of this unprecedented undertaking," and thus vulnerable to " 'show stopping' problems and delays that could lead to a further deterioration of public and scientific trust" (National Research Council, 1990:1). The board recommended, instead, a more flexible approach, permitting design and engineering changes as new information becomes available during repository construction and operation.

Perceptions of risk from radiation, nuclear power, and nuclear waste play a pivotal role in this story and need to be thoroughly understood if we are to make any progress in resolving the current impasse. Although we already know a good deal about perceptions in this domain (Slovic, 1990), most of our knowledge comes from rather general questions (e.g., "How great is the risk of a nuclear waste repository compared with the risks of X, Y, and Z?"). With some notable exceptions (Erikson, 1990; Lifton, 1967; Weart, 1988), there have been few attempts to penetrate the surface veneer of nuclear fear and provide insight into the nature and pervasiveness of people's concerns, the origins of these concerns, the emotions that underlie them, their legitimacy, and their likely stability. Analysis of the intense concerns associated with a nuclear waste repository is also important, we believe, for understanding the role that perceived risk plays in the opposition to many other unwanted facilities such as chemical-waste landfills and incinerators.

Attitude, Perception, and Opinion Surveys

There have been a number of surveys conducted recently to assess public attitudes, perceptions, and opinions regarding the management of high-level radioactive wastes. We shall focus here on the results from a series of surveys we conducted in 1988 and 1989 (Flynn et al., 1990).

Details of implementing these surveys are reported in Table 3-2. More than 3300 respondents were questioned by telephone with regard to their perceptions of the risks and benefits associated with a nuclear waste repository, their support or opposition for the DOE repository program, their trust in the ability of DOE to manage the program, and their views on a variety of other issues pertaining to radioactive waste disposal. In addition to a national survey, data were collected from three other populations of special interest:

Table 3-2   Survey Details

Survey and Location	Dates	Sample Size	Response Rate (%)
Phoenix	4/13–6/8/88	802	72
National	10/21–12/7/89	825	77
Southern California	12/6/89–1/1/90	801	77
Nevada			
Statewide	9/25–10/15/89	500	74
Nye County	9/25–10/15/89	204	74
Lincoln County	9/25–10/15/89	101	84
Esmeralda County	9/25–10/15/89	101	77
Nevada Total	9/25–10/15/89	906	77

residents of Nevada, the state selected as the site for the proposed national repository, and residents of southern California and Phoenix, Arizona, the two major sources of visitors to Nevada. The Phoenix survey was less extensive than the others and will be discussed in the next section. Respondents were selected by means of a random digit dialing procedure. When telephone contact was made with a household, the interviewer asked to speak to the person eighteen years or older who had the most recent birthday (to ensure random selection of respondents within each household). Response rates were high (72 to 84 percent) in each of the surveys.

When asked to indicate the closest distance they would be willing to live from each of ten facilities, the median distance from an underground nuclear waste repository was 200 miles in the national, Nevada,[1] and southern California surveys, twice the distance from the next most undesirable facility, a chemical waste landfill, and three to eight times the distances from oil refineries, nuclear power plants, and pesticide manufacturing plants. In response to the statement "Highway and rail accidents will occur in transporting the wastes to the repository site," the percentage of respondents who agreed or strongly agreed was 77.4 percent in Nevada, 69.2 percent in California, and 71.6 percent nationally. Similar expectations of problems were expressed with regard to future earthquake or volcanic activity at the site, contamination of underground water supplies, and accidents while handling the material during burial operations.

When asked whether a state that does not produce high-level nuclear wastes should serve as a site for a nuclear waste repository, 67.9 percent of the southern California and 76.0 percent of the national respondents answered 'no' (the question was not asked in Nevada). A majority of those polled in the southern California and national surveys judged a single national repository to be the least fair of five disposal options (including storage at each nuclear plant, in each state, and in each of several regions, and dual repositories in the East and West).

Strong distrust of the DOE was evident from the responses to questions such as "The U.S. Department of Energy can be trusted to provide prompt and full disclosure of any accidents or serious problems with their nuclear waste management programs." In southern California, 67.5 percent either somewhat or strongly disagreed with this statement. The corresponding rate of disagreement in the national survey was 68.1 percent.

Nevadans were asked whether or not they would vote in favor of a repository at Yucca Mountain; 69.4 percent said they would vote against it, compared to 14.4 percent who would vote for it. About 68 percent of the Nevadans surveyed said they agreed strongly with the statement "The state of Nevada should do all it can to stop the repository." Another 12.5 percent agreed somewhat with this statement; only 16.0 percent disagreed. When asked whether or not they favored Assembly Bill 222, which was passed in 1989 and made it illegal to dispose of high-level nuclear waste in Nevada, 74 percent were in favor and 18.4 percent opposed the bill. Finally, 73.6 percent of Nevadans said that the state should continue to do all it can to oppose the repository even if that means turning down benefits that may be offered by the federal government; 19.6 percent said the state should stop fighting and make a deal.

Follow-up surveys of Nevada residents in October 1990 and March 1991 suggest that opposition and distrust have continued to rise (Flynn, Mertz, and Slovic, 1991). The percentage of Nevadans who would vote against a repository at Yucca Mountain increased from 69.4 percent to 80.2 percent. In response to a request to indicate "how much you trust each of the following to do what is right with regard to a nuclear waste repository at Yucca Mountain," the governor of Nevada topped the list of officials, agencies, and institutions. DOE, the Nuclear Regulatory Commission (NRC), and the U.S. Congress were the least trusted entities (see Figure 3-1). Strong increases in trust were evident for the president, the governor of Nevada, and the Nevada state legislature.[2] In contrast, trust in DOE and NRC declined between 1989 and 1991.

Measures of trust in DOE, perceived risk, and opposition to a repository at Yucca Mountain were highly interrelated. Table 3-3 illustrates the link between trust in DOE to "provide full and prompt disclosure of any accidents or serious problems with a repository program" and perception of risk from highway or rail accidents during the transportation of wastes to a repository site. Table 3-4 illustrates the trust vs. perceived risk relationship with a rating scale measure of trust in DOE to "do the right thing with regard to a nuclear waste repository." In both tables, those who distrust DOE are

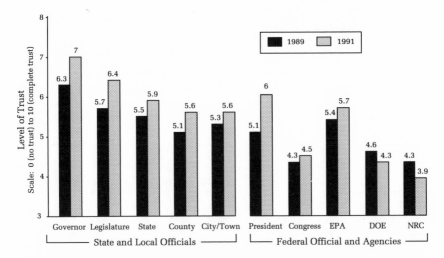

Figure 3-1   Responses of Nevada residents when asked to rate their trust in federal, state, and local officials and federal agencies to do what is right with regard to a nuclear waste repository at Yucca Mountain.

Table 3-3   Relationship Between Trust in DOE and
Perceived Risk of Transport Accidents

	Highway and Rail Accidents Will Occur[b]					
DOE Can Be Trusted[a]	Strongly Disagree	Somewhat Disagree	Neutral	Somewhat Agree	Strongly Agree	N
Strongly disagree	3.4	6.0	2.2	34.5	53.9	230
Somewhat disagree	5.7	13.6	5.0	51.4	24.3	140
Neutral	14.3	28.6	14.3	28.6	14.3	7
Somewhat agree	2.6	22.1	7.8	49.4	18.2	77
Strongly agree	25.0	25.0	10.7	14.3	25.0	28

$\chi^2 = 97.1$, $df = 16$, $p < .001$.
Note: Cell entries are row percentages based upon data from the Nevada survey.
[a] The DOE can be trusted to provide full and prompt disclosure of any accidents or serious problems with a repository program.
[b] Highway and rail accidents will occur in transporting the wastes to the repository site.

much more likely to agree that highway and rail accidents will occur than are those who trust DOE. Table 3-5 illustrates the relationship between trust in DOE to disclose problems and the respondent's response to the question "Would you vote for a repository at Yucca Mountain?" Among those who disagree that DOE can be trusted to disclose accidents or serious problems with a repository program, about 85 to 90 percent would vote against a re-

Table 3-4   Relationship Between Trust Rating of DOE and
Perceived Risk of Transport Accidents

| | Highway and Rail Accidents Will Occur[b] | | | | | |
Trust Rating[a]	Strongly Disagree	Somewhat Disagree	Neutral	Somewhat Agree	Strongly Agree	N
0	5.7	5.7	1.4	31.4	55.7	70
1–4	3.7	13.2	3.7	37.5	41.9	136
5	5.1	9.3	6.8	44.9	33.9	118
6–9	5.3	13.0	3.8	50.4	27.5	131
10	17.2	24.1	10.3	17.2	31.0	29

$\chi^2 = 40.7$, $df = 16$, $p < .001$.

Note: Cell entries are row percentages based upon data from the Nevada survey.
[a] Based upon rating trust in DOE to do what is right with regard to a nuclear waste repository (0 = no trust; 10 = complete trust).
[b] Highway and rail accidents will occur in transporting the wastes to the repository site.

Table 3-5   Relationship Between Trust in DOE and Response to the Question
"Would You Vote for a Repository at Yucca Mountain?"

| | DOE Can Be Trusted to Disclose Any Problems | | | | |
Yucca Mountain Vote	Strongly Disagree	Somewhat Disagree	Neutral	Somewhat Agree	Strongly Agree
Yes	8.3	14.4	33.3	34.9	66.7
No	91.7	85.6	66.7	65.1	33.3
N	204	118	6	63	21

$\chi^2 = 62.2$, $df = 4$, $p < .001$.

Note: Cell entries are column percentages. Responses are based upon the survey of Nevada residents.

pository at Yucca Mountain. The percentage of votes against a repository drops to 65.1 percent for those who somewhat agree that DOE can be trusted and falls further to 33.3 percent among the few who strongly agree that DOE can be trusted. Other questions assessing trust, perceived risk, and opposition to a repository produced relationships similar to those in Tables 3-3, 3-4, and 3-5.

Imagery and Perception

Prior to answering any of the attitude or opinion questions, respondents in the national, southern California, and Nevada surveys, along with the 802 respondents in Phoenix, were asked to free associate to the concept of a nuclear waste repository.

The potential for word associations to reveal the mental content of a person's subjective experience was recognized by Plato and has a long history in psychology, going back to Galton (1880), Wundt (1883), and Freud (1924). More recently, Szalay and Deese (1978) have employed the *method of continued associations* to assess people's subjective representative systems for a wide range of concepts. This method requires the subject to make repeated associations to the same stimulus, for example,

war: soldier
war: fight
war: killing
war: etc.

Szalay and Deese argue that the method of continued associations is an efficient way to determine the contents and representational systems of human minds without requiring those contents to be expressed in the full discursive structure of language. In fact, we may reveal ourselves through associations in ways we might find difficult to do if we were required to spell out the full propositions behind these associations through answers to questions. Evidence provided by Szalay and Deese and others demonstrates that responses produced by the method of continued associations are not erratic and whimsical but are stable and relate clearly and naturally to a person's experiences and preferences. They are organized and structured in much the same way as perceptions, beliefs, and attitudes.

A related view is provided by Fiske, Pratto, and Pavelchak, who describe an image as a "cognitive representation, a conception, or an idea, potentially containing both concrete and abstract impressions; . . . a mental picture, but not necessarily visual" (1983:42).

Cognitive images are often accompanied by affect and such affect-laden images have been found to have important behavioral consequences. Prejudicial images give rise to discrimination (Hamilton, 1981). Images of politicians affect voting behavior (Campbell et al., 1960). Images of nuclear war affect an individual's level of antinuclear political activity (Fiske, Pratto, and Pavelchak, 1983). Images of cities and states determine decisions about places to vacation (Slovic et al., 1991) or attendance at conventions (Kunreuther and Easterling, 1990).

The repository images were elicited using a version of the method of continued associations adapted for a telephone interview. The elicitation interview proceeded as follows:

My next question involves word association. For example, when I mention the word 'baseball,' you might think of the World Series, Reggie Jackson, summertime, or even hot dogs. Today I am interested in the first six thoughts or images that come to mind when you think of an *underground nuclear waste repository*.[3] Think about an underground nuclear waste repository for a minute. When you think about this underground nuclear waste repository, what is the first thought or image that comes to mind? What is the next thought or image you have when I say *underground nuclear waste repository?*
Your next thought or image?

This continued until six associations were produced or the respondent drew a blank.

The 3334 respondents in the four surveys produced a combined total of exactly 10,000 word association images to the repository stimulus. The associations were examined and were classified according to their content to 13 general or superordinate categories, one of which was a miscellaneous category. All but one superordinate category contained subordinate categories—in one case there were 17 subordinate categories that were judged to fit the theme of the major category. All in all, there were 92 distinct categories. Many of these contained multiple associations, judged to have similar meanings. For example, the subcategory labeled *Dangerous/Toxic*, within the superordinate category labeled *Negative Consequences*, included the terms danger, dangerous, unsafe, disaster, hazardous, poisonous, and so on.

The 13 superordinate categories and their 92 subcategories contained 9439 word association images (94.4 percent of the total). Some 561 associations were left uncategorized (5.6 percent of the total).[4]

Table 3-6 presents the 13 superordinate categories in order of their combined frequencies across all four samples. The one exception to this ordering is the relatively large miscellaneous category, which is presented last. The subordinate categories are also shown, ordered by frequency within their superordinate category. Table 3-7 presents an ordering of the subordinate categories without regard for the superordinate structure.

The most arresting and most important finding is the extreme negative quality of these images. The two dominant superordinate categories, *Negative Consequences* and *Negative Concepts*, accounted for more than 56 percent of the total number of images. The dominant subordinate category, *Dangerous/Toxic*, contained almost 17 percent of the total number of images. The five largest subordinate categories,

Table 3-6 Images of a Nuclear Waste Repository: Totals for Four Surveys by Superordinate and Subordinate Categories

	Number of Images			Number of Images
I. Negative Consequences		d	Nevada/Las Vegas	227
a Dangerous/toxic	1683	e	Waste/garbage/dumps	215
b Death/sickness	783	f	Isolated	107
c Environmental damage	692	g	Facilities and their	
d Leakage	216		construction	66
e Destruction	133	h	Bury it	30
f Pain and suffering	18	i	Locations—other	20
g Uninhabitable	7		Total	1390
h Local repository area		**IV. Radiation, Physical States**		
consequences	6	a	Radiation/nuclear	336
i Negative		b	Chemicals and physical	
consequences—other	8		states (liquids, gases)	55
Total	3546	c	Fire/hot	33
II. Negative Concepts			Total	424
a Bad/negative	681	**V. Safety, Security**		
b Scary	401	a	Safety	228
c Unnecessary/opposed	296	b	Facilities security	44
d Not near me (NIMBY)	273	c	Control, containment,	
e War/annihilation	126		and cleanup	32
f Societally unpopular	41	d	Caution	27
g Crime and corruption	40		Total	331
h Decay/slime/smell	39	**VI. Concerns**		
i Darkness/emptiness	37	a	Problems	119
j Negative toward		b	Questions	58
decisionmakers and		c	Health	25
process	32	d	Unsolvable	19
k Commands to not build		e	Family	18
or to eliminate them	24	f	Uncontrolled	14
l Wrong or bad solution	19	g	Controversy	13
m No nuclear, stop		h	Unpredictable	11
producing	15	i	Mistakes	8
n Unjust	14	j	Serious	7
o Violence	10	k	Skeptical	5
p Prohibited	5	l	Concerns—other	14
q Negative—other	15		Total	311
Total	2068	**VII. Societal Institutions**		
III. Locations		a	Government/industry	125
a Non-Nevada locations	245	b	Military/weapons	106
b Storage location/		c	Science, technology,	
containers	243		research, and progress	42
c Desert/barren	237	d	Political process	31
			Total	304

Table 3-6 *Continued*

		Number of Images			Number of Images
VIII. Ecology			c	Improved environment	9
a	Natural environment	124	d	Feasible	3
b	Food and water supply	25	e	Positive—other	1
c	Climate	9		Total	97
	Total	158	XIII. Miscellaneous		
IX. Necessary			a	Future/long lasting	85
a	Necessary	156	b	Energy/power	65
	Total	156	c	Transportation	38
X. Economics			d	Find alternatives	31
a	Cost	58	e	Natural disasters	
b	Employment	57		(potential or actual)	29
c	Money/income	29	f	Population	22
d	Economics—other	5	g	Degree of distance	21
	Total	149	h	Neutral/apathetic/mixed	
XI. Information, Knowledge				feelings	20
a	Uninformed	57	i	Supervison/	
b	Unsure/unknown	39		responsibility	14
c	Curiosity, interest, and		j	Public figures	12
	knowledge	24	k	Fiction	11
d	Media	9	l	Problem avoidance	9
e	Information,		m	Inevitability	8
	knowledge—other	2	n	Faith	5
	Total	131	o	O.K. If . . .	4
XII. Positive				Total	374
a	Positive, unconcerned	59	XIV. Uncategorized		561
b	Effective	25	TOTAL NUMBER OF ITEMS		10,000

Ia. Dangerous/Toxic	16.83 %
Ib. Death/Sickness	7.83 %
Ic. Environmental Damage	6.92 %
IIa. Bad/Negative	6.81 %
IIb. Scary	4.01 %

were thoroughly negative in affective quality and accounted for more than 42 percent of the total number of images. The four most frequent single associations were *dangerous* (n = 539), *danger* (n = 378), *death* (n = 306), and *pollution* (n = 276).

Positive imagery was rare. Category XII, *Positive*, accounted for only 1 percent of the images. Other generally positive concepts, *Necessary* (cate-

Table 3-7 Subordinate Categories Ordered by Decreasing Frequency

		Number of Images			Number of Images
I. a	Dangerous/toxic	1683	XIII. c	Transportation	38
I. b	Death/sickness	783	II. i	Darkness/emptiness	37
I. c	Environmental damage	692	IV. c	Fire/hot	33
II. a	Bad/negative	681	V. c	Control, containment, and cleanup	32
II. b	Scary	401			
IV. a	Radiation/nuclear	336	II. j	Negative toward decisionmakers and process	32
II. c	Unnecessary/opposed	296			
II. d	Not near me (NIMBY)	273			
III. a	Non-Nevada locations	245	VII. d	Political process	31
III. b	Storage location/containers	243	XIII. d	Find alternatives	31
			III. h	Bury it	30
III. c	Desert/barren	237	XIII. e	Natural disasters (potential or actual)	29
V. a	Safety	228			
III. d	Nevada/Las Vegas	227	X. c	Money/income	29
I. d	Leakage	216	V. d	Caution	27
III. e	Waste/garbage/dumps	215	VI. c	Health	25
IX. a	Necessary	156	VIII. b	Food and water supply	25
I. e	Destruction	133	XII. b	Effective	25
II. e	War/annihilation	126	II. k	Commands to not build or to eliminate them	24
VII. a	Government/industry	125			
VIII. a	Natural environment	124	XI. c	Curiosity, interest, and knowledge	24
VI. a	Problems	119			
III. f	Isolated	107	XIII. f	Population	22
VII. b	Military/weapons	106	XIII. g	Degree of distance	21
XIII. a	Future/long lasting	85	III. i	Locations–other	20
III. g	Facilities and their construction	66	XIII. h	Neutral/apathetic/mixed feelings	20
XIII. b	Energy/power	65	VI. d	Unsolvable	19
XII. a	Positive, unconcerned	59	II. l	Wrong or bad solution	19
VI. b	Questions	58	VI. e	Family	18
X. a	Cost	58	I. f	Pain and suffering	18
XI. a	Uninformed	57	II. q	Negative–other	15
X. b	Employment	57	II. m	Nuclear, stop producing	15
IV. b	Chemicals and physical states (liquids, gases)	55	VI. f	Uncontrolled	14
			II. n	Unjust	14
V. b	Facilities security	44	XIII. i	Supervision/responsibility	14
VII. c	Science, technology, research, and progress	42	VI. l	Concerns–other	14
			VI. g	Controversy	13
II. f	Societally unpopular	41	XIII. j	Public figures	12
II. g	Crime and corruption	40	XIII. k	Fiction	11
II. h	Decay/slime/smell	39	VI. h	Unpredictable	11
XI. b	Unsure/unknown	39			

Table 3-7 *Continued*

	Number of Images			Number of Images
II. o Violence	10		I. h Local repository area	
XII. c Improved environment	9		consequences	6
XI. d Media	9		X. d Economics–other	5
VIII. c Climate	9		II. p Prohibited	5
XIII. l Problem avoidance	9		XIII. n Faith	5
I. i Negative consequences–			VI. k Skeptical	5
other	8		XIII. o O.K. If . . .	4
XIII. m Inevitability	8		XII. d Feasible	3
VI. i Mistakes	8		XI. e Information,	
VI. j Serious	7		knowledge–other	2
I. g Uninhabitable	7		XII. e Positive–other	1

Note: Roman numerals indicate superordinate categories.

gory IX), *Employment* (category Xb), and *Money/Income* (category Xc) combined to total only 2.5 percent of the images. The response *safe* was given only 37 times (0.37 percent).

Other noteworthy features of the combined data are:

—There were 232 associations pertaining to war, annihilation, weapons, and things military (categories IIe and VIIb).

—There were 85 associations relating to the long duration of storage necessary for nuclear wastes or the transfer of risk and responsibilities to future generations (XIIIa).

—There were surprisingly few (38) transportation images (XIIIc).

—The famous NIMBY position ("not in my backyard") was expressed in 273 images (category IId).

—Nuclear waste repositories are sometimes referred to derisively as "dumps." Although dump imagery was definitely present, it was infrequent (40 associations).

—Studies of risk perception have found that the risks of nuclear reactors and nuclear wastes have a dread quality. There were definite signs of this in the images. Although the word *dread* was never mentioned specifically, many of the responses categorized as *Scary* (IIb) reflected this quality (e.g., *fear, horror, apprehension, terror*).

—Lack of trust in DOE or other governmental agents is a common finding in studies of public perceptions of nuclear waste management. Associations indicative of distrust appeared in category IIj, *Negative Toward*

Decisionmakers and Process, and categories IlI and VIi, dealing with mistakes. A number of images in the *Bad/Negative* category also seemed to reflect lack of trust (e.g., *stupid, dumb, illogical*).

—Jones et al. (1984) have attempted to characterize the key dimensions of stigma. Two of their major defining characteristics of stigma are peril and negative aesthetic qualities (ugliness, repulsion). These qualities dominate the repository images. Peril is pervasive throughout categories I and II and elsewhere, and negative aesthetics form the bulk of the subordinate categories *Bad/Negative* (IIa) and *Decay, Slime, Smell* (IIh).

The image frequencies were very similar from one survey to another. Demographic differences were also small. The negativity of repository images was remarkably consistent across men and women of different ages, incomes, education levels, and political persuasions.

After free-associating to the repository stimulus, each respondent rated the affective quality of his or her associations on a five-point scale ranging from extremely negative to extremely positive. These affective ratings were highly correlated with the respondent's attitudes and perceptions of risk. For example, Table 3-8 shows a strong relationship between a person's rating of the first image they produced and their response to the question "Would you vote for a repository at Yucca Mountain?" More than 90 percent of the persons whose first image was judged very negative voted against a repository at Yucca Mountain; more than half of the persons whose first image was judged positive voted in favor of the repository. A similarly strong relationship was found between affective ratings of images and a person's judgment of the likelihood of accidents or other problems at a repository. Negativity of

Table 3-8 Relationship Between Affective Rating of a Person's First Image of a Nuclear Waste Repository and Their Response to the Question "Would You Vote for a Repository at Yucca Mountain?"

| | Evaluation of the First Image | | | | |
Yucca Mountain Vote	Very Negative	Somewhat Negative	Neutral	Somewhat Positive	Very Positive
Yes	8.9	17.6	38.9	54.2	60.7
No	91.1	82.4	61.1	45.8	39.3
N	305	34	18	24	28

$\chi^2 = 81.4$, $df = 4$, $p < .001$.

Note: Cell entries are column percentages. Responses based on a survey of Nevada residents.

the image rating was also strongly related to support for the state of Nevada's opposition to the repository program.

What was learned by asking more than 3300 people to associate freely to the concept of a nuclear waste repository? The most obvious answer is that people don't like nuclear waste. However, these images (as well as the responses to the attitude and opinion questions) demonstrate an aversion so strong that to call it "negative" or a "dislike" hardly does it justice. What these responses reveal are pervasive qualities of dread, revulsion, and anger—the raw materials of stigmatization and political opposition.

Because nuclear waste is a by-product of an impressive technology capable of producing massive amounts of energy without contributing to greenhouse gases, one might expect to find associations to energy and its benefits—electricity, light, heat, employment, health, progress, the good life—scattered among the images. Almost none were observed.

Moreover, people were not asked to reflect on nuclear waste; instead, they were asked about a *storage facility* (Phoenix survey) or a *repository* (other surveys). One might expect, following the predominant view of experts in this field, to find a substantial number of repository images reflecting the qualities *necessary* and *safe* (see Table 3-1). Few images of this kind were observed.

It appears that the repository has acquired the imagery of nuclear waste, through some process of transference—guilt by association. The transference is so natural, so powerful, that one state official involved in nuclear safety, upon hearing of these imagery results, indignantly accused us of having biased our respondents by calling the facility a "nuclear waste repository."

Evidence that the quality of repository imagery has not heretofore been appreciated comes from exhortations by nuclear power proponents not to use the term *dump* when referring to the repository, because of the obvious negative connotations or imagery this word conveys (Carter, 1987). Not only is dump or garbage imagery relatively infrequent in the observed responses, such images would appear rather benign in comparison to the more prevalent responses.

How Did It Get This Way?

Imagery and attitudes so negative and so impervious to influence from the assessments of technical experts must have very potent origins. Weart's scholarly analysis of images shows that nuclear fears are deeply rooted in our social and cultural consciousness. He argues persuasively that modern thinking about nuclear energy employs beliefs and symbols that have been associated for centuries with the concept of transmutation—the passage through destruction to rebirth. In the early decades of the twentieth century, transmutation images became centered on radioactivity, which was associated with "uncanny rays that brought hideous death or miraculous new life; with mad scientists and their ambiguous monsters; with cosmic secrets of death and life; . . . and with weapons great enough to destroy the world . . . ." (Weart, 1988:421).

But this concept of transmutation has a duality that is hardly evident in the imagery we observed. Why has the destructive aspect predominated? The answer likely involves the bombing of Hiroshima and Nagasaki, which linked the frightening images to reality. The sprouting of nuclear energy in the aftermath of the atomic bombing has led Smith to observe: "Nuclear energy was conceived in secrecy, born in war, and first revealed to the world in horror. No matter how much proponents try to separate the peaceful from the weapons atom, the connection is firmly embedded in the minds of the public" (1988:62).

Research supports Smith's assertions. A study by Slovic, Lichtenstein, and Fischhoff (1979) found that, even before the accident at Three Mile Island (TMI), people expected nuclear-reactor accidents to lead to disasters of immense proportions. When asked to describe the consequences of a "typical reactor accident," people's scenarios were found to resemble scenarios of the aftermath of nuclear war. Replication of these studies after the TMI event found even more extreme "images of disaster."[5]

Fiske, Pratto, and Pavelchak (1983) studied public images of nuclear war and obtained results that were similar to our repository images. The dominant themes of nuclear war were physical destruction (long-term, short-term, and immediate), death, injury, weapons, politics, hell, oblivion, nothingness, pain, contamination, radiation, end of civilization, and genetic damage. Dominant emotional images included fear, terror, worry, and sadness, with anger, hate, helplessness, and peace mentioned somewhat less frequently.

The shared imagery of nuclear weapons, nuclear power, and nuclear waste may explain some of the surprising results that have come from surveys that have examined perceived risks for these various forms of nuclear hazards. A nuclear waste repository is judged to pose risks at least as great as a nuclear power plant or a nuclear weapons test site (Kunreuther, Desvousges, and Slovic, 1988). If asked to indicate the closest distance a facility could be built from one's home before one would want to move to another place or actively protest, people are far more averse to living near a nuclear waste repository than any other kind of facility studied, including a nuclear power plant, a chemical-waste landfill, or a pesticide-manufacturing facility (Flynn et al., 1990).

Further insights into the special quality of nuclear fear are provided by Erikson (1990), who draws attention to the broad, emerging theme of toxicity, both radioactive and chemical, that characterizes a "whole new species of trouble" associated with modern technological disasters. Erikson describes the exceptionally dread quality of technological accidents that expose people to radiation and chemicals in ways that "contaminate rather than merely damage; . . . pollute, befoul, and taint rather than just create wreckage; . . . penetrate human tissue indirectly rather than wound the surface by assaults of a more straightforward kind" (1990:120). Unlike natural disasters, these accidents are unbounded. Unlike conventional disaster plots, they have no end. "Invisible contaminants remain a part of the surroundings—absorbed into the grain of the landscape, the tissues of the body and, worst of all, into the genetic material of the survivors. An 'all clear' is never sounded. The book of accounts is never closed" (1990:121).

Another strong determiner of public perceptions is the continuing story of decades of mishandling of wastes at the nation's military weapons facilities operated by DOE (National Academy of Sciences, 1989). Leakage from these facilities has resulted in widespread contamination of the environment, projected to require more than $150 billion for cleanup over the next thirty years. The recent revelation of unprecedented releases of radiation from the Hanford, Washington, weapons plant in the 1940s and 1950s (Marshall, 1990) will certainly compound the negative imagery associated with a nuclear waste repository and further undermine public trust in government management of nuclear waste disposal.

A Crisis of Confidence

Analysis of these survey data provides insight into the remarkably negative attitudes toward radioactive waste disposal facilities and the impassioned opposition to government efforts to site high-level and low-level waste repositories. The negativity of perceptions and emotions associated with a repository are remarkable in light of the confidence that most technical analysts and engineers have in their ability to dispose of radioactive materials safely. Even the report of the National Research Council, though highly concerned about the difficulties of predicting the long-term performance of a repository, conceded that "these uncertainties do not necessarily mean that the risks are significant, nor that the public should reject efforts to site the repository" (1990:13).

Chauncey Starr, pointing to the public's lack of concern about the risks from tigers in urban zoos, has argued that "acceptance of any risk is more dependent on public confidence in risk management than on the quantitative estimates of risk . . . ." (1985:98). Public fears and opposition to nuclear waste disposal plans can be seen as a "crisis of confidence," a profound breakdown of trust in the scientific, governmental, and industrial managers of nuclear technologies.

Viewing the nuclear waste problem as one of distrust in risk management gives additional insight into its difficulty. Social psychological studies (Rothbart and Park, 1986) have validated "folk wisdom" by demonstrating that trust is a quality that is quickly lost and slowly regained.[6] A single act of embezzlement is enough to convince us that our accountant is untrustworthy. A subsequent opportunity to embezzle that is not taken does little to reduce the degree of distrust. Indeed, a hundred subsequent honest actions would probably do little to restore our trust in this individual.

In this light, the 1989 attempt by DOE to regain the confidence of the public, the Congress, and the nuclear industry by simply rearranging its organizational chart and promising to do a better job of management and science in the future (Moore, 1989) appears naive. Trust, once lost, cannot be so easily restored. Similarly naive is the aim professed by DOE officials and other nuclear industry leaders to change public perception and gain support by letting people see firsthand the safety of nuclear waste management. The nature of any low-probability, high-consequence threat is such that adverse events will demonstrate riskiness, but demonstrations of safety (or negligible risk) will require a very long time, free of damaging incidents. The intense scrutiny given to nuclear power and nuclear waste issues by the news media (Mazur, 1990) insures that a stream of problems, occurring all over the world, will be brought to the public's attention, continually eroding trust.

Where Next for Nuclear Waste Disposal?

Although everyone appreciates the sophisticated engineering required to store nuclear wastes safely, the political requirements necessary to design and implement a repository have not similarly been appreciated. As a result, notes Jacob, "while vast resources have been expended on developing complex and sophisticated technologies, the equally sophisticated political processes and institutions required to develop a credible and legitimate strategy for nuclear waste management have not been developed" (1990:164).

In the absence of a trustworthy process for siting, developing, and operating a nuclear waste repository, the prospects for a short-term solution to the disposal problem seem remote. The report of the National Research Council (1990) is quite sensitive to issues of risk perception and trust but makes the strong assumption that trust can be restored by a process that openly recognizes the limits of technical understanding and does not aim to "get it right the first time." It seems likely that such open admission of uncertainty and refusal to guarantee safety might well have opposite effects from those intended—increased concern and further deterioration of trust. Moreover, the NRC statement also assumes that DOE will continue to manage the nuclear waste program, thus failing to come to grips with the difficulties that DOE will face in restoring its tainted image.

The lack of a trustworthy process for siting, developing, and operating a nuclear waste repository has drawn a number of other comments and recommendations besides those of the NRC. Weinberg (1989) drew an analogy between fear of witches during the fifteenth through seventeenth centuries and today's fear of harm from radiation. He hypothesized that "rad-waste phobia" may dissipate if the intelligentsia (read "environmentalists") say that such fears are unfounded, much as eventually happened with fears of witches. Carter argued that "trust will be gained by building a record of sure, competent, open performance that gets good marks from independent technical peer reviewers and that shows decent respect for the public's sensibilities and common sense" (1987:416). He also recommended that the National Academy of Sciences undertake a study to determine how an independent and credible process of peer review could be established to increase public trust in repository siting and development and to determine how state and local governments can best be given a voice in siting investigations and in oversight of actual repository operations. Others have called for more radical changes, such as creating new organizations and developing procedures to ensure that state, local, and tribal governments have a much stronger voice in siting decisions and oversight of actual repository operations (e.g., Advisory Panel, 1984; Bella, Mosher, and Calvo, 1988; Bord, 1987; Creighton, 1990; Jacob, 1990). In this spirit, an official of the Canadian government has argued for making repository siting in that country voluntary by requiring public consent as an absolute prerequisite for confirming any decision (Frech, 1991).

Whatever steps are taken, it is unlikely that the current "crisis in confidence" will be ended quickly or easily. We must settle in for a long effort to restore the public trust. Krauskopf (1990) has noted that postponing the repository to an indefinite future can be defended on a variety of technical grounds, and points out that the choice between repository construction or postponement ultimately rests upon the shoulders of the public and their elected representatives. The problems of perception and trust described above imply that postponement of a permanent repository may be the only politically viable option in the foreseeable future.

In an address to the National Association of Regulatory Utility Commissioners in November 1990, Joseph Rhodes, Jr., himself a commissioner from Pennsylvania, pointed out the implications of the polls indicating that most Nevadans oppose the siting of a repository anywhere in Nevada and want state leaders to oppose such siting with any means available (Rhodes, 1990). "I can't imagine," said Rhodes, "that there will ever be a usable Yucca Mountain repository if the people of Nevada don't want it. . . . There are just too many ways to delay the program. . . ." (1990:6).[7]

What are the options in the light of dedicated public opposition to a permanent underground repository? Rhodes lists and rejects several:

> —Continuing on the present path in an attempt to site a permanent repository (which Rhodes refers to as the modern equivalent of "pyramids underground") is a costly and doomed effort.
> —Permanent on-site storage is unsafe.
> —Deploying a monitored retrievable storage (MRS) program is also politically unacceptable. Without a viable program to develop a permanent repository, the MRS would be seen, in effect, as the permanent site.
> —Reprocessing the spent nuclear fuel is also politically unacceptable because of concerns over nuclear weapons proliferation. Moreover, reprocessing reduces but does not eliminate high-level wastes, and the record of managing reprocessing residues at Hanford and other military sites is hardly encouraging.

Rhodes concludes that the only viable option is to delay the siting of a permanent repository for several decades and store the wastes on site in the interim—employing dry-cask storage that has been certified by NRC as being as safe as geological storage for 100 or more years (Nuclear Regulatory Commission, 1990). Technical knowledge would undoubtedly advance greatly during this interim period. Perceptions of risk and trust in government and industry might change greatly, too, if the problem of establishing and maintaining trust is taken seriously.

Beyond Yucca Mountain

The story of Yucca Mountain has implications for environmental decision-making that transcend the conflicts and concerns surrounding the disposal of radioactive wastes. People's perceptions of chemicals are almost as negative as their perceptions of radioactivity. Any major facility that produces, uses, transports, or disposes of chemicals will face similar problems originating from perceptions of risk that bear little resemblance to the risk assessments of technical experts. No one is happy about the current state of affairs. Industrialists, scientists, politicians, and the public are united only in their anger and frustration about the ways that environmental risks are currently managed.

Restoration and preservation of trust in risk management needs to be given top priority. A solution to the problem of trust is not immediately apparent. The problem is not due to public ignorance or irrationality but is deeply rooted in the adversarial nature of our social, institutional, legal, and political systems of risk management. Public relations won't create trust. Aggressive and competent government regulation, coupled with increased public involvement, oversight, and local control over decisionmaking might.

Notes

This chapter relies extensively on material from the authors' article by the same title published in *Environment*, 33, no. 3 (April, 1991): 6–11, 28–30, reprinted by permission from Heldref Publications. It also draws on material from P. Slovic, J. Flynn, and M. Layman, "Perceived Risk, Trust, and the Politics of Nuclear Waste," published in *Science*, 254 (December 13, 1991): 1603–7.

1   The Nevada results reported in this section are based upon the 500 respondents in the statewide survey.
2   The 1991 survey was conducted in the days following the conclusion of the Gulf War when President Bush's approval ratings had reached unprecedented levels.
3   Respondents in the Phoenix survey were asked to associate to the term *underground nuclear waste storage facility*.
4   A complete listing of all 10,000 images, including those that were not categorized, is available from the authors.
5   The fact that the earliest technical risk assessments for nuclear power plants portrayed "worst-case scenarios" of tens of thousands of deaths and devastation over geographic areas the size of Pennsylvania likely contributed to such extreme images (see Ford, 1977). These early projections received enormous publicity, as in the movie *The China Syndrome*.
6   Abraham Lincoln, in a letter to Alexander McClure, observed: "If you once forfeit the confidence of your fellow citizens, you can never regain their respect and esteem."
7   Rhodes's assertion echoes an earlier statement made by a former DOE official, John O'Leary, in an interview with Luther Carter: " 'When you think of all the things a determined state can do, it's no contest,' O'Leary told me, citing by way of example the regulatory authority a state has with respect to its lands, highways, employment codes, and the like. The federal courts, he added, would strike down each of the state's blocking actions, but meanwhile years would roll by and, in a practical sense, DOE's cause would be lost" (Carter, 1987: 185).

References

Advisory Panel on Alternative Means of Financing and Managing Radioactive Waste Facilities. 1984. *Managing Nuclear Waste: A Better Idea*. Report to the Secretary of Energy. Washington, D.C.: DOE.

Bella, David A., Charles D. Mosher, and Steven N. Calvo. 1988. "Technocracy and Trust: Nuclear Waste Controversy." *Journal of Professional Issues in Engineering* 114:27–39.

Bord, Richard J. 1987. "Judgments of Policies Designed to Elicit Local Cooperation on LLRW Disposal Siting: Comparing the Public and Decision Makers." *Nuclear and Chemical Waste Management* 7:99–105.

Campbell, Angus, Philip E. Converse, W. E. Miller, and D. E. Stokes. 1960. *The American Voter*. New York: John Wiley & Sons.

Carter, Luther J. 1987. *Nuclear Imperatives and Public Trust: Dealing with Radioactive Waste*. Washington, D.C.: Resources for the Future.

Cohen, Bernard L. 1983. *Before It's Too Late: A Scientist's Case for Nuclear Energy*. New York: Plenum.

Creighton, James. 1990. "Siting Means Safety First." *Forum for Applied Research and Public Policy* 5 (Spring): 97–98.

Erikson, Kai T. 1990. "Toxic Reckoning: Business Faces a New Kind of Fear." *Harvard Business Review* (January–February): 118–26.

Fiske, Susan T., Felicia Pratto, and Mark A. Pavelchak. 1983. "Citizen's Images of Nuclear War: Contents and Consequences." *Journal of Social Issues* 39:41–65.

Flynn, James H., C. K. Mertz, and Paul Slovic. 1991. *The 1991 Nevada State Telephone Survey: Key Findings*. Report 91–2. Eugene, OR: Decision Research.

Flynn, James H., Paul Slovic, C. K. Mertz, and James Toma. 1990. *Evaluations of Yucca Mountain*. Report 90-4. Eugene, OR: Decision Research.

Ford, Daniel F. 1977. *The History of Federal Nuclear Safety Assessment: From WASH 740 Through the Reactor Safety Study*. Cambridge, MA: Union of Concerned Scientists.

Frech, Egon R. 1991. "How Can We Deal with NIMBY in Nuclear Waste Management?" Paper presented at the International High-Level Radioactive Waste Management Conference, May, Las Vegas, NV.

Freud, Sigmund. 1924. *Collected Papers*. London: Hogarth.

Galton, Francis. 1880. "Psychometric Experiments." *Brain* 2:149–62.

Hamilton, David L. 1981. *Cognitive Processes in Stereotyping and Intergroup Behavior*. Hillsdale, NJ: Erlbaum.

Jacob, Gerald. 1990. *Site Unseen: The Politics of Siting a Nuclear Waste Repository*. Pittsburgh, PA: University of Pittsburgh Press.

Jones, Edward, Amerigo Farina, Albert Hastorf, Hazel Markus, Dale Miller, and Robert Scott. 1984. *Social Stigma: The Psychology of Marked Relationships*. New York: W. H. Freeman.

Kasperson, Roger E. 1990. "Social Realities in High-Level Radioactive Waste Management and Their Policy Implications." In *Proceedings, International High-Level Radioactive Waste Management Conference*, Vol. 1. LaGrange, IL: American Nuclear Society. 512–18.

Krauskopf, Konrad. 1990. "Disposal of High-Level Nuclear Waste: Is It Possible?" *Science* 249:1231–32.

Kunreuther, Howard, William H. Desvousges, and Paul Slovic. 1988. "Nevada's Predicament: Public Perceptions of Risk from the Proposed Nuclear Waste Repository." *Environment* 30 (8): 16–20, 30–33.

Kunreuther, Howard, and Douglas Easterling. 1990. *Imagery and Convention Decision Making*. Technical Report. Philadelphia: Wharton Risk and Decision Processes Center.

Lewis, H. W. 1990. *Technological Risk*. New York: Norton.

Lifton, Robert. 1967. *Death in Life*. New York: Basics.

Marshall, Eliot. 1990. "Hanford Releases Released." *Science* 249:474.

Mazur, Allan. 1990. "Nuclear Power, Chemical Hazards, and the Quantity-of-Reporting Theory of Media Effects." *Minerva* 28:294–323.

Moore, W. Henson. 1989. Remarks before the 1989 Nuclear Energy Forum, San Francisco, November 28.

National Academy of Sciences. 1989. *The Nuclear Weapons Complex: Management for Health, Safety, and the Environment*. Washington, D.C.: National Academy Press.

National Research Council. 1990. *Rethinking High-Level Radioactive Waste Disposal*. Washington, D.C.: National Academy Press.

Nealey, Stanley, and John A. Hebert. 1983. "Public Attitudes Towards Radioactive Waste." In Charles A. Walker, Leroy C. Gould, and Edward J. Woodhouse, eds., *Too Hot to Handle: Social and Policy Issues in the Management of Radioactive Wastes*. New Haven: Yale University Press. 94–111.

Nuclear Regulatory Commission. 1990. "Storage of Spent Fuel in NRC-Approved Storage Casks at Power Reactor Sites." *Federal Register* 55 (138): 29, 181–95.

Rhodes, Joseph Jr. 1990. "Nuclear Power: Waste Disposal: New Reactor Technology, Pyramids Underground." Paper presented at the 102nd Annual Meeting of the National Association of Regulatory Utility Commissioners, Orlando, FL, November 13.

Rothbart, Myron, and Bernadette Park. 1986. "On the Confirmability and Disconfirmability of Trait Concepts." *Journal of Personality and Social Psychology* 50:131–42.

Slovic, Paul. 1990. "Perception of Radiation Risks." In *Radiation Protection Today—The NCRP at Sixty Years: Proceedings of the Twenty-fifth Annual Meeting of the National Council on Radiation Protection and Measurements*, Vol. 11. Bethesda, MD: NCRPM. 73–97.

Slovic, Paul, Mark Layman, Nancy Kraus, James Chalmers, Gail Gesell, and James H. Flynn. 1991. "Perceived Risk, Stigma, and Potential Economic Impacts of a High-Level Nuclear Waste Repository in Nevada." *Risk Analysis* 11:683–96.

Slovic, Paul, Sarah Lichtenstein, and Baruch Fischhoff. 1979. "Images of Disaster: Perception and Acceptance of Risks from Nuclear Power." In Gordon Goodman and William Rowe, eds., *Energy Risk Management*. 223–45.

Smith, Kirk R. 1988. "Perception of Risks Associated With Nuclear Power." *Energy Environment Monitor* 4 (1): 61–70.

Starr, Chauncey. 1985. "Risk Management, Assessment, and Acceptability." *Risk Analysis* 5:97–102.

Szalay, Lorand, and James Deese. 1978. *Subjective Meaning and Culture: An Assessment through Word Associations*. Hillsdale, NJ: Erlbaum.

Technical Review Board. 1991. *Third Report to the U.S. Secretary of Energy*. Nuclear Waste Technical Review Board, Arlington, VA.

Weart, Spencer R. 1988. *Nuclear Fear: A History of Images*. Cambridge, MA: Harvard University Press.

Weinberg, Alvin M. 1989. "Public Perceptions of Hazardous Technologies and Democratic Political Institutions." Paper presented at *Waste Management '89*, Tucson, AZ.

Wundt, Wilhelm. 1883. "Über Psychologische Methoden." *Philosophische Studien* 1:1–38.

Part II

Public Reactions to

Preliminary Sites

4 Public Testimony in Nuclear Waste Repository Hearings: A Content Analysis

Michael E. Kraft and Bruce B. Clary

Numerous studies, including those reported in this volume, tell us that the American public considers radioactive waste to be among modern society's most dangerous technological risks. It ranks exceptionally high on two key dimensions of perceived hazards, the extent to which it is unknown and the degree to which it is dreaded (Slovic, 1987),[1] and public attitudes on the issue have proved difficult to change (Nealey and Hebert, 1983; Slovic and Fischhoff, 1983). These public perceptions and attitudes clearly pose major obstacles to siting a high-level nuclear waste repository, as mandated by the Nuclear Waste Policy Act of 1982 (NWPA) and its amendments in 1987.

Since adoption of the 1982 act, the Department of Energy (DOE) has been committed to rapid development of a permanent repository. The U.S. Congress has been highly supportive of that goal, but, in deference to public fears, it required in the 1982 act extensive involvement by citizens, the states, and Indian tribes in DOE's repository siting process. The logic of doing so was understandable enough: without some way to build public confidence in the siting process, any decision to locate the repository would likely be subject to vigorous public protest that could seriously delay or halt the program. Put more positively, successful implementation of the nuclear waste program depends upon building and maintaining a requisite level of public support (Mazmanian and Sabatier, 1983). The chief mechanism Congress chose to attain this objective was public hearings in those states being considered for the repository.

Given the general importance of public acceptance to policy implementation, the role of hearings in the site evaluation process merits close study. In this chapter, we examine public testimony during the second round siting process in four states (Wisconsin, Maine, North Carolina, and Georgia) to determine what the hearings held in those states can tell us about public perceptions and attitudes toward radioactive waste, especially aspects of public concerns that are difficult to measure through survey research. We want to ask as well about policy implications of the testimony in these states.

Public Hearings as a Form of Citizen Participation and as a Data Source

Hearings as Citizen Participation

Our focus here is on public attitudes toward nuclear waste rather than on the process or on the impact of citizen participation in the policy process. However, because we deal with testimony at public hearings, it is important to recognize that hearings typically are mandated by Congress as one mechanism for ensuring that governmental decisionmaking is responsive to public concerns. This is an ambitious objective, especially for technically demanding problems such as nuclear waste disposal. The literature is replete with trenchant criticisms of the potential of hearings to meet such expectations (Checkoway, 1981). The reasons commonly offered include assertions that hearings provide—at best—a weak opportunity for public influence on decisions, provide little opportunity for two-way communication or interaction, may result in poor and excessively technical presentation of information, may distort the general public's views because those testifying often are unrepresentative of the public, and may be used by administrators to provide evidence of public involvement while devoting little or no time to analysis of the information obtained (Arnstein, 1969; Heberlein, 1976).

Whatever truth there may be to such criticisms, hearings also function to legitimize agency decisions, defuse opposition, warn the agency of possible political obstacles, and satisfy legal or procedural requirements (Fiorino, 1989; Milbrath, 1981). They are not, and need not be, as inconsequential as described in the literature or as many administrators believe them to be (Aberbach and Rockman, 1978; Gormley, 1989). Indeed, it would seem that the efficacy of different approaches to citizen participation, including hearings, is heavily dependent on their design and execution as well as on the larger context of public perception of the problem, the consistency of participation with the agency's policy goals, and the agency's perceived credibility. We return to some of these considerations at the end of the chapter when we discuss policy implications of the repository siting hearings.

Hearings as a Data Source

Whatever one makes of the potential or limitations of public hearings, we believe that researchers can fruitfully explore hearing transcripts for evidence otherwise not available. They are a rich source of data on those individuals who are often the most concerned about an issue (the attentive public) and frequently the most informed. Thus, if used properly, hearings can promote understanding of citizen attitudes and their response to governmental policy decisions in ways that survey data cannot.

Analysis of hearings offers several advantages. First, in comparison to surveys, they are grounded in particular contexts. For example, public hearings are usually conducted in response to a specific report issued by the federal government, such as the draft *Area Recommendation Report for the Crystalline Repository Project*, to be discussed below. Second, hearings involve citizens who are likely to have greater, and more specific, knowledge about a problem in contrast to surveys of the general public's assessment of policy issues. The length of many statements provides detailed information about the many complexities and interrelationships of the issues that are addressed. Third, content analysis is an unobtrusive method; there is no direct interaction between the researcher and subject. The analyst does not dictate the form of response as in a structured questionnaire; coding categories are used, but the data are based on unprompted statements.

Hearing statements are not without their weaknesses as a form of policy information. Unlike surveys based on probability samples, there can be no claim to representativeness. Those testifying may be important political actors, such as state government officials, but we cannot generalize from them to the population at large. Additionally, statements at hearings are often made for political effect, and they may be more indicative of such considerations than of the attitudes or opinions of the individuals testifying.

Given our purposes, we should note also that criticism of hearings typically focuses more on how they are employed, especially when they are the only technique of public involvement used, than on the kinds of information they are capable of producing. In the specific case of NWPA, hearings were required, and we can make use of the transcripts to consider a segment of the population, the attentive public, which is not easily accessible in survey research (Key, 1963:265, 282–85).

Public Hearings in the Nuclear Waste Siting Process: 1984–1986

The NWPA called for nationwide screening of potentially suitable sites for two repositories, one in the West and another in the East. DOE was to conduct assessments of the sites that would address physical, social, and environmental impacts of site characterization (the actual testing of a potential site) as well as repository siting, construction, and operation. These assessments were to be subject to review and comment through a public hearing process, which DOE expected to encourage acceptance of the overall repository program.

DOE's implementation of the citizen participation provisions in the act occurred in two stages, reflecting the different tasks and circumstances in the western and eastern site evaluation process. In December 1984, the department issued nine draft environmental assessments (EAS) for public comment in what became known as the "first round" of repository siting. These involved sites in six states in the West and South: Washington, Nevada, Texas, Utah, Mississippi, and Louisiana. Over 20,000 comments were received on the EAS, an unmistakable sign of the salience of the nuclear waste issue in these states. Criticism was directed especially at omissions and other technical deficiencies in the data used, the methodology for site evaluation and comparison, and what was thought to be DOE bias in the evaluation process (Bryan, 1987; Clary and Kraft, 1988 and 1989). The result was an increase in state and public opposition to DOE's efforts and more determined actions to prevent the site evaluation process from going forward (Carter, 1987).

A similar pattern developed two years later in the East. A total of 235 sites were narrowed through regional and area surveys to 20 locations in seven states: Wisconsin, Minnesota, Maine, New Hampshire, Virginia, Georgia, and North Carolina. A preliminary environmental assessment, the draft *Area Recommendation Report for the Crystalline Repository Project* (ARR), was prepared for public review (DOE, 1986), and, during a three-month period in early 1986, DOE conducted some 39 briefings and 38 public hearings on the report. Some 18,000 people attended these sessions, and another 3200 individuals and organizations provided written comments; a total of some 60,000 comments (as DOE classifies them) were received (DOE, 1987:2).

The volume of public testimony and the highly negative reaction of the public and states to DOE's siting proposals make examination of the hearing process especially suitable for addressing several questions about public attitudes toward nuclear waste. First, what was the extent of opposition to the repository siting proposals, and how did the level of opposition vary by state and by the characteristics of those testifying? Second, what can we say about the nature of the opposition itself? Was the public's response thoughtful and well-informed? Were the issues seen in a broad context or in one that was narrow, parochial, and emotional in its outlook, as would be expected if opposition were a manifestation of the "not in my back yard" (NIMBY) syndrome (Kraft and Clary, 1991; Mazmanian and Morell, 1990)? Third, to what extent did those testifying express trust and confidence in the DOE and related governmental agencies? Fourth, what particular criticisms were directed at DOE and its report on the environmental effects of siting the repository?

Methodology

We selected the states of Wisconsin, Maine, North Carolina, and Georgia to provide what we believe to be a suitable sample of public reaction to DOE's plan in the seven eastern states being considered for a repository. By choosing one state in the upper Midwest, one in the Northeast, and two states in the South that differed on the issues, we hoped to tap concern about specific geological and environmental problems in repository siting that varied from state to state. Georgia was selected in part because the state's political leadership appeared less hostile to DOE than the others.

Within these four states, a total of 1045 individuals testified at DOE's formal hearings; 47 percent of the hearing statements were in Maine, 23 percent in Wisconsin, 20 percent in North Carolina, and 11 percent in Georgia. All hearings were held within the same three-month period, which lessens the influence of external time-related factors (such as new technical information or changes in the political climate) on variation across the hearings within each state and across the four states. DOE contracted with outside reporting services to transcribe tape recordings of the public testimony, and the Crystalline Repository Project Office in Argonne, Illinois, provided us with a full set of transcripts for the four states.[2] Following an initial review of selected transcripts, we developed an extensive coding scheme for the content analysis. We tried to answer fifteen questions, many with multiple subparts, for each statement at the hearings, the unit of analysis in this case. Some of these statements were fairly brief (the equivalent of a few paragraphs), while others were lengthy. Because of differences in the length of testimony and the scope of issues addressed, for the various analyses below we use only those cases where the individual's position could be coded.[3]

The Public Response to DOE's Repository Siting Proposals

The results of the content analysis confirm the widely held view that public response to DOE's recommendations (DOE, 1986) was extremely negative. Thirty percent of the 1045 people who testified showed evidence of having read some part of the ARR, and, of that 30 percent, 88 percent rejected the conclusions of the report. In 9 percent of the cases, the individual's position was unclear, but less than 1 percent spoke in favor of the report. If we exclude those whose position was not ascertainable, 96.5 percent (275 of the 285 who were familiar with the ARR) were in opposition to the siting proposals. The results are displayed in Table 4-1.

Data were collected on several other dimensions of the public's response to DOE's siting plans. All of the 1045 statements were coded for the general level of opposition to DOE's recommendations, whether specifically involving the ARR or not. Where a position was clearly discernable (which it was in 50 percent of the cases), 70 percent of those testifying voiced strong opposition to the department's plans. Twenty-six percent were categorized as showing moderate levels of disagreement, and only 4 percent were classified as having weak objections. The results are shown in Table 4-2.

On the question of whether a facility should be located within the state, our third measure of opposition, 57 percent of the persons testifying were strongly opposed, and 33 percent indicated moderate opposition (the latter position characterized by statements indicating the individual recognized the need to site a repository somewhere). For this variable, 10 percent were coded as having only weak opposition, and .4 percent as being unopposed to siting a repository within the state. These data are shown in Table 4-3.

For the second (general) and third (within states) measures of opposition

Table 4-1   Public Opposition to Repository Siting: Reaction to the Area Recommendation Report Among Those Having Read It

Reaction to the Area Recommendation Report	Percent[a]	N
Opposed	88	(275)
Neutral	3	(9)
Positive	0.3	(1)
Not ascertained	9	(29)
Total	100	(314)

[a] The ARR was addressed in 30 percent of the cases. These percentages refer to that subgroup of those testifying at the hearings. Rounding error is present in the percentages.

Table 4-2   Public Opposition to a Nuclear Waste Repository: Comprehensive Measure[a]

Position on Repository Siting	Percent	N
Strongly opposed	70	(366)
Moderately opposed	26	(135)
Weakly opposed	4	(20)
Total	100	(521)

[a] This measure refers to a coding of the entire statement, without regard to the ARR or an in-state repository location. A position was unambiguously ascertainable in 50 percent of the cases. The percentages listed are for that subgroup of those testifying.

Table 4-3   Public Opposition to Siting a Nuclear Waste Repository Within the State

Position on Repository Siting[a]	Percent	N
Strongly opposed	57	(591)
Moderately opposed	33	(343)
Weakly opposed	10	(103)
Not opposed	.4	(4)
Total	100	(1039)

[a] This measure of opposition refers to the individual's general stance on siting a repository in the state, aside from specific DOE recommendations contained in the ARR. Rounding error is present in percentages and figures.

Table 4-4   Public Opposition to DOE Siting Recommendations and In-State Repository Siting by State and Group Affiliation

State/Group Affiliation	DOE Siting Recommendations[a] Strongly Opposed		In-State Repository Siting Strongly Opposed	
	Percent	N	Percent	N
Wisconsin	81	(71)	56	(133)
Maine	65	(199)	58	(281)
North Carolina	74	(67)	70	(147)
Georgia	81	(29)	28	(30)
Total	70	(366)	57	(591)
Citizens (no stated affiliation)	73	(216)	62	(377)
Environmentalists	54	(13)	47	(16)
Interest groups (non-environmental)	64	(45)	61	(78)
Government officials	71	(49)	48	(59)
Indian tribes	85	(23)	49	(30)
Industry or utilities	30	(7)	24	(10)

[a] This comprehensive measure of opposition refers to a coding of the entire statement, without regard to the ARR or an in-state repository location. A position was unambiguously ascertainable in 50 percent of the cases. The percentages listed are for that subgroup of those testifying.

to DOE's proposals, we found, with a few exceptions, little variation across states or among individuals and groups who testified at the hearings.[4] The findings appear in Table 4-4. With the few exceptions to be noted, strong opposition was evident regardless of the state of residence and regardless of whether an individual spoke on his or her own behalf or represented a group. Georgia was much lower in opposition to an in-state site (28 percent), possibly because political leadership in the state did not take a strong position on the siting issue. Representatives of industry and utilities also showed markedly less opposition, as might be expected. We might also imagine that government officials (primarily state and local officials in this group) would react more positively to DOE than citizens in general; however, this was not the case. Yet, it must be remembered that the relations between DOE and state government offices, particularly in Wisconsin and Maine, were quite strained by 1986, after several years of "consultation and cooperation" had failed to bring about much agreement on repository siting (Shaefer, 1988).

A NIMBY Response?

A common explanation offered for public opposition to proposals for the siting of hazardous facilities is the NIMBY syndrome (Kraft and Clary, 1991). NIMBY refers to intense and often adamant resistance by the local population. In the classic formulation of this construct, public opposition is considered to be irrational; the public is pictured as poorly informed, interested primarily in avoiding local imposition of risks, and emotive rather than cognitive in its appraisal of the risk and in its response to siting proposals. To test whether the NIMBY explanation is helpful in understanding public opposition to DOE's repository siting efforts, we examined the level of technical knowledge displayed by those testifying at the hearings, the geographical orientation evident in their statements, and the level of emotionalism in the statements made.

Level of Knowledge. We measured the level of knowledge of those testifying by coding their statements in one of three categories of technical awareness: high, moderate, and low. Determination could be made for 87 percent of those testifying. Of these, 20 percent indicated a relatively high level of awareness of technical issues, 47 percent a moderate level, and 34 percent a fairly low level. In the high category, individuals used scientific terminology, criticized specific proposals, or indicated that they had read the ARR. Those coded as having a moderate knowledge level offered criticisms without detail, but they did exhibit a general understanding of at least some technical dimensions of nuclear waste disposal. Those in the low category showed little understanding of technical details, and often spoke in an emotional manner. Based on these criteria, the data indicate that those who attended and spoke at the public hearings could be described as relatively knowledgeable. Sixty-seven percent were identified as possessing at least a moderate level of information about nuclear waste issues as we measured it. Knowledge levels are displayed in Table 4-5.

Another finding is consistent with this evidence of a moderately well-informed public. Nearly one-fourth of those testifying offered suggestions for alternative ways to handle radioactive waste disposal, some of which

Table 4-5   Knowledge Level of Individuals Testifying at Public Hearings on Nuclear Waste Repository Siting by Group Affiliation (in percentages)

Knowledge Level[a]	Group Affiliation						
	Total	Citizens	Environ-mentalists	Government Officials	Interest Groups[c]	Industry & Utilities	Indian Tribes
High	20	16	28	35	18	31	24
	(179)	(82)	(8)	(39)	(21)	(11)	(13)
Moderate	47	48	55	42	47	46	33
	(422)	(248)	(16)	(47)	(54)	(16)	(18)
Low	34	37	17	24	35	23	43
	(305)	(191)	(5)	(27)	(40)	(8)	(23)
Total[b]	101	101	100	101	100	100	100
	(906)	(521)	(29)	(113)	(115)	(35)	(54)

[a] This variable was coded on the basis of a rater's estimate of a person's knowledge about and understanding of the dimensions of nuclear waste disposal based on the entire testimony.

[b] In 15 percent of the cases, no determination of the level of technical knowledge was possible. Cases were also dropped from the group affiliation section of the analysis if an affiliation was unclear. Hence, the total for the columns does not equal the total for the population as a whole. Totals exceed 100 percent due to rounding error.

[c] This category includes all interest groups other than environmental groups or those representing industry or utilities.

were quite specific. These suggestions concerned the production of nuclear waste itself (from reliance on nuclear energy), the possible use of a monitored retrievable storage (MRS) facility instead of permanent geologic disposal, and research into energy production and waste disposal alternatives. These citizens were not simply objecting to siting a facility in the local area; they were criticizing the design of the repository siting program and the reasons offered to justify it.

The differences across the states and among individuals and groups were limited, but some of them are of interest. For example, North Carolina had by far the greatest number of individuals in the lowest knowledge category, whereas Maine and Wisconsin had the largest proportions demonstrating a high level of knowledge on nuclear waste issues. One might also hypothesize that the relatively high level of knowledge among those testifying can be explained by the participation in these hearings of individuals with specialized knowledge, for example, government officials or representatives of environmental groups.

The data shown in Table 4-5 do not support this interpretation. There is little difference across the group categories as measured by an uncertainty coefficient of only .02.[5] Combining the high and moderate knowledge categories helps to explain this finding; the majority of people in all groups (57 to 83 percent) are highly or moderately knowledgeable. Environmentalists do indeed score higher than other groups, but not by very much, and government officials are not notably more knowledgeable than others.

One implication of these findings is that improved risk-communication processes will not necessarily alter opposition to repository siting. Opposition is not explained by a low level of understanding of the technical issues, and hence concentrating on building the public's knowledge of these issues offers no guarantee of reduced opposition. The greater task would seem to be to convince an already informed, attentive public that DOE's assessments of repository safety are scientifically valid and credible.

Geographic Orientation. The conventional view of the NIMBY response is that it is primarily a local, and often very personal, reaction against the siting of a feared or unwanted facility. There is indeed empirical support for the relationship between proximity to a proposed hazardous waste facility and reaction to it. Risks are perceived as greater and benefits as less by people who live in close proximity to a proposed site (Marks and von Winterfeldt, 1984). One might expect evidence of such a parochial outlook in hearings on nuclear waste repository siting.

Testimony at the hearings did reveal a local orientation. Sixty-eight percent of those making statements discussed their local communities and nearby areas in the context of other concerns raised. However, broader geographical perspectives also were evident. Sixty percent addressed the consequences of a repository for their state as a whole, 28 percent talked about difficulties other states and the nation faced in dealing with DOE, and 11 percent discussed the international implications of the nuclear waste question. Only 23 percent spoke exclusively in terms of local impacts. The only notable variation among groups is that representatives of industry and utilities and of Indian tribes exhibit a more pronounced local, as opposed to state-level, geographic focus. See Table 4-6.

Emotionalism. Finally, in most conceptions of the NIMBY syndrome, emotion is assumed to dominate over a thoughtful assessment of a facility's risks, costs, and benefits. The presumption is that people fear the risks associated with a facility and/or are angry at being forced to bear localized costs of a project for which they receive no special benefits.

Two coding categories were included in the content analysis to measure the extent to which persons at the hearing were emotive in their testimony. One variable measured whether nuclear waste was discussed in personal terms. As shown in Table 4-7, this theme was evident in 23 percent of the statements. Thus, over three-fourths of those testifying *did not* discuss nuclear waste issues in such a personal manner. The other variable was

Table 4-6 Geographic Focus of Public Comments

Geographic Focus	Percent[a]	N
Local	68	(712)
State	60	(623)
Other states or nation as a whole	28	(287)
International	11	(112)
Local only	23	(244)

[a] Percentages refer to those demonstrating a given geographical orientation. Since individuals held multiple orientations, the total exceeds 100 percent.

Table 4-7 Emotive Themes in Public Comments (in percentages)

Emotive Themes	Group Affiliation						
	Total	Citizens	Environ-mentalists	Government Officials	Interest Groups[b]	Industry & Utilities	Indian Tribes
Personalization of repository issues	23[a] (238)	26 (159)	12 (4)	15 (19)	23 (30)	17 (7)	21 (13)
Threats made over repository issues	14 (146)	12 (76)	9 (3)	18 (22)	12 (15)	15 (6)	26 (16)

[a] Numbers refer to the percentage of those testifying whose statements contained the emotive theme specified.

[b] This category includes all interest groups other than environmental groups or those representing industry or utilities.

whether emotional threats were made (to DOE and/or the federal government) about the consequences of building a repository in the area. These included statements about blocking access to the site and taking some unspecified action against the officials responsible for the repository program. Only 14 percent of those testifying made declarations of this kind. In short, the data indicate that emotive themes were present for only a relatively small number of those making statements; the vast majority did not make overly emotive statements of this kind in any of the four states. Thus, emotionalism is not as large a component of the public's reaction as would be predicted from the conventional view of the NIMBY syndrome.

One factor that might explain the likelihood of nuclear waste issues being viewed personally or a threat being made is an individual's background as measured by group affiliation. However, an uncertainty coefficient of .01 for both variables indicates that there are only minor differences across the groups as a whole. Emotionalism was somewhat more evident for Indian tribal spokespersons than other groups. They were more likely to make threatening remarks, with many viewing DOE's siting proposals as a challenge to the sovereignty of tribal lands. Similarly, unaffiliated citizens and representatives from public interest groups (other than environmentalists) were more likely to view radioactive waste in personal terms than were representatives of industry, government officials, or environmentalists.

Taken together, these findings provide little support for the conclusion that the public's reaction to DOE's repository siting plans was merely a NIMBY response. Certainly, elements of the conventional NIMBY construct were present, but there were clearly other reasons for the attitudes evident at these hearings. These and other dimensions of the testimony will be examined next.

Public Trust and Confidence in the Department of Energy
Trust and confidence in government is one of the most significant dimensions of siting controversies. Kasperson (1986) argues that the believability of risk information, such as the likelihood of leakage from a waste site, is closely related to the credibility of government and whether the public trusts what it says. He notes that the public's trust is based on judgments about whether the government agency is competent, unbiased, and responsive to the public's interest. Bella, Mosher, and Calvo (1988) have argued that a lack of confidence in the nuclear industry, and in DOE, underlies public opposition to nuclear installations, and that it has the potential to undermine the NWPA. The U.S. Office of Technology Assessment reached a similar conclusion in 1982 in saying "the greatest single obstacle that a successful waste management program must overcome is the severe erosion of public confidence in the Federal Government" (quoted in Kasperson, 1986:277).

Our coding categories included general perceptions of the credibility and competence of DOE as well as other actors in the nuclear policy arena. The results are given in Table 4-8. Fully 58 percent of those testifying raised questions about the credibility of DOE and/or indicated that they did not trust the department. Its technical competence was questioned by 35 percent, indicating a level of apprehension far above all other government agencies mentioned. These findings may reflect some generalized distrust of government, but DOE was clearly perceived as warranting less confidence than other institutions or experts. For comparison, a much smaller number viewed "government in general" as lacking credibility (31 percent) or technical competence (9 percent), and no other group mentioned (e.g., scientists, state government, Congress, or the Environmental Protection Agency) was singled out for criticism by more than a small number of people other

Table 4-8   Perceived Credibility and Competence of Government and Others

Group	Lacks Credibility		Lacks Technical Competence	
	Percent[a]	N	Percent[a]	N
Department of Energy	58	(603)	35	(361)
Government in general	31	(325)	9	(91)
Nuclear industry	13	(137)	11	(111)
Nuclear Regulatory Commission or Atomic Energy Commission	7	(66)	4	(39)
Environmental Protection Agency	2	(18)	1	(8)
Scientists	3	(33)	4	(41)
State government	1	(12)	.2	(2)

[a] The total percentage does not equal 100 because testimonies were coded for multiple responses.

than the nuclear industry (13 percent and 11 percent, respectively, on the credibility and competence questions).

Some variation by state was evident for these variables, especially for credibility. Nearly two-thirds of those testifying in Wisconsin and Maine viewed DOE as lacking credibility, compared to only 41 percent in Georgia and 48 percent in North Carolina. Industry representatives were more likely to express confidence in DOE than were other groups. And environmentalists were far more likely than other groups to challenge DOE on the grounds of competence (62 percent compared to an average for other groups of 33 percent); they stressed the technical limitations of site evaluation and hence the quality of DOE's work.

Of course, these findings about the credibility and competence of DOE in particular are not unexpected, given that the focus of the hearings was on a DOE report. Nevertheless, the importance of trust and confidence in the department should not be underestimated. The public's lack of confidence appears to be a significant influence on other attitudes that affect the level of opposition to DOE's siting plans. Other chapters in this volume reach a similar conclusion.

Public Concerns about a Nuclear Waste Repository

In addition to knowing that citizens objected to DOE's siting proposal, it would be instructive to learn about the particular issues that most concern them. Only with such knowledge can we understand the nature of public opposition to siting. Accordingly, we coded testimony for the nature of criticism offered. Three major sets of concerns were identifiable in hearing

Table 4-9   Technical Criticisms of the Department of Energy's Siting Analysis
Made at Public Hearings

Criticisms	Percentage of Total Number Making Criticisms[a]	N
Waste technology is unproven	32	(336)
Limited scientific understanding of the geology of rock formations	28	(295)
Analysis of transportation problems is incomplete	28	(292)
Site characterization involves many technical uncertainties	25	(256)
Level of risk posed by nuclear wastes is unacceptable	22	(230)
Little scientific knowledge about the movement of groundwater	21	(223)
Nuclear waste program characterized by inadequate planning	17	(182)
Isolation of radioactive material from the biosphere cannot be guaranteed	13	(140)
Need to explore alternative technologies such as breeder reactors or solar power	6	(65)
Nuclear waste repository should be located where the wastes are produced	4	(44)

[a] The total percentage does not equal 100 because testimonies were coded for multiple responses.

transcripts: technical, political and social, and concerns over the effects of a repository locally or across the state. The last includes both environmental and economic impacts, such as on agriculture or tourism. We examine each primarily by identifying the number of individuals offering specific complaints of these kinds.

Technical Criticisms of the Siting Process
Although not as critical of DOE's competence as of its political credibility, individuals who made statements at the four hearings voiced a great number of technical criticisms. These complaints focused on four basic dimensions: planning and analysis, risk and uncertainty, environmental assessment, and policy alternatives. Table 4-9 lists the percentage of those testifying who raised each of the issues, rank ordered by the frequency with which they were mentioned.

Most people at the hearings raised technical issues. Seventy-six percent addressed at least one technical issue and 53 percent addressed two or more.

On average, two technical issues were mentioned. Six of the ten criticisms included in the content analysis were made by more than 20 percent of those testifying. There was no marked variation across the four states or among the groups that participated in the hearings. The uncertainty coefficient for each of the ten technical variables for state of residence ranged from only .002 to .03. A similar range of low coefficients was evident for group affiliation (.001 to .02).

The most frequent technical criticism leveled at DOE (by 32 percent) was that the technology of waste disposal was unproven. This concern reflects a pervasive social attitude that too little is known about the risks from nuclear power and its by-products. A related point was made by 22 percent: the risks posed by nuclear waste are unacceptable. Other criticisms raised in more than one-fifth of the statements were specifically directed at DOE's site analysis. These focused on inadequacies in transportation analysis, limitations in the site characterization process, and limited baseline knowledge about geological and hydrological processes. Thus technical concerns revolved around two themes: the risks associated with nuclear waste disposal and the inadequacies of DOE's analysis of such risks. Overall, the findings indicate that despite DOE's extensive planning effort, as exemplified by its massive mission plan (DOE, 1985) and its elaborate environmental assessments, a widespread perception existed that the agency had not proceeded in a methodologically sound manner.

Political and Social Concerns about the Siting Process
As the previous discussion indicates, nuclear waste is a highly controversial issue with many political overtones. Statements made at the hearings often focused on political and social dimensions, and we used ten coding categories to measure the frequency with which these concerns were expressed. See Table 4-10. The mean number of political and social issues that were raised by those testifying was 1.9. In 80 percent of the statements, at least one argument of this kind appeared. This figure is a little higher than for either technical criticisms (76 percent) or for those dealing with the kinds of impacts people thought the facility might have (71 percent). This reaction by the public is not unexpected in light of the vocal censure of DOE by state officials, Indian tribes, and other groups involved in the siting process. As was the case with technical criticisms, there was no significant variation across the four states and the groups represented at the hearings. No correlation (uncertainty coefficient) exceeded .05 between either of the variables and the eight kinds of political concerns voiced.

Five of the ten political or social concerns included in the content analysis were mentioned by more than 20 percent of those who testified. The

Table 4-10   Political and Social Issues Raised at Public Hearings

Issues	Percentage of Public Statements at Hearings Where Issue Is Raised[a]	N
Burial of wastes is an unfair legacy to future generations	29	(298)
The government is violating political rights to life, liberty, and property	24	(251)
Production of nuclear waste linked to other aspects of nuclear power	23	(235)
Federal government (or specific agencies) motivated by political considerations	20	(207)
Society's responsibility to maintain the environment	20	(209)
Government is not representing our interests, lacks accountability, or is unresponsive to needs	18	(185)
Lack of time for citizens to adequately respond to the ARR	12	(126)
Public hearings do not provide adequate forum for citizen participation	8	(81)

[a] The total percentage does not equal 100 because testimonies were coded for multiple responses.

one raised most frequently (by 29 percent) was that the burial of nuclear wastes is an unfair legacy to future generations. Related value dimensions also were raised in many of the statements. Twenty percent emphasized that society has the responsibility to maintain the environment, and government violation of political rights was mentioned by 24 percent.

A frequent criticism of DOE by state and local officials was that DOE was motivated by political, not scientific, considerations (see Clary and Kraft, 1989). In particular, critics complained when, after DOE's site analysis ranked the Hanford location as the most expensive and least safe of nine sites evaluated in the first round, the department nevertheless recommended it as one of three for further study. Predictably, this theme was evident in the testimony. Twenty percent said the federal government was making its decision on political grounds.

The last sociopolitical theme that was mentioned (by 23 percent of those testifying) was the linkage of radioactive waste to the broader question of nuclear energy. In this respect, radioactive waste is different from other hazardous waste problems. It is embedded in a major political debate concerning the use of nuclear power plants and even nuclear weapons. The hearings confirm what survey data indicate (see chapter 2 by Rosa and Freudenburg in this volume): that attitudes toward nuclear energy influence public perception of waste disposal issues.

In sum, questions of political and social values figured prominently in the testimony. Intergenerational equity, social responsibility for environmental quality, political rights, and the role of nuclear energy in modern society were major themes. The data from this part of the content analysis underscore the importance of the nontechnical dimension of siting a high-level nuclear waste repository.

Concerns about Impacts of a Repository

One of the most frequent sets of concerns that appeared in the hearing statements dealt with environmental and economic impacts of the proposed repository. We list the ten major concerns in Table 4-11, along with the percentage of those testifying who raised them in each of the states. Despite the geographic diversity among the four states, there was generally little difference in perceived impacts, though Wisconsin does show inordinate concern with water resources. Uncertainty coefficients ranged from only .004 to .05 for relationships between state of residence and each of the ten impacts included in the analysis.

Perceived impacts figured prominently in the testimony, although slightly

Table 4-11   Perceived Impacts of a Nuclear Waste Repository by State

Impact (on)	Percentage of Public Statements at Hearings Where Issue Is Raised [a]				
	All Four States	Wisconsin	Maine	North Carolina	Georgia
Water resources	36	61	24	36	39
Economy	26	23	22	29	39
Public health	23	13	35	13	12
Tourism	22	27	22	25	2
Quality of life	18	6	24	19	17
Agriculture	13	25	8	13	7
Forests	10	18	9	7	2
Recreational opportunities	10	16	12	4	3
Sense of psychological well-being	8	5	9	10	10
Endangered species	6	7	8	3	5

[a] The total percentages do not equal 100 because testimonies were coded for multiple responses.

less than for technical and political/social criticisms. Seventy-one percent of those testifying raised issues of this kind, and almost one-half (44 percent) noted two or more impacts, with the mean value being 1.7.

Four of the ten impacts were cited in more than 20 percent of the statements. As Table 4-11 indicates, the issue of water resources was raised most frequently (by 36 percent). This is understandable, since all four states are water-rich, and much tourism and recreation is water-centered. In a technical sense, water is one of the major threats to the integrity of a repository. One of the probable ways that radioactive materials could reach the environment would be through water seeping into a repository and then migrating back to the biosphere. This contaminated water then could pollute aquifers and other water sources. Because of concern over these kinds of potential problems with a repository, it follows that public health would be among the chief impacts cited in the testimony (by 23 percent).

Negative impact on the economy appears second most frequently (26 percent). Many of the study sites were in areas where recreation (tourism was mentioned by 22 percent) is a very important part of the economy. DOE's analysis paid very little attention to this aspect of a repository's impact; because of regulatory requirements, more emphasis was given to geologic and other physical factors. Predictably, much of the criticism directed at DOE by public officials and the media focused on the limited weight given to socioeconomic elements.

Potential negative environmental and social impacts of a repository, then, were major concerns expressed at the hearings. There is little evidence to indicate that DOE was able to convince the public that environmental impacts would be minimal. Without confidence that the environment could be protected from the leakage of radioactive wastes, many of those testifying referred to negative social consequences, such as decreased tourism and loss of business, which they believed would result.

The Acceptability of Radioactive Waste Risks

A number of studies have confirmed that the public's perception of nuclear risks differs from what experts believe to be the actual risks and that the public considers these risks to be unacceptably high. We would expect public fear of the risks to be evident in testimony on repository siting and to play a major role in opposition to siting proposals. To determine the public's position on the acceptability of risks associated with repository construction and operation, we coded testimony for five aspects of risk evaluation. These are similar (though not identical) to dimensions of risk acceptability discussed by Slovic and Fischhoff (1983): whether the risk is perceived as voluntary or involuntary, the extent to which there may be catastrophic results, how much the risk is feared or dreaded, the level of familiarity with the risk, and the degree to which the risk is known. In Table 4-12 we provide the percentages of those who discussed each of our five dimensions.

At least one of these five dimensions appeared in 77 percent of the hearing statements. In 30 percent, two or more of them appeared. Four of the five dimensions of risk acceptability were mentioned by at least 23 percent of those testifying. One interpretation of the public's exceptionally negative response to the siting proposals, then, is that people believe the related risks to be unacceptable. For example, in 37 percent of the statements, concern was expressed about the involuntary nature of the risks imposed by a repository. Thirty-one percent explained their opposition in terms of the potential for catastrophic effects of siting. For 23 percent of those testifying, concern focused on the unacceptably high level of technical uncertainty. In addition, 23 percent spoke of the immediate consequences of locating a repository in the area (e.g., groundwater contamination; see Table 4-11). Of the dimensions we list in Table 4-12, the only one that was not evident in the testimony was a lack of familiarity; in almost all of the statements, people believed they understood the major risks of siting a repository locally.

To what extent were there differences in perceptions of risk among the individuals who testified? Earlier studies indicate that environmentalists and engineers differ in their views of the risks of nuclear waste disposal (Nealey and Hebert, 1983). Technical training, professional norms, personal attitudes, or location may affect risk perceptions and therefore may result in some differences across groups. However, as noted for other variables discussed above, we found little difference across groups represented at the hearings. For the five risk variables included in Table 4-12, there was little variation across groups (with uncertainty coefficients ranging from just .004 to .03). Nevertheless, some findings warrant brief mention.

The involuntary nature of radioactive waste risks is the dimension discussed most frequently by all groups, but it is especially notable among citizens, environmentalists, and particularly Indian tribal representatives. The second most frequently mentioned risk dimension is the potential for catastrophic effects, which was especially likely to be cited by citizens, environmentalists, and other interest groups and rarely by industry and utility representatives or Indian tribal representatives. The variation in risk perception across groups (and to some extent across states) is yet another indicator that important segments of the public apply different standards of judgment in evaluating risks of this kind than do technical experts in DOE and in the nuclear power industry.

Table 4-12   Acceptability of Nuclear Waste Risk by Group Affiliation

Percentage of Those Testifying Who Mention Risk Dimension

Risk Dimension	Total	Citizens	Environ-mentalists	Government Officials	Interest Groups[b]	Industry & Utilities	Indian Tribes
Involuntary exposure (repository forced on without consent)	37[a] (386)[c]	38 (230)	41 (14)	31 (38)	33 (42)	22 (9)	59 (36)
Catastrophic effects (fear of radioactive contamination)	31 (319)	34 (209)	32 (11)	21 (26)	34 (44)	12 (5)	6 (17)
Technical uncertainty (no technical solution to the problem)	23 (244)	23 (142)	35 (12)	24 (30)	28 (36)	12 (5)	16 (10)
Immediate consequences (repository impact in the local area)	23 (242)	21 (126)	35 (12)	32 (40)	26 (33)	39 (16)	18 (11)
Risk unfamiliarity (not familiar with risk of radioactive waste)	1 (14)	1 (8)	0 (0)	1 (1)	2 (2)	2 (1)	2 (1)

[a] The percentages refer to the number of individuals discussing the particular dimension of radioactive waste risks in his or her testimony. The risk dimensions are adapted from Slovic and Fischhoff (1983).
[b] This category includes all interest groups other than environmental groups or those representing industry or utilities.
[c] Totals do not equal the sum of the six group categories because group affiliation was not ascertained in 2.4 percent of the cases, or for 25 individuals.

Predicting Opposition to a Nuclear Waste Repository

Finally, another important question that can be asked about statements at the public hearings is what variables, if any, are useful in predicting the extent to which individuals oppose the DOE recommendations on repository siting. The dependent variable we use for this purpose is the public's overall assessment of DOE siting proposals. Recall that where a position was clearly stated, 70 percent were coded as strongly opposed, and 30 percent were coded as moderate or weak in disagreement. In order to determine what accounts for this admittedly limited variation in opposition to siting a waste repository, the dimensions of public testimony that were examined above and reported in Tables 4-5 to 4-12 were used here as independent variables in a regression analysis. The results of this analysis appear in Table 4-13. A relatively moderate amount of variance, 17 percent, is explained by these variables. Even though most of the variables in the equations have low explanatory power, several correlate with the dependent variable of degree of opposition to DOE's proposals.

The most important predictor is DOE's credibility, with strong opposition being related to low credibility of DOE. Of all the themes evident in the

Table 4-13   Multiple Regression of Public Opposition to Siting Proposals on Independent Variables[a]

Independent Variables	Bivariate r	Beta	Standard Error
DOE credibility	.34	.31	.06
Knowledge level	.17	.17	.04
Political/social concerns[b]	.22	.11	.02
NIMBY response[c]	.09	.08	.04
Technical criticisms[d]	.05	.07	.02
Repository impacts[e]	.07	.06	.01
DOE competence	.04	.04	.05
Risk acceptability[f]	.10	.04	.03
R^2	.17	—	—

[a] Due to listwise deletion, 602 cases of 1,045 are treated as missing values. Since the data are based on a population, not a sample, no statistical tests of significance are used.

[b] An index created by adding the number of items in Table 4-10 that were mentioned.

[c] A three-item scale based on whether a person (1) viewed the nuclear waste issue in personal terms, (2) made threats about stopping the repository project, and (3) viewed the siting proposal in local terms only.

[d] An index created by adding the number of items in Table 4-9 that were mentioned.

[e] An index created by adding the number of items in Table 4-11 that were mentioned.

[f] An index created by adding the number of items in Table 4-12 that were mentioned.

testimony, DOE's lack of credibility shows the strongest relationship to opposition ($r = .34$), and its effects are largely unaffected by inclusion of the other independent variables (beta $= .31$).

Two other variables have weaker but still substantial relationships with the opposition indicator—personal knowledge and political/social concerns. The higher the level of knowledge exhibited, the stronger the opposition expressed; also, the mention of political and social concerns was related to strong opposition. However, while the importance of knowledge is not affected when all of the other variables are examined simultaneously ($r = .17$, beta $= .17$), the importance of social/political concerns drops considerably from the bivariate level ($r = .22$) to the multivariate level (beta $= .11$).

Conclusions and Policy Implications

In this chapter, we have used content analysis of public hearings on repository siting in four eastern states to try to assess the extent, nature, and sources of public opposition to DOE's plans for nuclear waste disposal. Use of data from these hearings supplements the survey data found in the other chapters in this volume. Concentrating on hearings serves as a reminder that survey research usually reports on the views of the general public, for whom the issues may be low in salience, whereas individuals testifying at public forums such as the DOE-sponsored hearings may exhibit quite different characteristics. These statements provide a useful source of information about the attitudes, beliefs, and perceptions of a segment of the public that plays an especially important role in the policy process.

As we note above, however, content analysis of hearing data is not without weaknesses. Especially with regard to the NWPA hearings, statements are often made for political effect and may not accurately convey the individual's actual perception of an issue. Hearings, unlike general opinion surveys, are part of the political process and, as such, present both risks and benefits to the policymaker trying to understand public responses.

What did the testimony from four sets of hearings examined here say about the siting process? Clearly, there was overwhelming opposition to the DOE report on the candidate repository sites. Although the strength of opposition could be interpreted as a classic NIMBY response, the data suggest that this characterization does not adequately describe the public's reaction. In contrast to the prevailing conception of the NIMBY syndrome, the individuals who made statements at the hearings on the whole were reasonably well informed, nonemotional, and capable of viewing the nuclear waste issue in a broader context than simply that of their "back yard."

These findings have significant policy implications. The DOE faces a public which is fairly knowledgeable and which responds to siting proposals in a more sophisticated manner than commonly assumed. It will not likely be satisfied with DOE research it perceives as inadequate or politically motivated, and hearings will not suffice if they are the only means provided for citizen participation. A much more intensive and demanding process of public involvement is necessary. Free exchange of information is pivotal to such a process.

At a minimum, the public should have full access to scientific and technical information on which environmental risk decisions are based. There should also be appropriate institutional mechanisms for promoting two-way communication and democratic discourse on the issues. Evidence in this chapter and other studies of public participation in technical decisionmaking indicate that the public has the capacity to meet such expectations. Moreover, establishment of an improved process of public involvement would seem to be a prerequisite for building public trust for facilities of this kind even if there are no guarantees of success (National Research Council, 1989; Peelle and Ellis, 1987; Rosa, 1987).

A high degree of mistrust of DOE is a major finding of the content analysis. The attentive public represented at these hearings rated DOE low on credibility and raised significant doubts about DOE's technical competence in repository siting as well. The combination of great public fear of radioactive waste (see chapter 3) and lack of confidence in DOE hardly augurs well for the success of repository siting. Once lost, public trust and confidence in governmental institutions are difficult to rebuild, particularly in the short term. The implication is that somehow the federal government must find a more credible way to communicate information about repository risks in its further site characterization at Yucca Mountain and at other sites in the future.

The findings of the content analysis suggest that using good or acceptable science will be insufficient for solving the DOE's credibility problem. The public must be convinced that the department's motives are sound and that it is highly competent at its tasks. A long track record of successful management would help as well. All this will take far more than good public relations. That is where public participation can help. However, the common strategy of relying on public hearings, as noted above, will not suffice either. Hearings do not provide the public a genuine opportunity to understand nuclear waste issues from any perspective other than DOE's, and hearings provide no continuing basis for the agency to interact with citizens. If it is to be effective, citizen participation must involve more meaningful involvement by the public in repository planning and in oversight and program operations. Experience with hazardous waste siting indicates the potential of such approaches, especially for building public trust and a sense of citizen control over exposure to risk (Hadden, 1991; Lynn, 1987).

Finally, DOE needs to reconsider the factors that should be addressed in the evaluation of a candidate site. Its environmental assessments dealt largely with physical aspects of the environment, but the public was also concerned about other impacts. The economy, tourism, and public health were all mentioned by more than one-fifth of those testifying. DOE was limited in its ability to address such questions by the original wording of NWPA and the regulations governing its implementation. It is safe to assume, however, that the public cares less about legislative and bureaucratic requirements than about receiving credible reassurance that the full range of repository impacts is understood and will be addressed satisfactorily.

Notes

We wish to thank Tamara Crockett and Rebecca Spithill for their assistance with the content analysis reported here and Jolene Anderson for research assistance on nuclear waste issues related to this work. This research was supported in part by a grant from the Urban Corridor Consortium of the University of Wisconsin System and by the Herbert Fisk Johnson professorship at the University of Wisconsin-Green Bay.

1   According to Slovic (1987), unknown risk refers to new, unobservable risks that are unknown to science and to those exposed and whose effects are delayed. Dread risk refers to those that are uncontrollable, dreaded by the public, have potentially fatal or catastrophic consequences, pose a high and not easily reduced risk to future generations, and are involuntary.

2   We wish to thank the Crystalline Repository Project Office for their cooperation and generosity in supplying the transcripts for this analysis.

3   After extensive instruction from the investigators, one individual coded all 1045 comments. The coding was based on standard content analysis procedures as described in Krippendorff (1980) and Holsti (1969). To deal with the problems of reliability in using a single coder, a second individual coded 25 percent of the cases for the first state included in the analysis. Mean values were computed for 85 variables for each coder. The correlation between these values was .96. For the variables with the most discrepancy in coding, we reviewed coding decisions with the two coders and reached agreement on how the categories should be treated. Appropriate changes were made in the coding of all data to reflect the definitions agreed upon through this process.

4   We coded individuals according to their stated affiliation or how they identified themselves. Some 61 percent of those testifying identified themselves solely as private citizens; 40 percent, as a representative of a group. Among the more notable groups represented were environmentalists (3 percent), public interest groups other than environmentalists (13 percent), government officials (12 percent), Indian tribes (6 percent), and industry, utilities, or research firms (4 percent). Affiliation was used in the analysis only when an individual clearly fell into a single category (n = 1000).

5   Uncertainty coefficients are nonparametric statistics, ranging in value from zero to one, appropriate for nominal levels of measurement such as the variables in the con-

tent analysis. Similar to lambda, they are proportional reduction of error measures for nominal measurement and their interpretation is comparable to *gamma* and Pearson's *r*.

References

Aberbach, Joel D., and Bert A. Rockman. 1978. "Administrators' Beliefs About the Role of the Public: The Case of American Federal Executives." *Western Political Quarterly* 31:502–22.

Arnstein, Sherry. 1969. "A Ladder of Citizen Participation." *Journal of the American Institute of Planners* 35:216–24.

Bella, David A., Charles D. Mosher, and Steven N. Calvo. 1988. "Technocracy and Trust: Nuclear Waste Controversy." *Journal of Professional Issues in Engineering* 114:27–39.

Bryan, Richard H. 1987. "The Politics and Promises of Nuclear Waste Disposal: The View from Nevada." *Environment* 29:14–17, 32–38.

Carter, Luther J. 1987. *Nuclear Imperatives and Public Trust: Dealing with Radioactive Waste*. Washington, D.C.: Resources for the Future.

Checkoway, Barry. 1981. "The Politics of Public Hearings." *Journal of Applied Behavioral Science* 17:566–82.

Clary, Bruce B., and Michael E. Kraft. 1988. "Impact Assessment and Policy Failure: The Nuclear Waste Policy Act." *Policy Studies Review* 8:105–15.

———. 1989. "Environmental Assessment, Science, and Policy Failure: The Politics of Nuclear Waste Disposal." In Robert V. Bartlett, ed., *Policy Through Impact Assessment: Institutionalized Analysis as a Policy Strategy*. Westport, CT: Greenwood. 37–50.

Fiorino, Daniel J. 1989. "Environmental Risk and Democratic Process: A Critical Review." *Columbia Journal of Environmental Law* 14:501–47.

Freudenburg, William R., and Eugene A. Rosa, eds. 1984. *Public Reactions to Nuclear Power: Are There Critical Masses?* Boulder: Westview/American Association for the Advancement of Science.

Gormley, William T., Jr. 1989. *Taming the Bureaucracy: Muscles, Prayers, and Other Strategies*. Princeton: Princeton University Press.

Hadden, Susan G. 1991. "Public Perception of Hazardous Waste." *Risk Analysis* 11:47–57.

Heberlein, Thomas. 1976. "Some Observations on Alternative Mechanisms for Public Involvement: The Hearing, Public Opinion Poll, the Workshop, and the Quasi-Experiment." *Natural Resources Journal* 16:197–221.

Holsti, Ole R. 1969. *Content Analysis for the Social Sciences and Humanities*. Reading, MA: Addison-Wesley.

Kasperson, Roger E. 1986. "Six Propositions on Public Participation and Their Relevance for Risk Communication." *Risk Analysis* 6:275–81.

Key, V. O., Jr. 1963. *Public Opinion and American Democracy*. New York: Alfred A. Knopf.

Kraft, Michael E., and Bruce B. Clary. 1991. "Citizen Participation and the NIMBY Syndrome: Public Response to Radioactive Waste Disposal." *Western Political Quarterly* 44 (June): 29–328.

Krippendorff, Klaus. 1980. *Content Analysis: An Introduction to Its Methodology*. Beverly Hills: Sage.

Lynn, Frances M. 1987. "Citizen Involvement in Hazardous Waste Sites: Two North Carolina Success Stories." *Environmental Impact Assessment Review* 7:347–61.

Marks, Gary, and Detlof von Winterfeldt. 1984. "Not in My Back Yard: Influence of Motivational Concerns on Judgments About a Risky Technology." *Journal of Applied Psychology* 69:408–15.

Mazmanian, Daniel A., and David Morell. 1990. "The 'NIMBY' Syndrome: Facility Siting and the Failure of Democratic Discourse." In Norman J. Vig and Michael E. Kraft, eds., *Environmental Policy in the 1990s: Toward a New Agenda*. Washington, D.C.: CQ Press. 125–43.

Mazmanian, Daniel A., and Paul A. Sabatier. 1983. *Implementation and Public Policy*. Glenview, IL: Scott, Foresman.

Milbrath, Lester W. 1981. "Citizen Surveys as Citizen Participation Mechanisms." *Journal of Applied Behavioral Science* 17:478–96.

National Research Council. 1989. *Improving Risk Communication*. Washington, D.C.: National Academy Press.

Nealey, Stanley M., and John A. Hebert. 1983. "Public Attitudes Toward Radioactive Waste." In Charles A. Walker, Leroy C. Gould, and Edward J. Woodhouse, eds., *Too Hot to Handle? Social and Policy Issues in the Management of Radioactive Wastes*. New Haven: Yale University Press. 94–111.

Peelle, Elizabeth, and Richard Ellis. 1987. "Beyond the 'Not-In-My-Backyard' Impasse." *Forum for Applied Research and Public Policy* 2 (Fall): 68–77.

Rosa, Eugene A. 1987. "Namby Pamby and Nimby Pimby: Public Issues in the Siting of Hazardous Waste Facilities." *Forum for Applied Research and Public Policy* 3 (Winter): 41.

Schaefer, Jame. 1988. *State Opposition to Federal Nuclear Waste Repository Siting: A Case Study of Wisconsin 1976–1988*. Green Bay: Center for Public Affairs, University of Wisconsin-Green Bay.

Slovic, Paul. 1987. "Perception of Risk." *Science* 236:280–85.

Slovic, Paul, and Baruch Fischhoff. 1983. "How Safe is Safe Enough? Determinants of Perceived and Acceptable Risk." In Charles A. Walker, Leroy C. Gould, and Edward J. Woodhouse, eds., *Too Hot to Handle? Social and Policy Issues in the Management of Radioactive Wastes*. New Haven: Yale University Press. 112–50.

U.S. Department of Energy. 1985. Office of Civilian Radioactive Waste Management, *Mission Plan for the Civilian Radioactive Waste Management Program*, 3 vols. Washington, D.C.: GPO.

———. 1986. Office of Civilian Radioactive Waste Management. *Area Recommendation Report for the Crystalline Repository Project*, Vol. 1. Draft. DOE/CH-15(1). Washington, D.C.: GPO, January.

———. 1987. Office of Civilian Radioactive Waste Management, *Annual Report to Congress*. Washington, D.C.: GPO, April.

5 Sources of Public Concern About Nuclear Waste Disposal in Texas Agricultural Communities

Julia G. Brody and Judy K. Fleishman

From steak and potatoes to tortillas and beans, a good portion of America's pantry is stocked from food grown in Deaf Smith County, Texas. Located in the Texas Panhandle, Deaf Smith County was one of three sites named in 1986 by the U.S. Department of Energy (DOE) as finalists for the nation's first high-level nuclear waste repository. It's also farm country, with expansive fields of wheat, corn, sorghum, sugarbeets, and other crops.

The exceptionally fertile soil has sprouted a multibillion-dollar food processing industry, including Ralston-Purina, Frito-Lay, Holly Sugar, and Arrowhead Mills, a leading health food company. Deaf Smith is also the number two cattle county in the nation. Ninety percent of the *world's* sorghum seed is grown in the Deaf Smith impact area, and the proposed repository site includes a unique seed farm that maintains genetic stock for wheat grown in a seven-state region, working with Texas A&M University to distribute new varieties.

Deaf Smith's diverse agricultural cornucopia makes this area quite different from other proposed sites discussed in this volume. Yet public response to the nuclear waste program in Texas raises issues that are common to many communities facing hazardous facilities. Our purpose in this chapter is to address some of these issues, using the Texas experience as a laboratory for understanding the repository siting process nationally.

First, we offer an overview of what local residents think the repository would mean for their families and their communities. Our research provides extensive descriptive information about public opinion in communities surrounding the proposed Texas site and in other urban and rural communities in the region, documenting overall support and opposition toward the repository and focusing on expectations about both socioeconomic impacts and environmental impacts.

Second, we begin to explore how these expectations fit together. This analysis of environmental and socioeconomic expectations together is important because current legislation governing the nuclear waste program rests on the assumption that local residents weigh expected risks against expected economic benefits in deciding whether to favor or oppose a repository site nearby.

A third theme of our research is that perceptions of the repository may have significant social and economic impacts independent of the actual safety of the repository. A local banker raised this issue when DOE officials first met with citizens after identification of the Texas sites in 1984. He asked the DOE economist what he would do if he walked into a grocery where he had the choice of a steak raised next to a high-level nuclear waste repository or one that wasn't. While the importance of perceptions is highlighted in Texas because Deaf Smith's basic business is food, the underlying issues are similar in Nevada, where tourism is a concern, and, more generally, at all sites where citizens fear that perceptions of a repository will affect land values, investment in nonnuclear business, and the sense of home and community.

Similar issues—concerning both perceived and actual impacts—have been raised, too, in earlier studies of attitudes toward nuclear technologies. Numerous surveys of the general public and of host communities for nuclear power plants show substantial concern about the safety of nuclear facilities (Nealey, Melber, and Rankin, 1983; Freudenburg and Rosa, 1984; Rosa and Freudenburg in chapter 2 of this volume). Particularly since the accident at Three Mile Island, the public has seen risks to health and the environment as serious problems with nuclear power (Freudenburg and Rosa, 1984). In addition, Hughey et al. (1983) and Sundstrom et al. (1977) identify expectations about socioeconomic impacts—benefits such as increased employment and problems such as increased noise and traffic—as important factors in the attitudes of host communities toward a proposed power plant. Our research draws on this earlier work.

Design of the Texas Studies

We began investigating public responses to the repository in 1984 in a series of studies that assessed public opinion as part of a broader effort to use survey research to understand potential socioeconomic impacts of a repository in Texas. Because of the influence of perceptions on agricultural investments and marketing, Texas's elected officials, citizen activists, and business leaders were quick to identify public attitudes as a key element in evaluating the impact of a repository. Consequently, systematic opinion research began much earlier in Texas than in Nevada and Washington, the other "finalist" states described in this volume.

Sample

Our primary data source is a series of telephone surveys of randomly selected residents of the Texas Panhandle. The 1984 surveys included Deaf Smith and Oldham counties, which surround the finalist site; Swisher County, a preliminary site identified by DOE but dropped from consideration in 1986; and two comparison counties located outside the DOE-defined impact area but similar in socioeconomic composition to the site counties. The comparison counties are 65 to 150 miles from the Deaf Smith site. The 1984 interviews were conducted in late June and early July.

In October 1986, we again contacted our 1984 respondents. We also added participants from each sample area in order to assure adequate sample sizes for future years if Texas remained in the nuclear waste program, and we interviewed a random sample of residents of Amarillo, an urban area that would serve as a base for DOE activities associated with a Texas repository.

For the rural counties, current telephone books were used to select participant names and addresses. This procedure was chosen because previous research (Dillman, 1978) shows that response rates may be increased by sending an explanatory letter to research participants before the first telephone contact. Local telephone companies indicated that unlisted telephone numbers were rare in these counties. A random-digit dialing technique was used for Amarillo, where unlisted numbers are more common.

Participation rates for all of the telephone surveys were excellent. More than 90 percent of residents contacted in 1984 completed the survey. We reached 75 percent of these participants again in 1986, and 88 percent of those contacted completed the second survey. The 1984 results reported here are based on 605 interviews for the preliminary sites (Deaf Smith, Oldham, and Swisher counties) and 236 for the comparison counties. Results for 1986 are based on 340 interviews for the finalist site (Deaf Smith and Oldham counties) and 253 for comparison counties.

For Amarillo, 609 residents were interviewed. The participation rate was lower (63 percent) partly because the random-dialing technique meant that respondents did not receive an advance letter explaining the survey.

Instrument

Because of the intensely political atmosphere of the site selection process, we were cautious in designing survey instruments to avoid apparent bias, using items closely patterned on earlier research wherever possible. We included two questions assessing overall attitudes toward the repository:[1]

—"If it were up to you, would you allow construction of a high-level nuclear waste repository in Deaf Smith County?"

—"Do you think construction of the nuclear waste repository would be a good thing for your county?"

The five-point response scale ranged from "definitely yes," coded 1, to "definitely no," coded 5.

These items were taken from earlier research sponsored by Oak Ridge National Laboratories, which found that local citizens favored construction of a nuclear power plant (Sundstrom et al., 1977; Hughey et al., 1983). Those results indicated that the questions did not inherently produce antinuclear responses. Responses to these questions were highly correlated ($r > .80$ in our samples), so they were summed to create a scale with greater range and stability for multiple regression analyses.

Following these Likert-type questions, interviewers asked the open-ended question, "Why do you feel this way about the repository?" This question allowed respondents to state freely issues of concern to them before responding to issues raised by the interviewer.

Two other sets of questions asked about local residents' expectations about socioeconomic effects and environmental effects of a repository. The content of these items was based on issues raised in DOE hearings (DOE, 1986) and earlier research by Halstead et al. (1982) and Sundstrom et al. (1977).

Questions about the environment asked respondents to rate the likelihood of environmental problems at the repository on a four-point scale from "very unlikely," coded 1, to "very likely," coded 4. Items addressed the potential impact of the repository on air, soil, water, and food, and on the health of workers and local residents.[2]

Questions about potential socioeconomic impacts asked whether respondents expected variables to "go up," "stay the same," or "go down" during the next fifteen years if the repository were built in Texas. Questions addressed potential impacts on the respondents' family (e.g., household income), local community (e.g., crime rates), and the local economy (e.g., stores and businesses).

Both sets of questions form reliable scales. For socioeconomic effects, $alpha = .81$; for health and environmental risks, $alpha = .90$. It's important to note that the concepts of socioeconomic and environmental effects are interdependent, particularly since soil and water are critical economic resources in an agricultural community. The two scales are significantly correlated ($r = -.57$, $p < .01$).

Residents' knowledge about the nuclear waste disposal program was assessed with twelve true-false questions. For example:

—"High-level nuclear wastes are radioactive for thousands of years."
(True.)
—"All of the salt dug out of the repository during construction will be
put back into the repository eventually." (False.)

Knowledge items were verified in DOE-published documents, and they
were reviewed by state officials and knowledgeable citizen-advocates to assure consensus about their accuracy. They were also pilot-tested, as were
all survey items. Because of changes in DOE plans for the repository, the
1986 and 1984 knowledge questions are not identical. The average number
of correct responses was 6.7 in 1986 (standard deviation = 2.5, alpha = .66)
and 7.2 in 1984 (standard deviation = 2.5, alpha = .63).

An index of uncertainty was created to represent the number of times a
respondent answered "don't know" to survey questions. This procedure for
including "don't know" responses in multiple regression analyses is recommended by Cohen and Cohen (1975).

The 1984 surveys also included seven questions assessing attitudes toward other forms of industrial development, including nuclear facilities (a
power plant or low-level waste site); other energy facilities (for example, a
coal-burning or manure-burning power plant); and agricultural industries
(for example, a food-processing plant). Nealey, Melber, and Rankin (1983)
criticize many survey reports for interpreting responses as antinuclear without evaluating whether respondents oppose all forms of local industrial
development.

In addition to assessing attitudes, the survey asked respondents, "Have
you changed financial plans for your family or for your farm, ranch, or business because of the repository?" Those who answered yes were asked how
their plans had changed. A parallel set of questions asked about changes
in "personal plans." These questions are a beginning in the difficult task of
evaluating how attitudes toward the repository are reflected in behavioral
change.

Survey instruments also included questions about demographic characteristics, such as age, gender, and income. Survey design is described in
greater detail in Brody (1985a, 1985b) and Brody and Fleishman (1987).

Attitudes Toward the Repository
Results show that opposition to a Texas repository was strong and broad-
based throughout the region, and it persisted over the two-year study period.
About 70 percent of the residents in each geographic area said they "definitely" would not allow construction of the repository in Texas if it were
up to them. The margin of error is ± 3 to 4 percentage points for the 1984

preliminary site results and for Amarillo, and ± 5 percentage points for the 1986 finalist site. The margin of error is larger for the 1986 finalist site because of the smaller sample size. Results are shown in Table 5-1.

Responding to an open-ended question about why they favor or oppose the repository, many Panhandle residents mentioned environmental risks. Often their concerns focused on fears of water contamination, since construction of the repository would require drilling through the Ogallala and Santa Rosa aquifers, the primary source of water for drinking and irrigation. For example, respondents said:

—"Anything gets in the water and we're gone. The Ogallala is the lifeblood of this community."
—"If you think nothing would happen to the Ogallala, just remember the length of time for nuclear waste to become safe and remember Murphy's law."

In addition, about half of the respondents expressed fear that the repository would hurt agriculture directly, by contaminating soil and water, or indirectly, by stigmatizing local produce and devaluing farmland.

—"Leakage would ruin crops, water and soil. It would ruin our lifestyle. Our town would regress."
—"Someone is going to tell Mr. and Mrs. America that their beef is coming from a nuclear waste dump. . . . The antibeef people will have a field day."

Residents who favored the repository typically mentioned expected economic benefits.

—"I feel like [the repository] will get a little money circulating in the county."

Table 5-1 "Would You Allow Construction of a High-Level Nuclear Waste Repository . . . ?" (percentages)

	Definitely No	Probably No	Not Sure	Probably Yes	Definitely Yes
1986					
Finalist site	71	9	6	11	3
Comparison	75	10	6	7	3
Urban center	72	9	6	6	6
1984					
Preliminary sites	73	8	7	8	4
Comparison	66	17	7	4	5

Table 5-2   Percentages of Finalist Site Residents Who Mentioned Selected
Reasons for Supporting or Opposing the Repository, 1986[a]

Issue	Percentage
Harm to agriculture	49
Environmental hazards	
Water contamination	33
Transportation accidents	4
Other environmental and health risks	22
Economic harm, other than effects on agriculture	14
Economic benefits	
Increased employment	9
Other economic benefits	13
Need for more information	6

[a] Responses were coded from the open-ended question, "Why do you feel that way about the repository?" Since each survey participant may have mentioned none or more than one of these reasons, percentages do not sum to 100.

—"If they build one [repository] here, please send me a job application. It's bound to pay better than farming."

Table 5-2 shows reasons commonly cited in 1986 by residents of the finalist site counties for supporting or opposing the repository. Results are similar for other sample groups.

Expected Environmental Risks
Responses to other survey questions similarly indicate that concern about the potential environmental impact of the repository was widespread. Asked to rate the probability of six environmental hazards, a majority of the residents in each geographic area rated each risk above the midpoint. Again, residents expressed strong concerns about water contamination. More than three-fourths of the residents of each area believed radiation would leak into water supplies. Transportation accidents were of particular concern to rural residents: 80 percent rated them "very likely" or "somewhat likely." A majority of residents in each area said health problems and food contamination were likely in their own county if the repository were built in Deaf Smith County. Expected environmental impacts are summarized in Table 5-3.

Expected Socioeconomic Impact
Local residents were also pessimistic about the potential socioeconomic impact of the repository. In 1986, a majority of site-county residents said a repository would mean increases in taxes and the cost of living and de-

Table 5-3 Expected Environmental Risks, 1986[a] (percent rating each risk as "very likely" or "somewhat likely")

	Finalist	Comparison	Urban Center
Transportation accident	81	82	72
Radiation in water	77	80	75
Food contamination in "your county"	73	72	67
Health problems for repository workers	72	79	71
Radiation in air	67	70	57
Health problems for residents of "your county"	67	64	59

[a] Interviewers asked respondents to rate how likely it was that each type of environmental risk would occur because of the repository. "Somewhat unlikely," "very unlikely," and "don't know" responses are not shown in this table.

creases in agricultural production and the value of farmland. They expressed greatest concern about effects of the repository on agriculture. Almost 80 percent of site-county residents expected a decline in farmland values, and more than 60 percent of residents of the other study areas expected farmland values to fall in their own county if a repository were built in Deaf Smith. In Deaf Smith and Oldham counties, 24 percent expected the amount of industry in their county to increase, but twice as many expected local industry to decline because of the repository. Expected socioeconomic impacts are shown in Table 5-4.

Even in 1984, site-county residents reported behavioral effects of these expectations: 8 percent of site-county residents said they had already changed their personal or financial plans because of the repository. In 1986, the figure was 11 percent. Self-report data like these are difficult to evaluate, but some of the changes described were quite specific and were independently verified. For example, a farmer who had planned to build a sunflower-processing plant near the site built it elsewhere. Arrowhead Mills changed the lettering on peanut butter labels to de-emphasize their headquarters in Deaf Smith County. A family that planned to build a new house on their farm decided to stay in town. Other families that had planned to transfer farms to their children delayed the changes, fearing they would saddle their offspring with devalued land, and many residents reported delays in land purchases or farm improvements.

While residents anticipated negative effects of a repository on agriculture, they expected increases in other economic activity. Half of the Deaf Smith-Oldham residents expected an increase in the number of stores in their county. A majority of residents in the site counties and in Amarillo, the nearest urban area, thought the repository would mean more jobs.

At the same time, few Panhandle residents expected their own household to benefit. In Deaf Smith and Oldham counties, 16 percent expected their income to increase, while 29 percent expected it to decline if the repository were built there. These findings are typical of results throughout our research showing that local residents knew the repository would mean enormous increases in economic activity, but did not believe the money would go to citizens or businesses already located in the Panhandle.

Other Industrial Development

Although residents of the site counties and rural comparison counties saw the repository as a threat to local agricultural communities, they did not oppose all types of industrial development. The 1984 survey results show consistent opposition to nuclear facilities, including a power plant or low-level waste site, but support for other energy facilities. Respondents expressed strongest support for agricultural businesses, such as a feedlot or food-processing plant. These results indicate that local residents favored economic development projects they perceived as compatible with the existing agricultural base of the local economy. In the comparison counties, a substantial majority favored new industry related to local oil and gas resources, although these facilities would pose clear environmental hazards.[3] This perspective is consistent with findings by Fischhoff et al. (1978) that

Table 5-4   Expected Socioeconomic Impacts, 1986[a] (percentages)

Impact	Finalist		Comparison		Urban Center	
	Go Up	Go Down	Go Up	Go Down	Go Up	Go Down
Value of your home[b]	33	46	—	—	17	44
Your household income[b]	16	29	—	—	12	11
Farmland value	5	80	6	60	8	73
Crime rates	49	4	24	5	29	3
Cost of living	65	4	37	7	41	8
Number of jobs	58	23	18	31	54	15
Number of stores, businesses	50	26	10	32	32	25
Amount of industry	24	48	12	32	32	26
Amount of agriculture	2	78	4	48	2	65
Number of places to go for entertainment[b]	43	14	—	—	18	13

[a]Interviewers asked respondents whether a repository located in Deaf Smith County would mean changes in each variable in their own county during the next 15 years. "Stay the same" and "don't know" responses are not shown in this table.

[b]Interviews for the comparison counties used a shorter questionnaire that did not include these variables.

Table 5-5   Attitudes Toward Industrial Development,
Preliminary Sites, 1984[a] (percentages)

Type of Development	Favor	Oppose
Low-level nuclear waste disposal site	10	78
Nuclear power plant	13	75
Coal-burning power plant	51	28
Manure-burning power plant	52	27
Power plant burning other agricultural byproducts	66	14
Feedlot	68	15
Large number of solar cells for electric power	69	7
Large number of wind mills for electric power	85	2
Food processing plant	90	4

[a] Questions ask respondents how they would feel about these projects moving into their county. "Strongly favor" and "somewhat favor" responses are combined, as are "strongly oppose" and "somewhat oppose." "Neutral" responses are omitted.

the public sees nuclear hazards as qualitatively different from other risks. Results for the preliminary sites are shown in Table 5-5.

Predicting Expectations and Overall Attitudes
Results presented so far give a picture of how Panhandle residents see the repository affecting their communities. Understanding how these views develop and how they fit together is also important to policymakers seeking to break the stalemate between DOE and potential host communities.

The DOE nuclear waste program and other legislative proposals for payments to waste-site communities assume that citizens develop summary views of hazardous facilities by weighing expected economic benefits against possible risks. O'Hare, Bacow, and Sanderson (1983) describe the implications of this model for designing compensation in negotiations with potential host communities. The Nuclear Waste Policy Amendments Act of 1987, which offers $10 million a year during repository development and $20 million a year during operation to a state willing to forego its veto, is consistent with the risk-benefit approach.

Regression Model
One purpose of the Texas research was to explore the risk-benefit model empirically. If this conception of attitude formation is correct, then overall attitudes toward the repository (that is, support or opposition) can be predicted as a function of both positive expectations about socioeconomic benefits and negative expectations of environmental risk. These expectations, in turn, may come from a variety of sources. For example, knowl-

edge about the nuclear waste program may be a factor. In addition, previous research suggests that gender, age, and income may influence attitudes toward nuclear facilities to some degree (Nealey, Melber, and Rankin, 1983; Freudenburg and Rosa, 1984), and research after the accident at Three Mile Island suggests that families with young children evaluate risks differently (Kasl, Chisholm, and Eskenazi, 1981).

The Texas research differs from these earlier studies in giving greater emphasis to the possible role of individual economic stakes in shaping attitudes. For example, farmers—because of their reliance on environmental resources of soil and water—may be more sensitive to the environmental effects of a repository than business owners or employees, for whom socioeconomic factors may be more crucial.

Multiple regression analysis was used to test whether the relationships among attitudes and expectations observed in our data are consistent with a risk-benefit model of attitude formation. Overall attitude toward the repository (measured by the two-item scale described above) was regressed on scale scores for expectations about socioeconomic benefits and environmental risks, knowledge and uncertainty and individual characteristics of participants, such as farm income, age, gender, and education.

Results of this analysis can tell us whether socioeconomic and environmental impact expectations each contribute independently to overall attitudes and can assess the comparative importance of these factors in shaping overall attitudes. In addition, the results can indicate whether environmental and socioeconomic expectations have substantial independent explanatory power or whether direct effects of other factors, such as individual background characteristics, are better predictors.

The risk-benefit model implies that socioeconomic and environmental impact expectations will both be important predictors of roughly similar weight. It also implies that knowledge and background characteristics will have little independent effect on overall attitudes; rather, their impact will be mediated by their effect on expectations. Additional regression analyses explore these possible sources of expectations.

Overall Attitudes of Support or Opposition for the Repository
Results are consistent with the risk-benefit model for both survey years in each geographic area. About half the variance in overall support or opposition toward the repository can be explained by variables in the regression analysis, as shown in Table 5-6. Adjusted R^2 ranges from .46 for Amarillo to .53 for Deaf Smith and Oldham counties.

Looking at the regression analysis in more detail, the standardized regression coefficients (beta weights) show the independent effects of each

Table 5-6 Regression of Overall Attitudes Toward the Repository on Expected Impacts, Knowledge, Uncertainty, and Background Characteristics[a] (standardized regression coefficients, beta; unstandardized coefficients in parentheses)

Variable	Preliminary Sites (1984)		Finalist Site (1986)		Urban Center (1986)	
Environmental risks	.48**	(1.30)	.47**	(1.31)	.40**	(1.14)
Socioeconomic benefits	−.31**	(−1.78)	−.29**	(−1.79)	−.36**	(−2.35)
Nuclear waste knowledge	.03	(.03)	.12	(.11)	.07	(.07)
Uncertainty	−.01	(.00)	.03	(.02)	.03	(.02)
Economic stakes						
Farmer	.00	(.00)	.12**	(.67)	.00	(−.08)
Business owner	−.03	(−.15)	.02	(.12)	.03	(.17)
Wage earner	−.05	(−.24)	.11*	(.57)	.01	(.09)
Property ownership[b]	−.02	(−.10)	−.05	(−.26)	—	—
Socioeconomic status						
Income	−.02	(−.03)	−.01	(−.02)	.02	(.03)
Education	.00	(−.01)	.08	(.14)	−.02	(−.05)
Demographic characteristics						
Age	.10*	(.01)	.16**	(.02)	.09	(.01)
Gender[c]	.03	(.12)	−.04	(−.21)	−.04	(−.17)
Ethnicity[d]	−.03	(−.19)	.05	(.29)	.03	(.33)
Children at home	−.04	(.20)	.01	(.02)	.00	(.01)
(Constant)		(7.48)		(4.99)		(8.08)
Overall F	45.71**		25.72**		21.2**	
Adjusted R^2	.53		.52		.46	

**p<.01, *p<.05

[a] Higher scores indicate greater opposition to the repository.

[b] This question was not asked of residents of the urban area.

[c] Male is coded 1; female is 0.

[d] Hispanic is coded 1; Anglo and other are coded 0.

variable in the equation. That is, they show the effect of each variable on overall support or opposition toward the repository with the effects of all other predictors held constant.

Results show that expected socioeconomic benefits and expected environmental risks are the strongest independent predictors of overall attitudes, with somewhat greater weight given to possible risks. Expectations about the environment are particularly important in the rural counties, including both site counties and comparison counties.

In some samples, farmers, older residents, and wage earners expressed greater opposition to the repository, but, generally, effects of individual characteristics on overall attitudes are slight compared to effects of socioeconomic and environmental impact expectations. Knowledge and uncertainty are not significant independent predictors of support or opposition to the repository.

This pattern of higher coefficients for socioeconomic and environmental impact expectations, coupled with small or nonsignificant coefficients for other variables, is consistent with a model of overall attitudes toward the repository as a function of expected risks and benefits. As expected, other variables—knowledge, uncertainty, and background characteristics—have little independent effect on support or opposition toward the repository. However, looking at this initial regression alone leaves open the possibility that knowledge, uncertainty, and background characteristics may influence overall support or opposition indirectly by influencing expectations about benefits and risks. Additional regression analyses predicting expected socio-economic and environmental impacts explore this possibility.

Socioeconomic and Environmental Impact Expectations
Results show that knowledge, uncertainty, and background characteristics are significant predictors of expected impacts on communities. Although the patterns are slightly different for the different samples, gender, occupation, income, and knowledge emerge as the most consistent predictors. These regression results for the site counties and Amarillo are shown in Tables 5-7 and 5-8. Results for the comparison counties are similar.

As in previous research about nuclear energy, women rate environmental hazards at the repository as more likely than do men. In addition, the Texas studies indicate that women are more pessimistic than men in their expectations about socioeconomic benefits.

Farmers in the rural counties expect greater environmental risks and fewer socioeconomic benefits. These views may reflect farmers' special concerns that the repository will directly threaten their own livelihood by harming the soil and water or by ruining markets for their produce because consumers perceive food grown near the repository as unhealthy. Farmers may also differ because of the fundamental values of land stewardship they hold, which are inherent to farming as a way of life.

While farming is an important factor in predicting expectations about the repository, business ownership is not. Results show no significant effect of business ownership on expectations. Higher income, however, is generally associated with lower expected risk and higher expected benefits.

Relationships between knowledge and expectations are complex. Greater knowledge about the nuclear waste program is consistently associated with higher environmental risk estimates in all samples. This result is consistent with the National Academy of Sciences (1984) review suggesting that empirical evidence does not support the common assertion that opposition to nuclear facilities results from ignorance. For the 1986 finalist site counties, knowledge is also strongly associated with lower expectations about socio-economic benefits. At the same time, uncertainty—a variable representing the number of times a respondent said he or she didn't know the answer to an interview question—is strongly related to lower expected benefits for the finalist county sample. Neither of these variables is significantly related to socioeconomic expectations in the other samples. Possibly this difference reflects the impact in the finalist site counties of DOE's intensive public in-formation campaign. Further study of relationships between knowledge and attitudes would be helpful in understanding public response to the nuclear waste program.

Table 5-7 Regression of Expected Environmental Risk on Knowledge, Uncertainty, and Background Characteristics[a] (standardized regression coefficients, beta; unstandardized coefficients in parentheses)

Variable	Preliminary Sites (1984)		Finalist Site (1986)		Urban Center (1986)	
Nuclear waste knowledge	.17**	(.06)	.37**	(.12)	.20*	(.07)
Uncertainty	.03	(.01)	.19	(.04)	.12*	(.02)
Economic stakes						
Farmer	.15**	(.29)	.17**	(.34)	.12*	(.67)
Business owner	−.03	(−.05)	−.05	(−.10)	−.03	(−.07)
Wage earner	−.04	(−.07)	−.05	(−.10)	.00	(.01)
Property ownership[b]	.03	(.06)	.04	(−.09)	—	
Socioeconomic status						
Income	−.18**	(−.09)	−.22*	(−.11)	−.17**	(−.09)
Education	−.14	(−.08)	−.03	(−.02)	−.08	(−.06)
Demographic characteristics						
Age	−.11	(.00)	−.03		−.01	(.00)
Gender[c]	−.23**	(−.38)	−.25**	(−.41)	−.27**	(−.47)
Ethnicity[d]	.14**	(.31)	.14*	(.30)	.03	(.14)
Children at home	.00	(.00)	−.05	(−.08)	−.04	(−.08)
(Constant)		(3.55)		(2.77)		(3.16)
Overall F	8.46**		5.94**		4.79**	
Adjusted R[2]	.14		.15		.12	

**p<.01, *p<.05

[a] Higher scores indicate gender opposition to the repository.

[b] This question was not asked of residents of the urban area.

[c] Male is coded 1; female is 0.

[d] Hispanic is coded 1; Anglo and other are coded 0.

Table 5-8 Regression of Expected Socioeconomic Benefits on Knowledge, Uncertainty, and Background Characteristics[a] (standardized regression coefficients, beta; unstandardized coefficients in parentheses)

Variable	Preliminary Sites (1984)		Finalist Site (1986)		Urban Center (1986)	
Nuclear waste knowledge	−.03	(−.01)	−.33**	(−.05)	−.07	(−.01)
Uncertainty	.06	(.00)	−.33**	(−.03)	−.03	(.00)
Economic stakes						
Farmer	−.19**	(−.18)	−.17**	(−.15)	−.08	(−.20)
Business owner	.02	(.02)	.06	(.05)	.09	(.08)
Wage earner	.08	(.06)	.12	(.10)	−.04	(−.04)
Property ownership[b]	−.03	(−.03)	.02	(.02)	—	
Socioeconomic status						
Income	.11*	(.03)	.13	(.03)	.22**	(.05)
Education	−.03	(−.01)	−.01	(.00)	−.04	(−.01)
Demographic characteristics						
Age	.00	(.00)	.06	(.00)	−.04	(.00)
Gender[c]	.14**	(.11)	.13**	(.11)	.21**	(.16)
Ethnicity[d]	.08	(.09)	−.02	(−.02)	−.02	(−.03)
Children at home	.03	(.02)	.09	(.07)	.03	(.02)
(Constant)		(1.60)		(1.95)		(1.75)
Overall F	4.46**		4.39**		3.56**	
Adjusted R^2	.07		.11		.08	

**p<.01, *p<.05

[a] Higher scores indicate gender opposition to the repository.

[b] This question was not asked of residents of the urban area.

[c] Male is coded 1; female is 0.

[d] Hispanic is coded 1; Anglo and other are coded 0.

While some significant predictors are identified in this analysis, much of the variance in socioeconomic and environmental expectations remains unexplained. For environmental risks, only 12 to 15 percent of the variance is explained; for socioeconomic expectations, only 7 to 11 percent is explained. Future research to investigate the sources of local residents' expectations about the impact of a nuclear waste repository would be useful. However, it is important to remember that this kind of analysis focuses on efforts to understand how and why residents in the site communities differ from each other in their expectations. Future research to understand these differences must not lose sight of the very substantial consensus within Texas communities in opposing the repository.

Comparisons Across Rural and Urban Communities

While the discussion so far has focused on associations within each research sample, as represented by standardized regression coefficients, comparisons across samples are also informative. These comparisons are best understood by looking at the unstandardized coefficients reported in Tables 5-6 through 5-8 (shown in parentheses).

In general, the unstandardized coefficients are similar for each variable across all samples. However, some differences are seen, consistent with the different economic stakes of the urban center and the site communities. Expectations about socioeconomic benefits play a stronger role in predicting overall attitudes toward the repository (Table 5-6) in the urban center than in the site communities, perhaps reflecting expectations that the urban economy will be better able to absorb benefits of the repository. Being a farmer has a stronger effect on perceived environmental risk within the urban sample than in the other two (Table 5-7), probably due to a greater degree of shared dependence on agriculture among respondents in the rural samples. For factors that contribute to expected socioeconomic benefits (Table 5-8), there is no noticeable variation in the unstandardized coefficients across the samples.

Summary of Regression Results

Looking at the regression analyses as a whole, the results suggest a pattern of relationships that is generally stable. Expected socioeconomic benefits and environmental risks are strong predictors of overall support or opposition toward the repository in all geographic areas for both survey years, with expected environmental risks consistently outweighing possible socioeconomic benefits. Farmers, who have a special stake in the environment both as an economic resource and as a way of life, consistently express strong opposition. As in other research, women expect greater risks; and in Texas, they also expect fewer benefits. Knowledge about the nuclear waste program is not independently associated with overall support or opposition toward the repository. Knowledge *is* associated with greater expected risks, but the relationship with socioeconomic expectations is complex. Finally, while environmental and socioeconomic expectations play an important role in overall support or opposition to the repository, the sources of these expectations are not well explained by variables included in this research.

Conclusion

Issues in Common
This chapter complements other chapters in this volume by focusing on themes that are common to all the nuclear waste repository sites. First, the Texas research demonstrates the importance of a longitudinal strategy for developing systematic information about representative groups of local residents. It shows the value of using repeated measures and comparison groups outside the repository impact area to test the generality of research models and to identify trends that are unique to the impact area.

Second, the basic issues—concerns about water and air, health, and changes in the local economic base—are widely relevant to communities facing hazardous facilities. In Texas, several findings on these issues emerge consistently from all sample groups:

(1) Representative samples of Panhandle residents express deep, broad-based fear that the repository will not be safe.
(2) Residents worry that even if the repository doesn't leak, public fears and the stigma of radioactive waste will threaten their communities by undermining the local economy.
(3) At the same time, residents don't believe they will benefit from enormous expenditures on the repository.
(4) In weighing risks and benefits, residents give greater weight to environmental risks.

Farm Country Differences
Beyond the issues common to all repository sites, each site raises unique concerns as well. We can learn about the nuclear waste program as a whole by being sensitive to variations in public reactions stemming from unique local concerns. In Texas, concerns about agriculture are paramount.

While other researchers have talked about fear and dread as dominant emotions for communities facing nuclear facilities (Fischhoff et al., 1978), the rich agricultural abundance of the Panhandle may be the source of another key to the response of Texas farm families: dismay. Our research did not set out to build a standardized "dismay scale," but this emotion comes through in open-ended comments about the nuclear waste program.

The sense of these comments is not simply that the repository may bring fewer benefits or greater risks than DOE projections but rather that the entire site selection process and its elaborate methodology are outside the realm of everyday logic. Comments repeatedly refer to the inherent worth—beyond dollars—of fertile land, the responsibility of the region to produce food, and the pride of being part of one of the most productive agricultural communities in the world. Comments reflect the belief that the Texas counties never would have been chosen if these values were included in the site selection process. Some examples from the interviews illustrate the sense of dismay about the Texas Panhandle being considered as a potential nuclear waste repository site:

—"It is beyond me why the government would even consider placing the repository in one of the most agriculturally productive areas of the United States."
—"I believe common sense should take the effect here. Any person who has common sense can see it is foolish to contaminate land and then attempt to raise the very food you have to eat to survive over the contaminated place."
—"Deaf Smith County, where our family lives and farms, has the most productive rich land whose vegetables, grain, and beef, etc. feeds so many people in this nation and the world. To put a nuclear waste site in this area is insane. This chancy thing should not even be thought about for this area."
—"I realize a waste dump is a national need. But this is some of the most fertile soil in the USA, and before we jeopardize this we need to think of feeding people further down the line in 100 or 200 years. Right now our nation has a surplus of food and this county or area's production may not be needed, but this situation will not last forever—think about it!"

Comments by these farmers and other survey participants reflect deeply held values and beliefs that were not adequately addressed by the nuclear waste program in Texas. Indeed, agriculture was *not* a substantial factor in the site-selection process, and federal documents describe farming as merely a transitory activity at the Texas sites (DOE, 1986).

Although agriculture is not an issue at other sites, the dismay registered by Texas residents signals a fundamental failure by the DOE to grasp the importance of local concerns in the siting process. This failure to address local values is significant beyond the particular concerns at this particular site.

Moving Forward
Although Deaf Smith County is not now being considered as a site for the repository, results of the Texas research offer some valuable guideposts for future efforts to manage nuclear wastes. Systematic research in the site communities and comparison communities over a two-year period shows the breadth and consistency of public opposition to the repository, and results

identify sources of opposition, indicating fundamental issues that must be addressed in an effective nuclear waste program.

First, planners must recognize that communities have genuine fears about repository safety. Research findings indicate that these fears do not stem from ignorance, so there is no "quick fix" via improved public information and risk communication. The existing nuclear waste program, with existing technologies and—just as important—existing strategies for *managing* technology, is not likely to be accepted in agricultural communities like those in the Texas Panhandle.

In addition to recognizing the depth and consistency of public concerns, public expectations about the repository must be taken seriously as potential impacts of the siting process. Local residents report that some expectations have already been translated into actions, including concrete changes in agricultural investments.

Finally, concerns about equity must be addressed, and survey results indicate that current policy and legislation are not adequate to meet these concerns. Results suggest that some aspects of mitigation policy are not perceived as fair. For example, the Nuclear Waste Policy Act provides for mitigation for communities in the DOE-defined impact area, but it does not provide benefits for communities outside the impact area. Yet residents of the Texas comparison communities believe they, too, face economic and environmental risks. Similarly, site-county residents do not expect that benefits from the repository will be targeted at those who face losses.

More importantly, research results raise fundamental questions about the equity of repository siting. Present policy—reflected most baldly in the Nuclear Waste Policy Amendments Act of 1987, which offers payments to the state accepting the repository—rests on the premise that any risks associated with the repository will be balanced by benefits, including economic development and mitigation. Results of multiple regression analyses are consistent with this model. However, the greater weight given to risk estimates over benefits estimates suggests the possibility that economic benefits, no matter how large or how carefully targeted, may not be perceived as an equitable balance to risks to health and the environment. That is, maybe you cannot pay people enough to accept certain kinds of risk.

While these results suggest fundamental problems in current nuclear waste policy, identifying and facing the issues raised by Texas residents who participated in this research is an important step in moving forward. The challenge of managing high-level nuclear waste is not only a problem of deep geology but also of understanding human communities above-ground.

Notes

This research was conducted at the Texas Department of Agriculture in cooperation with the Office of the Governor. Funding was provided by the Texas Nuclear Waste Programs Office (TNWPO), under a grant from the Nuclear Waste Fund. The authors wish to thank Steve Frishman, formerly director of TNWPO, for his innovative support for this project, and Mary Alice Davis, for her many valuable comments.

1   Wording of these questions varies slightly to make them applicable to each of the sample groups. Complete information about the wording of all survey items and introductory materials is available from the first author.
2   For example, "How likely do you think it is that radioactive wastes would escape into the air outside the repository?"
3   For the comparison counties, 75 percent favor an oil refinery in their county, and 69 percent favor secondary recovery of oil and gas. Additional details about attitudes toward industrial development are shown in Brody (1985b).

References

Brody, Julia G. 1985a. "New Roles for Psychologists in Environmental Impact Assessment." *American Psychologist* 40:1057–60.

———. 1985b. *Panhandle Residents' Views of High-level Nuclear Waste Storage.* Austin: Texas Department of Agriculture.

Brody, Julia G., and Judy K. Fleishman. 1987. *Effects of a High-level Nuclear Waste Repository on Local Communities.* Austin: Texas Department of Agriculture.

Cohen, J., and P. Cohen. 1975. *Applied Multiple Regression/Correlation Analysis for the Behavioral Sciences.* Hillsdale, NJ: Erlbaum.

Dillman, Don A. 1978. *Mail and Telephone Surveys: The Total Design Method.* New York: John Wiley & Sons.

Fischhoff, Baruch, Paul Slovic, Sarah Lichtenstein, S. Read, and B. Combs. 1978. "How Safe is Safe Enough? A Psychometric Study of Attitudes Towards Technological Risks and Benefits." *Policy Sciences* 9:127–52.

Folkman, Susan, and Richard Lazarus. 1980. "An Analysis of Coping in a middle-aged community sample." *Journal of Health and Social Behavior* 21:219–39.

Freudenburg, William R., and Rodney K. Baxter. 1984. "Host Community Attitudes Toward Nuclear Power Plants: A Reassessment." *Social Science Quarterly* 65:1129–36.

Freudenburg, William R., and Eugene A. Rosa, eds. 1984. *Public Reactions to Nuclear Power: Are There Critical Masses?* Boulder: Westview/American Association for the Advancement of Science.

Halstead, John M., F. Larry Leistritz, D. G. Rice, D. M. Saxowsky, and Robert A. Chase. 1982. *Mitigating Socioeconomic Impacts of Nuclear Waste Repository Siting.* Report prepared for the Office of Nuclear Waste Isolation. North Dakota State University, Fargo.

Hughey, Joseph B., John W. Lounsbury, Eric Sundstrom, and Thomas J. Mattingly, Jr. 1983. "Changing expectations: A longitudinal study of community attitudes toward a nuclear power plant." *American Journal of Community Psychology* 11:655–72.

Kasl, Stanislav V., Rupert F. Chisholm and Brenda Eskenazi. 1981. "The impact of the accident at the Three Mile Island on the behavior and well-being of nuclear workers.

Part II: Job tension, psychophysiological symptoms, and indices of distress." *American Journal of Public Health* 71:484–95.

National Academy of Sciences. 1984. *Social and Economic Aspects of Radioactive Waste Disposal: Considerations for Institutional Management.* Washington, D.C.: National Academy Press.

Nealey, Stanley M., Barbara D. Melber, and William L. Rankin. 1983. *Public Opinion and Nuclear Energy.* Lexington, MA: Lexington Books.

O'Hare, M., L. Bacow, D. Sanderson. 1983. *Facility Siting and Public Opposition.* New York: Van Nostrand Reinhold.

Sundstrom, Eric, John W. Lounsbury, C. Richard Schuller, James R. Fowler, and Thomas J. Mattingly, Jr. 1977. "Community Attitudes Toward a Proposed Nuclear Power Generating Facility as a Function of Expected Outcomes." *Journal of Community Psychology* 5:199–208.

U.S. Department of Energy, Office of Civilian Radioactive Waste Management. 1986. *Environmental Assessment, Deaf Smith County Site, Texas.* DOE/RW-0069.

6   Local Attitudes Toward Siting a High-Level

Nuclear Waste Repository at Hanford, Washington

Riley E. Dunlap, Eugene A. Rosa, Rodney K. Baxter,

and Robert Cameron Mitchell

The selection of the Hanford Nuclear Reservation in May of 1986 as one of
the three "finalist sites" for the nation's permanent high-level nuclear waste
repository came as no surprise, especially not to the residents of the state
of Washington. Washingtonians were long accustomed to nuclear activities
at the reservation, dating back to the 1940s, when Hanford was chosen as
the nation's first plutonium production facility. Hanford produced the plutonium used for the world's first atomic explosion, the Trinity bomb test at
Alamogordo, New Mexico, in July of 1945. One month later, an atomic bomb
with Hanford-produced plutonium was dropped on Nagasaki, Japan, ending
World War II. Although Hanford continued to serve as a defense facility, its
range of nuclear activities broadened over the years to include a wide range
of commercial nuclear activities. Among its broadened functions was the
temporary storage of low-level nuclear wastes in burial trenches and high-
level nuclear wastes in underground storage tanks. Because of its history of
nuclear operations in general, and waste storage in particular, Hanford was
often mentioned as a "natural" choice for a permanent high-level nuclear
waste repository (Impact Assessment, Inc., 1987).

Despite being accustomed to nuclear activities at the reservation, and notwithstanding their typical support for expanding Hanford's nuclear operations, Washington residents were far from enthusiastic about the selection
of Hanford for a permanent repository. In fact, opposition to the repository,
which had been growing steadily since 1983, when Hanford was selected
as one of the original nine semifinalist sites, had become substantial by the
time the three finalists were announced in 1986. The partial exception was
the Tri-Cities area of the state, consisting of the cities of Richland, Kennewick, and Pasco, adjacent to Hanford. The home of a majority of Hanford
employees, and heavily dependent on the nuclear industry, the Tri-Cities

area had historically been extremely pronuclear. Indeed, its presumed favorable political climate was acknowledged in the site selection process, and the weight given to this factor presumably led to Hanford being a finalist, despite its relatively unfavorable geological characterization (Keeney, 1986). Interestingly, over the next couple of years the atmosphere seemed to change somewhat, to the point that the repository became surprisingly controversial even in the Tri-Cities area.

Drawing on several sources, this chapter will examine the attitudes of residents of the Tri-Cities area toward the proposed nuclear waste repository. Making use of a series of polls sponsored by major newspapers in the state, we will first compare the views of those living near the Hanford reservation with those of citizens statewide, highlighting trends toward increased opposition over time in both locales. Then we will present the results of a detailed study of attitudes toward the repository among residents of the two counties constituting the Tri-Cities area, Franklin and Benton counties, noting that by late 1987 there was a surprising lack of support for the repository even in this traditionally pronuclear area. The remainder of the chapter will examine a wide range of factors, ranging from demographic characteristics to perceptions of the likely impacts of the repository, that help explain variation in repository attitudes among Tri-Cities residents.

The Research Setting

Why a desert site in southeastern Washington on the Columbia River was chosen for the nation's first plutonium production facility is easy to understand. The Columbia, the largest river flowing into the Pacific Ocean in North America, could provide the vast quantities of water needed to cool the cores of "single-pass" reactors—so named because cooling water was pumped into the reactor cores and then directly back into the river (Stenehjem, 1990). The river served another crucial need for the facility, as its big dams provided the large quantities of electricity needed to power Hanford. Finally, being remote and sparsely populated, mostly by scattered dairy farms, also made the site attractive for military production purposes. There were few people to relocate, few eyes, ears, and lips to compromise the high degree of secrecy that needed to be maintained, and few potential casualties in the event of an accident (U.S. Department of Energy, 1991a).

The city of Richland was constructed by the federal government to house Hanford employees, and over time the community of Kennewick (just across the Columbia) also developed as a home for many Hanford employees, as did the nearby agricultural community of Pasco. While these three communities varied in their ties to Hanford, with Richland having the strongest and

Pasco the weakest, together the "Tri-Cities" became both strongly identified with, and economically dependent on, the Hanford reservation (Impact Assessment, Inc., 1987). The pronuclear orientation of the area is legendary; indeed, nuclear power of any kind was a source of great community pride. The Richland High School's nickname of "Bombers" and its insignia of a nuclear blast (painted, for example, on football helmets) is one of the most publicized illustrations, reflecting local pride in having played such a critical role in the construction of the nation's nuclear weapons (see Loeb, 1982 for a more detailed discussion of the pronuclear orientation of the Tri-Cities).[1]

For the first three decades of operation, the Hanford Nuclear Reservation generally escaped scrutiny from the rest of the state. The Cascade Mountains separated it from the state's main population center, the Seattle-Tacoma area, and from the state capitol in Olympia. The only metropolitan area in the eastern part of the state was Spokane, the commercial center of the "Inland Empire" comprising the eastern two-thirds of Washington, northern Idaho, and western Montana. A bastion of social and political conservatism, Spokane paid little attention to the nuclear activities being conducted 150 miles or so to the southwest. The inattention given to Hanford outside of the Tri-Cities gradually began to shift, however, as nuclear phenomena began to attract increasing attention by the media and public interest groups in the 1970s and, especially, the 1980s.

A number of factors seem to have contributed to the growing salience of Hanford outside the Tri-Cities. First, the large-scale leakage of a waste storage tank in 1973 received considerable media coverage within the state as well as nationally and served to nudge nuclear waste onto the public agenda (Rankin and Melber, 1980). The leak also received considerable attention in Portland, Oregon, the Northwest's second largest metropolitan area, located downstream on the Columbia River. The emergence of a visible antinuclear movement and the related decline nationwide in public support for nuclear power in general (see chapter 2 of this volume by Rosa and Freudenburg) surely affected the state of Washington as well. Finally, the financial collapse of the Washington Public Power Supply System's overly ambitious plans for constructing several nuclear power plants, and the negative publicity attending the unprecedented bond forfeiture, further served to blacken the nuclear eye within the state (Impact Assessment, Inc., 1987).

Insofar as these events contributed to the salience of the repository and to growing citizens' anxieties, they were reinforced by other concerns: a repository's unique risk characteristics—especially the dangerous, long-lived radioactivity of nuclear waste—and the stigmatization associated with becoming the nation's "nuclear dump." The cumulative result was that by the early 1980s there were already signs of concern around the state, a concern heightened by the assumption that the Tri-Cities would openly welcome the repository (Impact Assessment, Inc., 1987). This changing mood was reinforced by the emergence of the "Downwinders," a loose coalition of rural residents (mostly farmers) in the Tri-Cities area who began to claim that their families and neighbors suffered disproportionate rates of serious maladies, such as thyroid cancer, due to operations at Hanford. Singled out as key culprits were radioactive releases—large, deliberate releases in the early days of Hanford and other, less dramatic releases due to faulty storage facilities. Information on both types of releases, long treated as top-secret, had only recently become available to the public.

The Downwinders' claims were given additional credibility after a powerful, award-winning series of investigatory articles appeared in the Spokane *Spokesman-Review*, the largest-circulation newspaper east of the Cascade Mountains (Steele, 1985). The series stimulated noticeable concern among the heretofore apathetic residents of Spokane and eastern Washington more generally. Concerns over Hanford activities in general were given increased credibility by the fact that the area had experienced substantial ashfall (several inches deep) from the Mount St. Helens' eruption in 1980. The Spokane area was in the path of a prevailing northeasterly wind flow, making it "downwind" of Hanford just as it had been downwind of Mount St. Helens.[2] Further magnifying these concerns was the realization that Spokane lay on both the most likely rail and highway routes for transporting nuclear waste to Hanford from the rest of the nation. The convergence of these phenomena resulted in the emergence of a highly visible "anti-Hanford" movement within the historically conservative community of Spokane as well as in a few other eastern Washington and northern Idaho communities (including the twin border towns of Pullman, home of Washington State University, and Moscow, home of the University of Idaho). Most prominent in the movement was the Spokane-based "Hanford Education Action League," or HEAL, which published a newsletter entitled *The Hanford Journal* and issued several reports highly critical of Hanford activities, emphasizing the health threats posed by radioactive contamination (see, for example, Benson and Shook, 1985). Even the Spokane chapter of the mainstream League of Women Voters (1985) issued a critical report on Hanford's nuclear operations, and the implications for Spokane of having a repository located there.

Despite an initial reception of near-ridicule, at least in the Tri-Cities area, the claims of the Hanford Downwinders received another boost in credibility, this time from a most unexpected source: the U.S. Department of Energy itself. In response to growing pressure, the department initiated a Dose Reconstruction Project (still ongoing), whose goals are to estimate the

amount and dispersion of radioactive releases from Hanford, using historical records, and to trace the incidence of cancer among those living in the
Hanford area at the time of the releases.

Trends in Repository Attitudes: Tri-cities and the State

That the proposed repository was becoming a salient political issue for
Washingtonians was further evidenced by the actions of the major newspapers in the state. To gauge the climate surrounding the proposed repository, they sponsored several public opinion polls over a three-year period.
Some were statewide polls and some were limited to the Tri-Cities area, and
on occasion they involved both state and Tri-Cities samples. In combination, the poll results allow us to determine levels of support and opposition
for a repository in the entire state as well as in the local Tri-Cities area and
to make comparisons between the two populations. The results also permit
an examination of trends in opinions over time, both statewide and in the
Tri-Cities.

Five statewide surveys and four Tri-Cities surveys were conducted between early 1985 and late 1987. While not employing identical questions,
all included a relatively comparable item that appears to provide an unam-
biguous indicator of basic attitude toward the repository (favor, oppose, or
unsure). Consequently, despite the variation in wording, reasonably meaningful comparisons across surveys and among populations are possible. In
addition, a November 1986 state ballot measure, Referendum 40, on the
potential nuclear waste repository also provides an indicator of public sentiment toward the repository.

The results for all of the surveys as well as the ballot measure are presented in Table 6-1. Focusing first on the state as a whole, two points stand
out: there was overwhelming opposition by the time of the first poll in early
1985, and this opposition grew rather consistently over the next two-and-a-
half years. For the Tri-Cities area—sometimes defined as all of Benton and
Franklin counties and sometimes as just the three cities (which contain the
vast majority of residents of the two counties)—support was consistently
higher than for the entire state, though it was not as overwhelming as widely
assumed. Even in the early 1985 poll, "only" 59 percent favored the repository; after that, the support declined to the point where it barely exceeded
opposition. By the time of the in-depth survey of Benton and Franklin counties reported in this chapter, it was obvious that overwhelming local support
for a repository, once taken for granted, had withered away.

Also worth noting is the remarkable similarity between the results from

Table 6-1 Summary of Washington State Polls on Nuclear Waste Repository Siting at Hanford: Statewide and Tri-Cities (percent)

		State	Tri-Cities
1. *Spokane Spokesman-Review*[a]			
Statewide: N = 400, Feb. 11, 1985	Yes	21	59
Tri-Cities: N = 385, Jan. 21–26, 1985	No	68	26
	Don't know	12	14
2. *Seattle Times*[b]			
N = 500 reg. voters	Favor	19	
Feb. 1–6, 1986	Oppose	69	
	No opinion	12	
3. *Seattle Times*[c]			
Statewide: N = 409 voters	Favor	10	43
Tri-Cities: N = 250 voters	Oppose	78	37
July 27–29, 1986	No opinion	12	20
4. *Tacoma News Tribune*[d]			
Statewide: N = 1000 voters	Favor	16	47
Tri-Cities: N = 200 voters	Oppose	72	32
Sept. 22–28, 1986	Undecided	12	21
5. *Tacoma News Tribune*[e]			
N = 600 voters	Favor	12	
Oct. 26–28, 1986	Oppose	76	
	Undecided	12	
6. *WSU/Impact Assessment, Inc.*[f]			
Tri-Cities: N = 658 RDD Sample	Favor		46
Dec. 3–21, 1987	Oppose		43
	Unsure		11
7. *Referendum 40—Nov. 1986*[g]	Yes	83	46
	No	17	54

[a] "Should the government locate the nation's first permanent nuclear waste disposal site at Hanford?"
[b] "Based on what you know now, do you favor or oppose having a national nuclear waste repository at Hanford?"
[c] Same question as in 2 above.
[d] "Do you favor or oppose location of a nuclear waste repository at the Hanford nuclear reservation in eastern Washington?"
[e] Same question as in 4 above.
[f] Same question as in 2 above.
[g] "Shall state officials continue challenges to the federal selection process for high-level nuclear repositories, and shall a means be provided for voter disapproval of any Washington site?"

Referendum 40 and those from the polls conducted before and after the election. Deleting the "undecided" category from the October 1986 statewide poll by the *Tacoma News Tribune* shows 86 percent opposed to a repository, close to the 83 percent of the voters who expressed opposition by voting in favor of Referendum 40. Similarly, deleting the undecideds in the December 1987 survey of Benton and Franklin county residents reported in this chapter results in 48 percent opposed to the repository, just 2 percent above the 46 percent of the voters in these two counties who voted for Referendum 40.[3] These results provide strong testament to the validity of the survey results in terms of accurately gauging Washingtonian opinions on the repository.

In sum, the available surveys show that, by mid-decade, the residents of Washington were strongly opposed to the repository and became even more so over time. The residents of the Tri-Cities were an exception to this general pattern, but, even in this traditionally pronuclear area, the support was far from overwhelming. Still, the fact that support for a repository exceeded opposition in the Tri-Cities, in stark contrast to the situation in the rest of the state, is a very interesting departure from the now familiar NIMBY ("not in my back yard") syndrome—strong *local* opposition to the siting of any noxious facility (Another exception is reported by Krannich, Little, and Cramer in chapter 10 of this volume). This finding points to the importance of understanding historical experiences, economic dependence, and local culture as they relate to the proposed facilities in those communities where significant NIMBY reactions fail to materialize. In fact, as illustrated by the historical receptivity of the Tri-Cities to nuclear facilities, such phenomena can create a quite different reaction. Apparently stemming from lack of fear of the relevant technology (based on long-term experience with it and trust in those responsible for its operation), plus the perceived economic benefits of the facility, such an orientation might be termed a "PIMBY," or "put in my back yard," orientation (Rosa, 1987).[4] PIMBY provides an apt description of the history of nuclear facilities in the Tri-Cities (Loeb, 1982), particularly the efforts of the Tri-Cities Nuclear Industrial Council to attract such facilities to Hanford (Fleischer, 1974).

The Current Study

The original survey data reported in this chapter were collected in December of 1987, as part of the state of Washington's assessment of the possible socioeconomic impacts of having a permanent high-level nuclear waste repository (HLNWR) located at Hanford. The state's Office of High-Level Nuclear Waste Management contracted with Impact Assessment, Inc., of La Jolla, California, for an in-depth study of these impacts, which in turn subcontracted with Washington State University's Social and Economic Sciences Research Center for the survey portion of the study. The basic purpose of the survey component was to assess Washington residents' views of nuclear waste and of repository siting—particularly differences between residents of the Tri-Cities area and citizens statewide—and to identify the key factors (besides geographical location) that account for variation in these views.[5] The remainder of this chapter reports the results of a survey of residents of Benton and Franklin counties (the officially designated "impact zone"); the statewide survey that was to have been conducted for comparative purposes was canceled due to the congressional decision in late December 1987 to remove Hanford from consideration as a site for a repository.

Since we wanted to be able to make comparisons between Benton and Franklin residents, and since the population of the former is over three times that of the latter, we employed a disproportional stratified sample in which Franklin residents were over-sampled. Using random-digit-dialed samples provided by Survey Sampling, Inc., we completed 426 telephone interviews with residents of Benton County and 232 with residents of Franklin County between December 3 and 21 of 1987. Results of comparisons between the two counties are based on these two samples, but results for the total "Tri-Cities sample" (the two counties combined) were weighted to reflect accurately the appropriate sizes of the two county populations.[6]

Comparisons of the characteristics of our respondents with the population characteristics of the two counties reveal that both samples are representative in terms of age and gender but that minorities are underrepresented in Franklin County and that in both counties the samples are above-average in educational level (Dunlap and Baxter, 1988:6–11). Such biases are almost inevitable in survey research but are doubtless of lower magnitude than in most other techniques of obtaining citizen input into policymaking (e.g., formal hearings and letters to public officials). Furthermore, our samples are likely to be more representative of the voting public than of the total public and should therefore provide reasonably accurate readings of the citizens' views that policymakers are most likely to consider when making decisions.[7]

Descriptive Results

A frequent criticism of survey research is that it often elicits "opinions" that have little meaning, as respondents are being asked questions on topics about which they have little personal awareness, concern, or knowledge. In contrast, it is believed that opinions expressed on topics which are salient to the public have more meaning and stability (i.e., validity and reliability). Consequently, we began our interviews with an open-ended question concerning what respondents saw as the "two or three most important issues" facing the state—a standard technique for measuring salience, since respondents are most likely to mention issues that are "on their minds." That nuclear energy was an enormously salient issue in the Tri-Cities at the time of our survey was clearly reflected by the volunteered responses to this item.

Although respondents were not aware of the focus of our survey at the outset,[8] various types of nuclear issues were mentioned (first, second, or third) by 62 percent of the Benton residents, 50 percent of the Franklin residents, and 58 percent of the total (weighted) sample. This was more than twice that mentioning "environmental issues," the second most-cited topic with 27 percent total, and triple or more the percentages mentioning issues such as taxes (20 percent), jobs (18 percent), government (15 percent), and education (11 percent). Since it is rare for any noneconomic issue to be mentioned by over half of the public in response to this type of question (Smith, 1985), these results indicate that nuclear issues were indeed a topic of considerable interest to our respondents. It is therefore very unlikely that our interviews elicited superficial responses to later questions that focused specifically on nuclear power, nuclear waste, and repository siting.

Because volunteered responses to open-ended questions about most important issues or problems is a very stringent measure of salience, we next presented the respondents with an alternative measure of salience (Dunlap, 1987)—asking about their concern over the repository relative to other issues in the state. Specifically, we gave respondents "a short list of issues facing our state that some people are concerned about," and asked them to indicate their personal level of concern "regardless of whether you favor or oppose" each of the four: "promoting trade with Japan and other nations," "constructing a nuclear carrier base in Everett," "attracting more high-tech industries," and "locating a national nuclear waste repository at Hanford." Again, the results indicate that a nuclear waste repository had become a very salient issue, as 58 percent of the Benton-Franklin County residents said they were "very concerned" about it. The repository was followed by promoting trade with Japan (53 percent "very concerned"), attracting more high-tech industries (46 percent "very concerned"), and constructing a nuclear carrier base (only 28 percent "very concerned"). While the latter was no doubt of more salience to Washingtonians west of the Cascades (where it had become a major source of controversy), the fact that the waste repository elicited twice the level of strong concern among our respondents again reflects the salience of repository siting to Tri-Cities residents.

Attitudes Toward Siting a Repository at Hanford

In addition to examining the salience of nuclear issues at the outset of our interviews, we took two more steps to ensure that we obtained a good indicator of respondents' global attitude toward having a permanent, HLNWR located at Hanford: First, we used three separate measures of repository attitudes, a standard procedure for improving the validity and reliability of attitudinal measures. Second, we followed a less conventional procedure: during the interviews, respondents were provided with considerable information on the siting process in order to insure that they were providing well-informed opinions. The three items and the responses to them for both counties and for the total weighted sample are shown in Table 6-2. When examining responses to these items, it must be kept in mind that the attitudes of the Tri-Cities residents were no doubt more positive toward the repository than would have been the case for citizens in the rest of the state (as clearly suggested by the poll results reported in Table 6-1 above).

The first of the three items, presented about a fourth of the way into the interview, was the first question in the interview to focus specifically on a waste repository.[9] After explaining the site selection process mandated by Congress, it asked respondents if they "favor or oppose having a national nuclear waste repository at Hanford." As shown in Table 6-2, in Benton County those in favor exceeded those opposed, but in Franklin the opposite was found; overall the proponents exceeded the opponents of a repository by only 2 percent (45.6 vs. 43.3 percent). The fact that less than half of the sample favored a repository, and nearly as many opposed it, clearly indicates that even in the traditionally pronuclear Tri-Cities area the repository had become a controversial issue.

Next we presented a series of questions (to be discussed below) concerning respondents' knowledge about the repository siting process; respondents were provided the correct information after giving their answers. Then, the testing program envisioned for Hanford was described, and respondents were asked whether they favored or opposed it. As shown in Table 6-2, there was considerably more support for the testing program itself than for the actual siting of a repository at Hanford (as measured by the prior item). While support was again noticeably higher in Benton County than in Franklin County (81.7 vs. 70.7 percent), the principal finding is the strong support for the testing program in the overall sample. Those in favor outnumbered those opposed by over four to one (79 vs. 18.3 percent). Given the obvious economic benefits of the proposed testing program for the Tri-Cities economy, and the clear stipulation that we were asking only about the testing program, these results are not surprising. In fact, various degrees of testing

Table 6-2   Attitudes Toward Siting a High-Level Nuclear
Waste Repository at Hanford (percent)

		Benton County	Franklin County	Weighted Total
Attitude toward	Favor	47.9	38.8	45.6
a repository[a]	Oppose	41.1	50.0	43.3
	Unsure/don't know	7.8	8.2	7.9
	No answer	3.3	· 3.0	3.2
Attitude toward	Favor	81.7	70.7	79.0
testing program[b]	Oppose	15.5	26.7	18.3
	Unsure/don't know	2.1	1.7	2.0
	No answer	.7	.9	.7
Intended vote if	Yes	63.9	48.7	60.1
Hanford safest site[c]	No	24.2	38.4	27.7
	Wouldn't want to vote	6.3	6.5	6.4
	Unsure/don't know	4.7	5.2	4.8
	No answer	.9	1.3	1.0
	N	(426)	(232)	(567)

[a] "As you may know, Congress has decided the nation needs a permanent storage place or repository for high-level radioactive wastes, and has directed the Department of Energy to find the best site for it. The department has selected three sites for thorough study, one of which is the Hanford Nuclear Reservation in Washington. Based on what you know now, do you favor or oppose having a nuclear waste repository at Hanford?"

[b] "If Hanford is chosen, the wastes will be buried in basalt rock 3,000 feet underground. Over the next several years, the Department of Energy plans to spend about a billion dollars at each of the three sites (Texas, Nevada, and Hanford) to determine if the underground rock formations can safely store radioactive wastes. Only after these tests have been completed will a final decision be made. Thinking just about the testing program, do you favor or oppose allowing the Hanford site to undergo these tests?"

[c] "Suppose that after all the tests are completed the federal government said the Hanford site was the safest place for the nation's first nuclear waste repository, but it wouldn't be located there unless state residents voted in favor of it. If this were the case, would you vote for it, against it, or wouldn't you vote on this issue?"

for the appropriateness of burying nuclear waste at Hanford had begun via the Basalt Waste Isolation Program prior to the 1982 Nuclear Waste Policy Act. Thus, the area was already used to testing programs, and putting an end to testing might have been interpreted by some respondents as meaning a loss of existing jobs.

The third item designed to measure attitudes toward the repository came near the end of the interview, just before the demographic questions. Also shown in Table 6-2, it asked respondents to indicate how they would vote if tests showed Hanford to offer the safest site, but approval by state voters was required before the repository would be located there. The caveat that Hanford be assumed the safest site clearly affected some respondents, as this item elicited a much more favorable response than did the first item—48.7 percent of Franklin residents, 63.9 percent of Benton residents, and 60.1 percent overall would vote for a repository under these conditions.

Taken together, responses to these three items indicate the complex nature of the repository and the ability of residents, owing to their familiarity with nuclear technology, to discriminate among features of the complexity. Our initial inquiry concerning the repository revealed nearly as much opposition as support, a surprising finding given the strong pronuclear orientation of the Tri-Cities area. However, a testing program per se, with no commitment for siting, was strongly supported. There was also considerable support for a repository given the crucial caveat that Hanford be found to provide the safest site for it. Responses to these three items thus revealed different aspects of Tri-Cities residents' attitudes toward a repository and the siting process. Fortunately, we need not debate which one offers the best indicator of respondents' views of locating a repository at Hanford. Since our primary interest is in understanding the factors which affect repository attitudes, the use of three separate questions, each tapping different aspects of these attitudes, serves us well. We combined the responses to the three items into a single variable measuring overall attitude toward siting a HLNWR at Hanford (ranging from "strongly opposed" to "strongly in favor").[10] This composite variable, termed *Attitudes toward Repository*, forms the dependent variable for our subsequent analyses.

Attitudes Toward Nuclear Power, Nuclear Waste,
and Nuclear Accidents
Among the factors which we expected to be useful in explaining variation in citizens' attitudes toward repository siting were their overall views of nuclear energy—including their attitudes toward nuclear power, their perceptions of the risk posed by high-level radioactive waste, and their views of the likelihood of accidents occurring in the transportation and burial of nuclear waste. We expected that examination of these factors would also be useful in explaining differences in the reactions of Tri-Cities residents and those of other Washington residents to a repository. While the cancellation of the proposed statewide survey limits our ability to examine the latter issue, we can nonetheless gain additional insight into the uniquely pronuclear orientation of the Tri-Cities region by comparing the results of our Benton-Franklin counties survey with those of existing surveys at the national and state level. In the process we will be giving an overview of

some of the important variables to be examined in our explanatory analyses reported later.

A standard way of measuring overall attitude toward nuclear power is to ask respondents the following question: "In general, do you favor or oppose the building of more nuclear power plants in the United States?" We inserted this item between general attitude questions toward science and technology and the first item on repository attitudes. A comparison of our results with those from national surveys using this item at about the same time again reflects the strong pronuclear orientation of the Tri-Cities area. Whereas 65 percent of our Tri-Cities respondents (surveyed in December 1987) favored the construction of more nuclear power plants, and only 28 percent were opposed, results reported in Figure 3 of chapter 2 by Rosa and Freuden-burg reveal that nationwide the results were very different: A fourth quarter 1987 survey found 22 percent in favor and 65 percent opposed, while a first quarter 1988 survey found 24 percent in favor and 65 percent opposed—a perfect inverse of the Tri-Cities attitudes. These comparative results reaffirm the long-standing pronuclear mood of the Hanford area.

Another stark contrast between Tri-Citians' nuclear attitudes and those of the rest of the nation emerges when we compare results from our survey with those from the national survey reported in chapter 7 by Desvousges, Kunreuther, Slovic, and Rosa. In both surveys respondents were "read a list of several sources of pollution," and then told: "on a scale from 1 to 10, with 1 meaning not at all serious and 10 very serious, please tell me how serious a problem you think each source of pollution is for the United States as a whole." For three of the pollution sources, the responses of the Tri-Cities respondents were very similar to those of the national survey, as reflected by the mean ratings given to each source: 5.9 for our sample and 6.1 for the national sample for "garbage from city or county landfills;" 7.2 and 7.1, respectively, for "air pollution from cars and factories;" and 7.7 and 7.9, respectively, for "water pollution from toxic chemicals." However, the ratings of the fourth source, "radioactive wastes from nuclear power plants," differed dramatically: only 5.4 for Tri-Cities sample but 7.4 for the national sample.[11] The fact that the national sample rated nuclear waste as second only to toxic chemicals, whereas the Tri-Cities residents rated it as much less serious—indeed, less serious than garbage—is another vivid indicator of the relative lack of concern over nuclear phenomena in the Hanford area.

Our third indicator of global attitude toward nuclear energy, focusing more specifically on repository siting, deals with the perceived likelihood of accidents occurring in the transport and storage of nuclear wastes at Hanford. Here we employed an item used in the February 1986 statewide survey sponsored by the Seattle Times (no. 2 in Table 6-1). In both surveys, respon-

Table 6-3 Likelihood of Accidents Associated with Siting a Repository at Hanford[a]

	Percent Responding "Major Problem"			
Potential Accidents	Benton County	Franklin County	Weighted Total	Statewide Sample[b]
Accidents may occur in transporting the wastes to the site	37.6	41.4	38.5	72
Future generations may accidentally dig into the site	19.3	22.8	20.1	39
An earthquake may release the materials	28.6	40.1	31.5	63
The waste may leak out of the containers and contaminate underground water supplies within the next 50 years	39.9	50.0	42.4	72

[a]"Now I am going to read a list of concerns that have been expressed about putting a nuclear waste repository at Hanford. For each one I read, please tell me whether you think it would be a major problem, a minor problem, or not a problem at all. The first one is, 'Accidents may occur in transporting the wastes to the site.' Would this be a major problem, a minor problem, or not a problem at all?"
[b]Results from a February 1986 survey for the *Seattle Times*.

dents were read "a list of concerns that have been expressed about putting a nuclear waste repository at Hanford," and then asked to indicate for each one whether they thought it "would be a major problem, a minor problem, or not a problem at all." The four types of accidents examined are shown in Table 6-3, along with the percentages indicating "major problem" for each type in our Benton-Franklin county sample and in the statewide sample. In every case, the residents of the Tri-Cities area were far less likely to rate the potential accident as a major problem than those in the statewide sample, often being only half as likely to give such a response.

In sum, the three sets of results just reviewed—overall attitude toward nuclear power, comparative seriousness of nuclear waste and other pollutants, and likelihood of accidents in transport and burial of nuclear waste—again reveal the striking difference between the nuclear views of residents of Benton and Franklin counties and those of citizens statewide and nationwide. On all three aspects of nuclear power assessed, Tri-Cities residents were far more pronuclear than the rest of the state and the nation. Because the items tap important aspects of nuclear power, responses to them should provide a good indication of orientation or predisposition toward nuclear power generally. We therefore expect that these responses will be useful in predicting citizens' receptivity to the siting of a nuclear waste repository at their locale. Consequently, in the analyses that follow, we include each as a potential explanatory variable: the first item as a measure of overall "attitude toward nuclear power," the numerical rating of the seriousness of nuclear waste as a measure of "perception of nuclear waste," and a composite of responses to the four accident items as an indicator of perceived "likelihood of accidents." [12]

Distinguishing NIMBY from PIMBY Reactions

The foregoing results provide a reasonably good idea of how residents of Benton and Franklin counties viewed having the nation's first permanent HLNWR located at the Hanford Nuclear Reservation. While support for a repository was not as strong in the Tri-Cities area as widely assumed, and therefore cannot accurately be termed a PIMBY (as opposed to NIMBY) response, comparisons of Tri-Cities residents with statewide and national samples in terms of their overall perceptions of nuclear phenomena provide insight into why the proposed repository did not generate a strong NIMBY response in the two counties. The results also indicate why, historically, most proposals for nuclear operations at Hanford—ranging from new power plants for the Washington Public Power Supply System program, continuing operation of the N-Reactor for plutonium production, and development of a fast-flux reactor—were warmly endorsed by local residents. Presumably, a combination of familiarity with nuclear operations, economic dependence on such operations, perception of the safety of such operations, and trust in those responsible for supervising the operations all seem to have fed a widespread and generally uncritical acceptance of nuclear phenomena at Hanford (i.e., rating nuclear waste as less serious than garbage).

We argued earlier that understanding why some proposed projects provoke a NIMBY reaction and others a PIMBY reaction requires a detailed examination of local experiences with and economic dependence on related technologies as well as the local culture which developed as a result of such experiences and dependencies. Our survey results, especially those comparing results from statewide and national surveys, supplement Loeb's (1982) previously cited documentation of the pronuclear orientation of the Tri-Cities area. Yet, although a relatively strong pronuclear orientation emerged in our study, and accounts for the lack of a strong NIMBY response (entailing deep and widespread opposition) to the repository, our results also show that the strong PIMBY reaction among Tri-Cities residents expected by the rest of Washington State did not emerge either. Why so many residents of an area that had historically welcomed all types of nuclear facilities balked at the idea of having it become the home of the nation's first permanent HLNWR represents something of an enigma. Thus, the Tri-Cities' reaction to a nuclear waste repository provides neither a clearcut case of a NIMBY nor of a PIMBY reaction but rather reflects a surprising amount of ambivalence.

An understanding of this ambivalence can be obtained by looking at a final set of descriptive data, the perceived likelihood of various impacts—both positive and negative—from having a repository located at Hanford. Data on expected impacts have frequently been used to explain attitudes toward siting nuclear power plants (e.g., Hughey, Sundstrom, and Lounsbury, 1985) and were also examined in the studies of Texas residents' views of a nuclear waste repository reported by Brody and Fleishman in chapter 5 of this volume. Consequently, we provided our Tri-Cities respondents with a list of fifteen potential impacts (both positive and negative) from siting a repository at Hanford and asked them to indicate the likelihood of each one occurring—"very likely," "somewhat likely," "somewhat unlikely," or "very unlikely." Even though we do not have comparative data from either statewide or national surveys, the descriptive results presented in Table 6-4 provide insight into the ambivalence of Tri-Cities residents toward having a repository located at Hanford.

As can be seen in Table 6-4, the creation of more jobs is the outcome of having a repository at Hanford that Tri-Cities residents saw as most likely, as 93.6 percent saw it as a likely outcome (either "very likely" or "somewhat likely"). Large majorities also thought that a repository would make residents of the area more certain about its economic future (85.3 percent) and that it would significantly increase tax revenues for the state (81.0 percent). Nearly three-fourths thought that a repository would bring more stores and entertainment to the Tri-Cities, while more than seven out of ten thought it would lead the U.S. Department of Energy (DOE) to clean up existing contamination from the defense wastes already stored at Hanford. The only positive impact seen as likely by less than a majority was that the rest of the country would respect Washington for meeting an important national need if it accepts the repository, and very close to half saw this as likely.

Although the presumed negative impacts were generally viewed as less likely than the positive ones, several were viewed as likely by over half of the Tri-Cities sample. Seen as most likely—by over three-fourths of the respondents—was that a repository would keep the area's economy too dependent on nuclear energy, a major concern after years of boom and bust cycles resulting from changing federal nuclear policies (Impact Assessment, Inc., 1987). Also seen as likely by over 70 percent was that a repository would increase fears about the dangers of nuclear operations at Hanford among state residents and that the rest of the nation would tend to see the state as a "dump site." Over 60 percent thought it likely that siting the repository would lead to a tax increase for local services and increase prejudice against

Table 6-4   Expected Impacts of a HLNWR at Hanford[a] (percent)

Impacts	Benton County	Franklin County	Weighted[b] Total
Positive Impacts			
a. Create many more jobs in the Tri-Cities area	94.0	92.7	93.6
b. Make Tri-Cities residents feel more certain about the area's economic future	86.9	80.6	85.3
c. Significantly increase tax revenues for the state	82.4	76.8	81.0
d. Bring more stores and entertainment facilities to the Tri-Cities area	76.1	68.1	74.0
e. Cause the DOE to clean up existing contamination from the storage of defense wastes at Hanford	69.5	74.2	70.7
f. Lead the rest of country to respect the state for meeting an important national need	48.9	50.4	49.3
Negative Impacts			
g. Keep the economy of the Tri-Cities too dependent on nuclear energy	74.9	78.0	75.7
h. Increase state residents' fears about the dangers of nuclear operations at Hanford	73.2	72.0	72.9
i. Cause the rest of the country to view Washington as a "dump site"	73.7	71.1	71.8
j. Create a need in the Tri-Cities to raise taxes for more teachers, policemen, and firemen	62.5	64.6	64.1
k. Increase prejudice toward Tri-Cities residents by the rest of the state	62.5	59.5	61.8
l. Make people hesitant to buy agricultural products grown in or near the Tri-Cities area	48.3	59.0	51.0
m. Lead to dangerous accidents in the shipping of nuclear wastes to Hanford	47.9	53.8	49.4
n. Lead to radioactive contamination of the Columbia River	39.4	43.5	40.5
o. Expose Tri-Cities residents to dangerous levels of radioactivity	27.7	29.8	28.2

[a] "Having a high-level nuclear waste repository at Hanford could bring both benefits and problems. I am going to read a list of possible benefits and problems, and I'd like you to tell me whether you think each one is likely or unlikely to be a result of having a repository located at Hanford. First, how likely do you think that having the repository at Hanford will 'create many more jobs in the Tri-Cities area?' Is this very likely, somewhat likely, somewhat unlikely, or very unlikely to occur?"

[b] Percentages are the sum of "very likely" and "somewhat likely" responses and are weighted by county.

Tri-Cities residents by the rest of the state. Approximately half of the Tri-Cities sample thought it likely that a repository would make people hesitant to buy agricultural products grown in the area and that dangerous accidents could occur in the transportation of nuclear waste to Hanford. Seen as less likely to occur was radioactive contamination of the Columbia River (by 40 percent) and, least of all, the exposure of Tri-Cities residents to dangerous levels of radioactivity (by only 28 percent).

Overall, we see that, although the Tri-Cities sample thought that siting a permanent high-level nuclear waste repository at Hanford would likely produce several positive impacts, particularly economic ones, a variety of negative impacts were also seen as quite likely—that a repository would increase the area's dependence on nuclear power, increase state residents' fear of Hanford activities, and lead the rest of the nation to see the state as a "dump site." Coupled with the majority belief that a repository would further increase prejudice against the Tri-Cities residents within the state (perhaps even making the area's agricultural products less marketable), these findings do suggest some elements of the typical NIMBY response. While the great fear of health impacts often associated with NIMBY is not in evidence among Tri-Citians—perhaps due to their familiarity with nuclear activities—there was nonetheless considerable concern about becoming stigmatized by the rest of the state and nation (Mitchell, Payne, and Dunlap, 1988). The negative imagery that the public associates with nuclear waste, documented in chapter 3 by Slovic, Layman, and Flynn and the stigma often experienced by residents of communities contaminated by toxins of any kind (Levine, 1982; Edelstein, 1988) suggest that this concern is neither groundless nor trivial.

Thus, on the one hand, our sample clearly expected the economic benefits of a repository to be far more likely than any environmental and health threats it would pose. On the other hand, respondents were very cognizant of the fact that a repository would likely exacerbate an "image problem" for the Tri-Cities—both within the state as well as nationwide. We suspect that a concern with the "dump-site" image may well have counteracted the historical tendency for residents of the area to see the economic benefits of nuclear operations as far outweighing the environmental and health risks of such operations and thereby contributed to the surprising degree of ambivalence with which our sample viewed the potential repository. This gives credence to the argument that there are indeed "special impacts" associated with nuclear waste that do not come into play in controversies over siting most other technological facilities (e.g., Freudenburg, 1985; Mitchell, Payne, and Dunlap, 1988).

Explanatory Results

The foregoing results should provide a general picture of how the possibility of having the nation's first permanent HLNWR located at Hanford was viewed by citizens in the area. We now turn to the task of explaining variation in overall attitude toward the repository, as measured by the previously described three-item index. In addition to examining attitude toward nuclear power, the perceived risk of nuclear waste, the perceived likelihood of accidents, and the expected impacts of a HLNWR (all of which were described in the previous section), we will examine a wide range of other variables in an effort to explain variation in citizens' attitudes toward repository siting. We begin with demographic characteristics, then examine some basic cognitive orientations, such as political ideology, and finally turn to a range of attitudes and beliefs related to nuclear power and waste—including those already examined in the descriptive results.

Demographic Characteristics
Past research on a wide variety of nuclear attitudes has generally found demographic characteristics to be of only limited utility in accounting for differing attitudes. Women have typically been found to be significantly, albeit modestly, less supportive of nuclear power than are men, and, to a lesser degree, pronuclear attitudes have been found to be positively related to education and income (Nealey, Melber, and Rankin, 1983:43–57). The relationship with age is ambiguous, but studies finding age to be negatively related to concern about environmental problems (Jones and Dunlap, 1992) suggest that young adults may be more concerned about nuclear waste disposal as well. Likewise, a few studies have found that parents are more likely to be concerned about environmental problems than are nonparents (Hamilton, 1985), and thus we might expect parenthood to be related to negative views toward the repository.

We also examined two variables that seem particularly relevant to the Hanford setting, employment at Hanford and involvement in farming. It seems logical to expect that those who are themselves employed at Hanford, or who have a family member employed there, would be more likely to hold favorable views toward the repository, both out of economic self-interest and out of a greater trust in the Hanford authorities' ability to dispose of radioactive wastes safely. Conversely, as already noted, the farmers played a leading role in the "Downwinders" group that formed in opposition to Hanford based on their perception of health problems (for their families as well as their animals) from radioactive releases at Hanford. Similarly, many farmers and vineyard owners expressed concern that having a nuclear

Table 6-5   Regression of Attitudes Toward Repository on Demographic Characteristics: Bivariate Correlation Coefficients (rs), Standardized Regression Coefficients (betas) and Multiple Correlation Coefficient (R^2)

	r	beta
Gender	−.15***	−.10**
Children	−.01	.01
Age	.02	.04
Education	.18***	.10*
Income	.21***	.11*
Farming	−.09*	−.05
Hanford employment	.27***	.22***
All variables		
R^2		.13
F value		11.96
Probability		.001

*p<.05, **p<.01, ***p<.001

waste repository at Hanford might raise questions about the safety of the agricultural produce and the increasingly popular wines of the area.

Table 6-5 shows the results of both bivariate and multivariate regression analyses, where attitudes toward the repository (with high scores representing prorepository attitudes) are regressed on these demographic variables both singly and in combination.[13] The bivariate results show that Hanford employment is most strongly related to repository attitudes, followed by income, education, and gender. Thus, having someone in the family employed at Hanford, higher education, higher income, and being male are all related to holding positive attitudes toward the repository. Neither age nor having children is related to repository attitudes, and those involved in farming are only slightly more negative toward the repository than those in other industries. When the demographic variables are used simultaneously to predict repository attitudes, a similar pattern emerges. As apparent from the standardized regression coefficients or betas, which show the effect of each variable while the effects of all of the other demographic variables are taken into account, Hanford employment remains the best predictor, and income, education, and gender remain significant predictors, of attitudes toward the repository. Farming joins age and having children as insignificant predictors. Finally, it is also apparent that, in general, the demographic variables are fairly weak predictors of repository attitudes, as all seven in combination account for only 13 percent of the variance in such attitudes (although this is highly significant).

Table 6-6 Regression of Attitudes Toward Repository on Cognitive Characteristics: Bivariate Correlation Coefficients (rs), Standardized Regression Coefficients (betas) and Multiple Correlation Coefficient (R^2)

	r	beta
Political ideology	.09*	.01
Faith in science and technology	.24***	.05
Attitude toward nuclear power	.53***	.35***
Perception of nuclear waste	−.55***	−.37***
All variables		
R^2		.40
F value		95.14
Probability		.001

*$p<.05$, **$p<.01$, ***$p<.001$

Cognitive Variables

We next examined a range of cognitive variables as possible predictors of attitudes toward having a nuclear waste repository sited at Hanford, beginning with two broad orientations—political ideology and faith in science and technology. Several studies have found political conservatism to be related to pronuclear attitudes (Webber, 1982; Dunlap and Olsen, 1984), and we therefore expected that those expressing a conservative political ideology would be more favorable toward the repository.[14] Although not as carefully studied, the widely assumed link between holding a very positive view of science and technology and pronuclear attitudes has been documented in a few studies (e.g., Kuklinski, Metlay, and Kay, 1982; Mitchell, 1984). These findings, and the fact that differing views of nuclear waste disposal are often attributed to divergent perspectives on science and technology, lead us to expect that the expression of a high degree of faith in science and technology will be related to prorepository attitudes.[15]

Although less fundamental or broadscale than either political ideology or faith in science and technology, basic attitudes toward nuclear power and perceptions of the risk of nuclear waste per se (relative to risk of other hazards), appear also to be underlying factors likely to influence attitudes toward repository siting. Thus, these two variables (for which the wording and distribution of responses were both presented in the descriptive results section) were examined along with political ideology and orientation in science and technology.

In Table 6-6 we see that each of the variables is significantly related to repository attitudes in the expected direction at the bivariate level, although political ideology is barely so. Thus, political conservatism and a high degree of faith in science and technology are both associated with prorepository attitudes. Not surprisingly, attitudes toward nuclear power in general and perceptions of nuclear waste are much more strongly related to repository attitudes. As expected, those who favor the construction of more nuclear power plants and those who rate the seriousness of nuclear waste relatively low (on the 1-to-10 scale described previously) are much more likely to hold favorable attitudes toward siting a HLNWR at Hanford.

When we turn to the multivariate results, we see that attitude toward nuclear power and perceptions of the risk posed by nuclear waste remain very significant predictors of repository attitudes but that the effects of political ideology and faith in science and technology are reduced to insignificance. This should not be surprising, given that overall view of nuclear power and the perceived risk of nuclear waste are much closer to repository attitudes in what we might think of as a "funnel of causality" that begins with broad social forces, such as demographic characteristics, moves on to basic cognitive orientations, such as faith in science and technology, and then to general views of nuclear power and nuclear waste. It is also not surprising that these cognitive variables are, in general, much more strongly related to repository attitudes than are the demographic variables, as obvious from the fact that together they explain 40 percent of the variance in attitudes toward repository siting.

Trust in DOE and Knowledge about Nuclear Waste

Two other cognitive variables frequently mentioned in debates over nuclear waste repository siting, and nuclear issues in general, are levels of trust and confidence in the authorities responsible for nuclear energy and waste management and knowledge about nuclear waste and nuclear power in general. In particular, it seems widely agreed by both proponents and opponents of permanent nuclear waste disposal that the degree of trust in the DOE, the agency with final authority for waste management, is a critical factor affecting public attitudes toward repository siting (see the editors' introduction and chapter 3 by Slovic, Layman, and Flynn). Although proponents tend to see the public's general distrust of DOE's waste management capabilities as misplaced, while critics see it as clearly warranted, both agree that a low level of confidence in DOE currently undermines public support for a HLNWR.

In contrast, proponents and critics of repository siting hold very different views of the effects of the public's knowledge about repository siting on repository acceptance. Proponents of nuclear power have long believed that public opposition to nuclear power was a simple matter of ignorance—the public did not understand the technology, in large part because of its negative treatment in the media (e.g., Rothman and Lichter, 1982). For them the solution to this problem has long been clear and straightforward: a "knowledge fix." If the public could be educated, if only it "had the scientific facts," it would doubtless recognize the necessity of nuclear power and, therefore, support it. This persistent belief in a knowledge fix has permeated the debate over a HLNWR as well, with proponents arguing that the public has an exaggerated view of the risks associated with a repository. Indeed, based upon this reasoning, the DOE in late 1991 embarked upon two initiatives to inform the public: its draft of the plan to revise the siting of the HLNWR in Nevada featured programs of education (DOE, 1991b), and it launched a concerted public information and public relations campaign in the state of Nevada (Schneider, 1991; also see chapter 11, the concluding chapter of this volume, for a more detailed description of this campaign). Despite widespread acceptance of the need and possibility of a knowledge fix within the nuclear industry, the DOE, and elsewhere (National Research Council, 1989), and despite recent efforts to employ this strategy, it rests on painfully little empirical evidence. The several available studies have found at best only limited evidence of a relationship between levels of nuclear knowledge and support for nuclear power (e.g. Kuklinski, Metlay, and Kay, 1982; Mitchell, 1984). Based upon earlier, but similar, results, some observers (Nelkin, 1977) and critics of nuclear power have argued that education programs designed to increase support for nuclear power are unlikely to succeed.[16]

In view of the foregoing, our survey measured both trust in DOE and knowledge about nuclear waste, with the expectation that trust would be the better predictor of repository attitudes. Interestingly, we found a more modest level of trust and confidence in DOE than might be expected in the traditionally pronuclear Tri-Cities area, again pointing (and likely contributing) to the degree to which the repository program had become negatively viewed even in this unique region. Specifically, after briefly describing the proposed repository plan, we asked respondents, "All things considered, how confident are you that the Department of Energy can safely manage a nuclear waste program like this?" The responses from the Tri-Cities sample were as follows: 29 percent were "very confident," 40 percent were "somewhat confident," 16 percent were "not very confident," and 13 percent were "not at all confident" (with 2 percent unsure or not giving an answer). We then asked respondents, "How much trust do you have in the following people or agencies to tell you the truth about whether or not nuclear waste could be stored safely at Hanford?" and included "U.S. Department of Energy" among a list of five potential sources of information. A modest 26 percent indicated that they trusted DOE "a great deal," 46 percent "some," 14 percent "a little," and 12 percent "not at all" (with 2 percent unsure or not answering).[17]

Table 6-7   Regression of Attitudes Toward Repository on Confidence in Department of Energy and Knowledge about Nuclear Waste: Bivariate Correlation Coefficients (rs), Standardized Regression Coefficients (betas) and Multiple Correlation Coefficient (R^2)

	r	beta
Confidence in Department of Energy	.63***	.63***
Knowledge about nuclear waste	.20***	.10**
All variables		
R^2		.43
F value		216.80
Probability		.001

*$p < .05$, **$p < .01$, ***$p < .001$

In combination, responses to the above two items reveal only a modest level of confidence and trust in the DOE to manage a safe repository program at Hanford, lower than might be expected given the region's history of strong ties to DOE and its predecessors. Of more significance for the present study, a variable constructed by combining responses to these two items was found to be very strongly related to attitudes toward having a HLNWR at Hanford.[18] As shown in Table 6-7, trust in DOE correlates .63 with repository attitudes, indicating that those with a high degree of trust tend to hold strongly prorepository attitudes, while those with a low degree of trust tend to hold the opposite attitudes.

To measure knowledge about nuclear waste, we used three items. First, after indicating that Hanford was one of the three finalist sites, we asked, "Do you happen to know the names of the other two states that were chosen by the Department of Energy for further study?" and coded responses 2 if both Texas and Nevada were mentioned, 1 if either was mentioned, and 0 for neither. Second, after describing existing nuclear operations at Hanford, we stated that, "Hanford is also being considered as a place where high-level wastes from commercial nuclear power plants can be stored permanently," and then asked, "Do you happen to know the storage method the Department of Energy is considering for Hanford?" Those who chose "in tunnels deep underground" were coded as 1 while those choosing "in specifically designed storage tanks at ground level," "in old nuclear power plants," or who were "unsure" or gave no answer were coded 0. Finally, we noted that, "Most of the waste will be used fuel rods from nuclear reactors. They will need to be stored until their radioactivity has decayed and they are no longer dangerous," and then asked, "Do you happen to know how long experts say this will take?" We gave the respondents three choices, coding "1000 years or more" as 1 and "less than 1000 years," "about 500 years," and "unsure" or "no answer" as 0.[19] The resulting knowledge index, ranging from 0 to 4, had a mean of 2.3 and a standard deviation of 1.2.

Looking at the second row in Table 6-7, we see that knowledge about nuclear waste is significantly, but fairly modestly, related to attitudes toward repository siting at both the bivariate and multivariate levels. The bivariate relationship is less than a third of that observed for trust in DOE, while controlling for the latter variable reduces the effect by half. Thus, it is apparent that trust in DOE has a much stronger effect on attitudes toward repository siting than does level of knowledge about nuclear waste. This is further evidence that the knowledge-fix approach of DOE's public education programs may be misplaced. It appears that it might be more promising to try to increase the public's trust in DOE's handling of repository siting. It should be emphasized, however, that either task is likely to prove very difficult. It is also important to note that the two variables combined account for 43 percent of the variance in attitudes toward the repository, more than the combined effect of the four cognitive variables examined in Table 6-6. Trust in DOE alone accounts for 36 percent of the variance in repository attitudes, reinforcing the pattern of evidence in chapters 3, 7, 9, and 10 showing that trust is pivotal to the siting of a HLNWR.

Perceived Impacts and Likelihood of Accidents
Previously we pointed out that a good deal of research on attitudes toward various aspects of nuclear energy, especially nuclear power plant construction, has examined the effect of perceived impacts of plant construction on overall attitudes toward construction (e.g., Hughey, Sundstrom, and Lounsbury, 1985; Hughey and Sundstrom, 1988; Eiser, Van Der Plight, and Spears, 1988). This approach was applied to the public acceptance of a HLNWR in Texas, as described in chapter 5 by Brody and Fleishman, with considerable success, and we likewise employed it in the current study. As already described, we examined respondents' views of the likelihood of a wide range of potential impacts, presented in Table 6-4 above, that might result from siting the repository at Hanford. However, unlike most studies of perceived impacts, which focus only on negative environmental, health, and social impacts and on generally positive economic impacts, we also investigated the potential impacts of stigma associated with a HLNWR.

Previous research has shown that the contamination of a community with toxic substances can also contaminate the social identity of community residents with stigma (Levine, 1982; Edelstein, 1988). On logical grounds we expected that residents living in a community with a HLNWR might, likewise, become stigmatized—after all, high-level nuclear wastes are, literally, some of the most toxic substances known to humankind. On empirical grounds, our expectation stemmed from the widespread stigmatization that resulted from the accidental radioactive contamination (with cesium 137) of 249 people in Goiania, Brazil, in 1987 (Petterson, 1988; also see chapter 8 of this volume by Easterling and Kunreuther). Our initial exploration of the possibility that stigma would be associated with a HLNWR at Hanford consisted of focus groups with Tri-Cities residents in August of 1987 (Mitchell, Payne, and Dunlap, 1988). The group discussions revealed that many residents had already experienced some degree of stigmatization, having been subjected to comments such as "Do you glow in the dark?" when outsiders learned they lived near Hanford. Focus-group participants also expressed concern that a "waste dump" nearby might further magnify the stigma already associated with Hanford. Added to these concerns with personal stigma were concerns among some residents that the region's agricultural products would become stigmatized, as had occurred in Goiania.

In order to examine systematically the effect of perceived impacts on attitudes toward repository siting, we first factor analyzed (specifically, we used a principal components analysis with varimax rotation) the fifteen impacts listed in Table 6-4. This analysis yielded four discernible dimensions, and the results are generally consistent both with prior research on expected impacts (including that reported by Brody and Fleishman in chapter 5) and with the notion that stigma is a significant, additional impact associated with nuclear waste.

The items loading most heavily on the first factor consist of a combination of environmental and health impacts, including radioactive contamination of the Columbia River (n), transportation accidents (m), exposure of residents to radioactivity (o), as well as the item designed to measure stigmatization of agricultural products (l)—which is at least indirectly related to environmental impacts. The second factor consists primarily of items dealing with economic benefits, including, on the positive side, more jobs (a), increased tax revenues (c), more stores and entertainment (d), and create more certainty about the future (b), and, on the negative side, a need to raise taxes (j). The third factor clearly deals with issues surrounding nuclear stigma, including items dealing with increased fear about nuclear operations by state residents (h), the rest of the nation viewing Washington as a dump site (i), and increased prejudice toward Tri-Citians by the rest of the state (k). Finally, a more ambiguous fourth factor is composed of the rest of the nation respecting Washington for meeting an important national need (f), keeping the Tri-Cities too dependent on nuclear energy (g), and causing DOE to clean up existing contamination (e). Because of the heavy loading of item f on this dimension, we have labelled it "nuclear reputation," and it

Table 6-8   Regression of Attitudes Toward Repository on Perceived Repository Impacts and Likelihood of Accidents: Bivariate Correlation Coefficients (rs), Standardized Regression Coefficients (betas) and Multiple Correlation Coefficient (R^2)

	r	beta
Environmental and health impacts	−.70***	−.52***
Economic benefits	.28***	.16***
Nuclear stigma	−.31***	.02
Nuclear reputation	.34***	.14***
Likelihood of accidents	−.60***	−.20***
All variables		
R^2		.57
F value		153.76
Probability		.001

*p<.05, **p<.01, ***p<.001

might be seen as somewhat the opposite of stigma. We constructed indexes for these four dimensions by combining responses to the items loading most heavily on each one, so that high scores indicate that the impacts are seen as likely.[20]

Because accidents associated with a HLNWR are clearly a type of potential impact, we included the perceived likelihood of accidents as another dimension of expected impact. To represent this dimension we combined the four items described earlier in Table 6-3 into a "likelihood-of-accidents index."[21] High scores on the index indicate that such accidents are seen as a likely impact of siting a repository at Hanford.[22]

Attitudes toward a repository were then regressed on these five impact dimensions and the results are shown in Table 6-8. Not surprisingly, all five are found to be significantly related to repository attitude, in the expected directions, at the bivariate level. Most influential on attitudes toward a repository, and producing negative attitudes, are perceived environmental and health impacts (r = −.70) and belief that repository accidents are likely (r = −.60). Considerably lower in magnitude, but still highly significant, are the influences of the area's reputation enhancement (r = .34) and economic benefits (r = .28) on repository acceptability. Fear of stigmatization, consistent with our expectation, has a negative effect on repository attitudes, with the magnitude of the effect (r = −.31) similar to those of the positive economic influences.

When the effects of all five of these perceived impact variables are examined simultaneously, however, the situation changes somewhat. Not surprisingly, the effect of each one is reduced when the others are taken into account. Nonetheless, environmental and health impacts remains the best predictor of repository attitudes, and likelihood of accidents remains the second best. Although the effect of the latter is reduced substantially, this is likely due to its high level of shared variance with the former.[23] Similarly, the effects of economic benefits and nuclear reputation are also reduced considerably, but they continue to have significant effects on repository attitudes. In contrast, the effect of nuclear stigma disappears when the other variables are controlled, suggesting that, while fear of stigma may be real, it has no effect once the other impacts (especially environmental and health) are taken into account.

What is also notable about Table 6-8 is that these five variables together explain a very large amount of variance, 57 percent, in attitudes toward siting a repository at Hanford. While this is not surprising, given that these variables all deal with repository related impacts, the results reemphasize the importance of examining the expected impacts of nuclear projects if one is to understand the roots of support and opposition to them. In this case, seeing the repository as bringing economic benefits or as entailing risks to the environment and human health (especially via accidents) clearly have major effects on respondents' views of having one sited at Hanford. The importance of the "special impacts" stemming from issues such as stigmatization or, conversely, local pride over meeting an important national need, remain less clearcut (Freudenburg, 1985; Mitchell, Payne, and Dunlap, 1988). The bivariate results indicate that they may pale in comparison to issues of risk and economic gain. Since this is the first attempt to examine these special impacts, there is clearly a need for additional research before firm conclusions can be drawn about their influence on attitudes toward nuclear facilities.

Examining Predictive Models

The foregoing analyses have examined the relationships between attitudes toward siting a repository at Hanford and four general sets of variables: (1) demographic variables, which were found to be fairly poor predictors of repository attitudes; (2) general cognitive characteristics, of which only overall attitude toward nuclear power and perception of nuclear waste were strong predictors; (3) level of trust in DOE and knowledge about repository siting, with the former being far more strongly related to repository attitudes than the latter; and (4) perceived impacts (including the likelihood of accidents) of siting a repository at Hanford, which were found to be most strongly related to repository attitudes. Having examined these four sets of variables separately, we conclude our data analysis by employing various combinations of them in an effort to develop more comprehensive predictive models of repository attitudes.

It will be recalled that of the cognitive variables examined in Table 6-6, overall attitude toward nuclear power and perception of seriousness of nuclear waste relative to other forms of pollution were the best predictors of repository attitudes. Adding this set of variables to the demographic variables, and thereby controlling for the latter, produces little change in these results. As seen in the first column of Table 6-9, attitude toward nuclear power and perception of nuclear waste continue to have very significant effects on repository attitudes even when demographic variables are taken into account. These effects easily exceed those of employment at Hanford, the only demographic variable to have a significant effect in Model 1. In fact, adding the demographic characteristics to the cognitive variables produces virtually no improvement in explained variance in repository attitudes (41 percent in Model 1 of Table 6-9 vs. 40 percent in Table 6-6).

Not surprisingly, however, when we add the next set of variables—trust in DOE and knowledge about nuclear waste—to create Model 2 we observe a marked improvement in predicting repository attitudes, with the amount of variance increasing to 54 percent. Also not surprising is that this increase appears to be due entirely to trust in DOE, as in this model the modest effect observed for knowledge in Table 6-7 completely disappears. And, while the effects of attitude toward nuclear power and perception of nuclear waste are diminished somewhat in Model 2, they remain highly significant.

Finally, Model 3 incorporates all of the potential predictive variables examined in this chapter. Adding the five impact dimensions to the previous variables again results in a significant increase in the predictability of repository attitudes—to 63 percent of the variance. Two points are especially noteworthy about this full model: first, environmental and health impacts continue to have the most effect, and nuclear stigma the least effect, among the impact dimensions, as was true in Table 6-8; second, even though the effects of all of the previously significant variables are diminished somewhat, they continue to be significant predictors of repository attitudes. Thus, attitudes toward nuclear power, perception of nuclear waste, and trust in DOE continue to have a significant effect, with trust second in importance only to environmental and health impacts. These three, along with all of the expected impacts except stigma, have significant effects on attitudes toward a repository. Interestingly, the modest effect originally observed for Hanford employment declines to insignificance, while farming barely attains significance, when all of the cognitive, perceptual, and attitudal variables are taken into consideration.

Table 6-9 Regressions of Repository Attitudes on Sets of Predictor Variables: Standardized Regression Coefficients (betas) and Multiple Correlation Coefficients (R^2)

	Model 1[a]	Model 2[b]	Model 3[c]
Gender	−.01	−.01	−.01
Children	.01	.02	.03
Age	−.04	.03	−.02
Education	−.01	.02	.01
Income	.00	−.01	−.01
Farming	−.04	−.04	−.06*
Hanford employment	.09**	.07*	.03
Political ideology	.02	.01	.02
Faith in science and technology	.04	−.03	−.04
Attitude toward nuclear power	.34***	.21***	.12***
Perception of nuclear waste	−.35***	−.24***	−.12***
Trust in DOE		.44***	.23***
Knowledge about nuclear waste		.01	−.03
Environmental and health impacts			−.31***
Economic benefits			.14***
Nuclear stigma			.01
Nuclear reputation			.08**
Likelihood of accidents			−.12**
All variables			
R^2	.41	.54	.63
F value	36.34	51.79	54.28
Probability	.001	.001	.001

*$p<.05$, **$p<.01$, ***$p<.001$

[a] Includes independent variables from Tables 5 and 6.

[b] Includes independent variables from Tables 5, 6, and 7.

[c] Includes independent variables from Tables 5, 6, 7, and 8.

In sum, our most exhaustive multivariate analysis reveals that among residents of the Tri-Cities area, environmental and health risks and trust in DOE are the best predictors of repository attitudes: those seeing little threat to the environment or their health, and those who are trusting of DOE, are most likely to favor the repository. Both are much more important than expected economic benefits from a repository, a variable that is frequently believed by policymakers to be the most important factor in siting acceptability. Although we found a positive relationship between expected economic benefits and repository attitudes, the effect is similar to three other of the model's significant variables: overall attitude toward nuclear power (the more pronuclear, the more prorepository); perception of nuclear waste as a serious form of pollution (the higher the seriousness, the less favorable the attitude toward the repository); and perceptions of accident likelihood (the greater the likelihood, the less favorable the attitude).

The only other variables having a significant effect when all of the independent variables are examined simultaneously are nuclear reputation and farming (but just barely). The fact that the former has a greater effect than stigma suggests two observations: although fear of increased stigmatization resulting from a HLNWR being located at Hanford may exist among Tri-Cities residents (recall the results in Table 6-4), this fear does not (at least not yet) play an important role in generating opposition to the repository; and it may be the case that a certain degree of community pride about nuclear technology may still exist. That stigma is far less important than factors such as the expected environmental and health impacts is, in retrospect, understandable. Tri-Cities respondents would be directly, perhaps significantly, affected by health-related problems, including those stemming from accidents. They might also directly benefit from an improved economy, but the effects of stigma would only come indirectly—from their interacting with others outside the area troubled by the presence of the repository. It seems logical that respondents would place more emphasis on the probable direct impacts, as opposed to those mediated through other people, in forming their opinions about the repository.

Conclusion and Implications

A number of results have been reported in this chapter that warrant highlighting. First, in terms of the descriptive results, we found that, despite the fact that the Tri-Cities area has historically been staunchly supportive of nuclear operations at Hanford (indeed our sample was far more favorable toward constructing nuclear power plants than are citizens nationwide), there was a good deal of ambivalence about having a HLNWR located there. This ambivalence was widely spread across various segments of the population, as opposed to being anchored in specific sectors such as women or farmers. Presumably, the ambivalence stemmed from a variety of factors, including a surprisingly low level of trust in DOE's ability to manage the repository. Although residents saw some economic benefits associated with a repository, far more important in shaping their attitudes were threats to the environment or their health.

Thus, even in a uniquely pronuclear region like the Tri-Cities, low levels of trust in DOE and the fear of the environmental and health hazards posed by a repository, appear to have strong effects on attitudes toward a repository. These findings, which are similar to those reported in other chapters in this volume, suggest that arguments about the pivotal role played by lack of trust in generating opposition to the nation's effort to develop its first permanent HLNWR are quite robust and not limited to hostile regions such as Texas and Nevada (see Slovic, Flynn, and Layman, 1991). Likewise, the results also reinforce findings from several other studies in this volume that the perceived impacts of a repository, including the likelihood of accidents, are the other key factors in generating opposition (see chapter 9 by Mushkatel, Nigg, and Pijawka and chapter 5 by Brody and Fleishman). Both trust and expected impacts are far more important than is actual knowledge about nuclear waste, a finding that calls into question the accuracy of the nuclear industry's efforts to "educate" the public about the benign nature of nuclear waste and waste repository. Only if such educational efforts are able to build up trust in DOE (as well as in the nuclear industry and other relevant agencies) and change the public's perceptions of the possible impacts of a waste repository will they be successful.

Building, and especially regaining, trust is a Herculean task, however, as noted by Slovic, Flynn, and Layman (1991) and in chapter 3 of this volume. This point is underscored by our results that show a low level of trust in DOE even among residents of a region generally supportive of and economically dependent upon nuclear energy. If a region such as the Tri-Cities that depends so heavily on the nuclear industry and prides itself on having played a central role in producing our nation's nuclear weapons is not hospitable to a HLNWR, where will a suitable host site be found? The options do not appear promising.

Notes

This research was supported by the Washington State Office of Nuclear Waste Management via a contract with Impact Assessment, Inc., of La Jolla, California, which in turn subcontracted with Washington State University's Social and Economic Sciences Research Center to conduct the survey reported herein.

1   While quite negative in his assessment, Loeb (1982) nonetheless provides an excellent portrayal of the degree to which the local culture was permeated by an unquestioned acceptance of nuclear energy. He notes, for example, the popularity of references to nuclear power in the names of Tri-Cities businesses—e.g. a bowling alley named "Atomic Lanes." In part this pronuclear orientation reflected the enormous impact of Hanford on the local economy, an impact encouraged and promoted by local influentials. For example, the publisher of the Tri-City Herald, an exceptionally friendly voice for the nuclear industry, played a prominent role in the development of the Tri-City Nuclear Industrial Council, a booster organization that lobbied heavily—and often successfully—for additional nuclear facilities at Hanford (see Fleischer, 1974 for an historical analysis of "TRICNIC").

2   This fact, and comparisons of the effects of a nuclear accident with the eruption, were frequently emphasized by Hanford critics in eastern Washington during the 1980s.

3   Not surprisingly, the opposition was lower in Benton County, which includes both Richland and Kennewick (and the Hanford reservation) than in Franklin County, home of Pasco. In fact, Benton was the only county in the state in which Referendum 40 failed, as only 43.5 percent voted for it. In contrast, in Franklin County it passed with 57.2 percent. The figure reported in Table 1 for the two counties combined is much closer to the Benton results than to those for Franklin because the former had over four times as many voters as did the much smaller Franklin County.

4   Similar reactions were observed in many of the western U.S. "boomtowns" stimulated by coal mining in the 1970s and 1980s, as well as in a range of rural communities offered opportunities for economic development. This points to the necessity of improving our understanding of how some proposed developments become defined as hazardous or noxious, and therefore generate opposition, while others are primarily judged by their economic benefits—i.e., the different dynamics that generate NIMBY rather than PIMBY reactions. While still poorly understood, these discrepant reactions highlight the critical role that risk perceptions play within our society.

5   For a more detailed discussion of the objectives of the survey, see Dunlap and Baxter (1988).

6   Specifically, Franklin County responses are weighted by .61 in order to create a correctly proportioned total sample of 567. For more information on the sampling and other methodological details, see Dunlap and Baxter (1988).

7   Recall that our results differed by only 2 percent from the vote on Referendum 40, as reported in Table 1.

8   To provide an accurate statement of the purpose of the interview, while avoiding direct cues concerning our interest in the topic of nuclear waste, the following introduction was used by the interviewers: "Hello, my name is ———, and I'm calling from Washington State University in Pullman. We're calling homes throughout the state and asking people's opinions on several important environmental and energy issues facing our state."

9   The only prior reference to a nuclear waste repository in the interview was in the list of four issues designed to measure salience, described above, where respondents were not asked to give their own opinion on repository siting at Hanford.

10   Respondents answering "favor" or "oppose" to the first two items in Table 2 were then asked whether they were "strongly" or "moderately" in favor (or opposed), yielding a five-point measure—5 ("strongly favor"), 4 ("moderately favor"), 3 ("unsure"/"don't know"), 2 ("moderately oppose") and 1 ("strongly oppose"). In order to give the third item equal weight, it was coded as 5 for "yes," 3 for "unsure" or "don't know," and 1 for "no." The resultant summated rating scale thus ranges from a low of 3 to a high of 15, with high scores indicating strong favorability toward the repository. The actual scores range from 3 to 15, and have a mean of 10.8 and a standard deviation of 3.9.

11   Results from the national survey were provided by William Desvousges and Hillary Rink of the Research Triangle Institute. See chapter 7 of this volume for a description of the national survey.

12   Specifically, responses to the four potential accidents were coded as 1 for "not a problem," 2 for "a minor problem," and 3 for "a major problem," resulting in a variable ranging from a low of 4 to a high of 12, with a mean of 8.3 and a standard deviation of 2.3.

13   Gender was coded as male = 0, female = 1; children as 1 if children under eighteen are in the household, 0 if not; age on a six-point scale ranging from under-20 = 1 to 65-

and-over = 6; education on a seven-point scale ranging from less than high school = 1 to professional/graduate degree = 7; income (total family income for 1987) on an eight-point scale ranging from less than $10,000 to $100,000 or more; farming as 1 if respondent or "any member of immediate family currently owns or works on a farm, ranch, orchard, or vineyard" or had worked on one within the "past ten years" and 0 if no one in the family was so employed; and Hanford employment as 1 if respondent "currently works for the Department of Energy or a Department of Energy contractor" or had done so in the past ten years and 0 if otherwise.

14 Political ideology was measured by asking respondents, "Which of these terms best describes your usual stand on political issues?" and coding their responses as "liberal" = 1, "slightly liberal" = 2, "middle of the road" = 3, "slightly conservative" = 4, and "conservative" = 5.

15 Faith in science and technology was measured by combining responses to the following two items: "Scientists can solve any problem we might face if they are given enough time and money" and "Science and technology do as much harm as good." Responses to the first item were coded as "strongly agree" = 5, "somewhat agree" = 4, "unsure"/"don't know"/"missing data" = 3, "somewhat disagree" = 2, and "disagree" = 1, while those for the second were coded in reverse direction. The resulting index ranges from 2 to 10, with a mean of 6.7 and standard deviation of 2.1.

16 Bisconti (1991) discusses evidence suggesting that intensive, small-scale educational efforts can produce more favorable attitudes toward nuclear waste disposal. However, we suspect that efforts to reach large segments of the public, which inevitably will be less intensive, are likely to be less successful, and may even produce the opposite effect. See, for example, Liu and Smith's (1990) findings that a national "risk-communication program" concerning the need for a new nuclear power plant in Taiwan not only failed to increase support for the plant, but seems to have had— at least to a modest degree—the opposite effect.

17 DOE was ranked third, behind "scientists who work at Hanford" (trusted a "great deal by 45 percent) and "scientists from Washington's universities" (37 percent) but ahead of the "U.S. Environmental Protection Agency" (22 percent) and "Washington State's Office of Nuclear Waste Management" (only 17 percent). These responses reflect the relatively greater trust placed in Hanford authorities, especially relative to the state agency established to represent the state's interest in repository siting (an Olympia-based agency that was widely regarded as being opposed to the repository).

18 Responses to the first item were coded as "very" = 4, "somewhat" = 3, "not very" = 2, and "not at all" = 1, while those for the second item were coded as "a great deal" = 4, "some" = 3, "a little" = 2, and "not at all" = 1. The resulting index can range from 2 to 8, and its mean is 5.7 and standard deviation is 1.7.

19 Given that most technical experts usually mention 10,000 years or more as the necessary isolation time for high-level nuclear wastes, the wording of this question improves the respondent's chances of answering successfully.

20 The environment and health impacts index was created by adding together responses to the four items, all coded as "very unlikely" = 1, "somewhat unlikely" = 2, "unsure"/no answer = 3, "somewhat likely" = 4, and "very likely" = 5; thus, high scores indicate that these impacts are seen as quite likely. The nuclear stigma index was constructed similarly, so that high scores again indicate that stigma is seen as quite likely. For the economic benefits index, item j was reverse coded, so that high scores indicate that *positive* economic benefits are seen as quite likely. Finally, item g in the

nuclear reputation index was reverse coded so that high scores indicate that positive aspects of having a nuclear reputation are seen as quite likely.

21   These were coded as "major problem" = 3, "minor problem" = 2, and "not a problem" = 1. The index has a range of 3 to 12, with a mean of 8.3 and a standard deviation of 2.3.

22   The content of this index obviously overlaps somewhat with that of the environmental and health impacts index, especially due to item m in the latter, but we will analyze it separately for two reasons. First, because its items have very different response categories, we are reluctant to add them into the factor analysis; second, the possibility of accidents per se was a crucial issue in debates over siting a repository at Hanford, so it seemed appropriate to examine this set of items separately.

23   Because environmental and health impacts and the likelihood of accidents are highly correlated with one another, .71, we checked for potential problems in our regression model stemming from multicollinearity. Deleting likelihood of accidents from the multivariate regression equation increased the beta for environmental and health impacts (to $-.66$), but did not affect the betas for the other three impact variables (all of which remained identical rounded to the second decimal), and reduced the amount of variance in repository attitudes only slightly, from .57 to .55. Thus, except for attenuating the relative impact of environmental health impacts to a modest degree, inclusion of likelihood of accidents does not alter our basic findings concerning the relative importance of the four dimensions of expected repository impacts.

References

Benson, Allen B., and Larry Shook. 1985. *Blowing in the Wind: Radioactive Contamination of the Soil Around the Hanford Reservation*. Spokane: Hanford Education Action League.

Bisconti, Ann S. 1991. "Public Opinion and Communication on Nuclear Waste." Thirteenth Annual Low-Level Waste Management Conference, Atlanta, November 19.

Dunlap, Riley E. 1987. "Public Opinion and Environmental Policy," In James P. Lester, ed., *Environmental Politics and Policy: Theories and Evidence*. Durham: Duke University Press. 87–134.

Dunlap, Riley E., and Rodney K. Baxter. 1988. *Public Reaction to Siting a High-Level Nuclear Waste Repository at Hanford: A Survey of Local Area Residents*. Pullman, WA: Social and Economic Sciences Research Center.

Dunlap, Riley E., and Marvin E. Olsen. 1984. "Hard-Path Versus Soft-Path Advocates: A Study of Energy Activists." *Policy Studies Journal* 13:413–28.

Edelstein, Michael R. 1988. *Contaminated Communities: The Social and Psychological Impacts of Residential Toxic Exposure*. Boulder: Westview.

Eiser, J. Richard, Joop Van Der Plight, and Russell Spears. 1988. "Local Opposition to the Construction of a Nuclear Power Station: Differential Salience of Impacts." *Journal of Applied Social Psychology* 18 (8): 654–63.

Fleischer, Christian Calmeyer. 1974. *The Tri-City Industrial Council and the Economic Diversification of the Tri-Cities, Washington, 1963–1974*. Unpublished M.A. Thesis, Department of History, Washington State University, Pullman, WA.

Freudenburg, William R. 1985. "Waste Not: The Special Impacts of Nuclear Waste Facilities." In Roy G. Post, ed., *Waste Isolation in the U.S.* Vol. 3, *Waste Policies and Programs.* Tucson: University of Arizona Press. 75–80.

Hamilton, Lawrence C. 1985. "Concern about Toxic Wastes: Three Demographic Predictors." *Sociological Perspectives* 28:463–86.

Hughey, Joseph B., and Eric Sundstrom. 1988. "Perceptions of Three Mile Island and Acceptance of a Nuclear Power Plant in a Distant Community." *Journal of Applied Social Psychology* 18(10): 880–90.

Hughey, Joseph B., Eric Sundstrom, and John W. Lounsbury. 1985. "Attitudes Toward Nuclear Power: A Longitudinal Analysis of Expectancy-Value Models." *Basic and Applied Social Psychology* 6:75–91.

Impact Assessment, Inc. 1987. *Socioeconomic Impact Assessment of the Proposed High-Level Nuclear Waste Repository at Hanford Site, Washington.* La Jolla: Impact Assessment.

Jones, Robert Emmett, and Riley E. Dunlap. 1992. "The Social Bases of Environmental Concern: Have They Changed Over Time?" *Rural Sociology* 54:28–47.

Keeney, Ralph L. 1986. *An Analysis of the Portfolio of Sites to Characterize for Selecting a Nuclear Repository.* Report in the Decision Analysis Series. Los Angeles: Systems Science Department, University of Southern California.

Kuklinski, James H., Daniel S. Metlay, and W. D. Kay. 1982. "Citizen Knowledge and Choices on the Complex Issue of Nuclear Energy." *American Journal of Political Science* 26:615–42.

League of Women Voters. 1985. *Radiation in Eastern Washington: Following the Trail of Nuclear Waste.* Spokane: League of Women Voters of Spokane.

Levine, Adeline G. 1982. *Love Canal: Science, Politics, and People.* Lexington, MA: Lexington Books.

Liu, Jin Tan, and V. Kerry Smith. 1990. "Risk Communication and Attitude Change: Taiwan's National Debate Over Nuclear Power." *Journal of Risk and Uncertainty* 3:331–49.

Loeb, Paul. 1982. *Nuclear Culture: Living and Working in the World's Largest Atomic Complex.* New York: Coward, McGann & Geoghegan.

Mitchell, Robert Cameron. 1984. "Rationality and Irrationality in the Public's Perception of Nuclear Power." In William R. Freudenburg and Eugene A. Rosa, eds., *Public Reactions to Nuclear Power: Are There Critical Masses?* Boulder: Westview. 137–79.

Mitchell, Robert Cameron, Barbara Payne, and Riley E. Dunlap. 1988. "Stigma and Radioactive Waste: Theory, Assessment, and Some Empirical Findings from Hanford, WA." In Roy G. Post, ed., *Waste Management '88: Proceedings of the Symposium on Waste Management.* Tucson: University of Arizona Press. 95–102.

National Research Council. 1989. *Improving Risk Communication.* Washington, D.C.: National Academy Press.

Nealey, Stanley M., Barbara D. Melber, and William L. Rankin. 1983. *Public Opinion and Nuclear Energy.* Lexington, MA: Lexington Books.

Nelkin, Dorothy. 1977. *Technological Decisions and Democracy: European Experiments in Public Participation.* Beverly Hills: Sage.

Petterson, John S. 1988. "Perception vs. Reality of Radiological Impact: The Goiania Model." *Nuclear News* 31 (14): 84–90.

Rankin, William L., and Barbara D. Melber. 1980. *Public Perceptions of Nuclear Waste Management Issues.* BHARC/411-80-004. Seattle: Battelle Human Affairs Research Centers.

Rosa, Eugene A. 1987. "Namby Pamby and Nimby Pimby: Public Issues in the Siting of Hazardous Waste Facilities." *Forum for Applied Research and Public Policy* 3 (Winter): 41.

Rothman, Stanley, and S. Robert Lichter. 1982. "The Nuclear Energy Debate: Scientists, the Media and the Public." *Public Opinion* 5:47–52.

Schneider, Keith. 1991. "Nuclear Industry Plans Ads to Counter Critics." *New York Times*, November 13, A1.

Schneider, William. 1986. "Public Ambivalent About Nuclear Power." *National Journal* 18:1562–63.

Slovic, Paul, James H. Flynn, and Mark Layman. 1991. "Perceived Risk, Trust and the Politics of Nuclear Waste." *Science* 254:1603–7.

Smith, Tom W. 1985. "The Polls: America's Most Important Problems—Part I: National and International." *Public Opinion Quarterly* 49:264–74.

Steele, Karen Dorn. 1985. "'Downwinders'—Living with Fear." Spokane *Spokesman-Review*, July 28, 1.

Stenehjem, Michele. 1990. "Indecent Exposure." *Natural History* (September): 2–22.

U.S. Department of Energy. 1991a. *Overview of the Hanford Cleanup Five-Year Plan*. Richland, WA: U.S. Department of Energy.

———. 1991b. *Draft Mission Plan Amendment*. Washington, D.C.: U.S. Department of Energy, Office of Civilian Radioactive Waste Management.

Webber, David J. 1982. "Is Nuclear Power Just Another Environmental Issue? An Analysis of California Voters." *Environment and Behavior* 14:72–83.

Part III

Public Reactions to the

Yucca Mountain, Nevada, Site

7 Perceived Risk and Attitudes Toward Nuclear Wastes: National and Nevada Perspectives

William H. Desvousges, Howard Kunreuther,

Paul Slovic, and Eugene A. Rosa

Risk Perceptions

Introduction

Fundamental to a sound understanding of people's viewpoints toward a high-level nuclear waste repository (HLNWR) is an understanding of their perceptions of the risks associated with such a facility. Perceptions, representing sensory experiences that have become recognized or that have gained meaning, stand between simple, reflexive responses and complex behavior. As such, they are *at* the foundation of, if not themselves *the* foundation of, beliefs, values, opinions, attitudes, and behaviors—all the human responses relevant to the siting of a repository. The principal focus of this chapter is on risk perceptions associated with a HLNWR: their seriousness, their most important characteristics, how they are conditioned by institutional factors and personal characteristics, and their acceptability.

The goal of the research presented here is not only to deepen our understanding of perceptions associated with complex, risky technologies such as waste repositories but also to provide useful input for public policy decisions. Research such as this, representing a conjuncture between basic and applied goals, presents unusual challenges in research design. On the one hand is the need to address fundamental processes so that findings are robust and generalizable to a wide variety of settings. On the other is the need to take into account the unique history and other special features of the repository setting because these factors, doubtless, influence perceptions of repository risks, too.

Preliminary Activities

The approach taken to address the two design challenges of this research comprised three complementary activities: (1) development of an inventory of concerns; (2) focus group sessions; and (3) implementation of a survey. As a first step toward taking into account the historical and political context of the repository siting, we developed an inventory of concerns. The inventory was developed by closely monitoring the popular press, other news media, and public comments on the repository. As a second step, focus group sessions were held with groups in three locales expected to be impacted by the repository. The results from the focus groups were combined with the step one inventory concerns in order to develop the instrument for the survey. The crafting of the first two steps' combined results into the survey instrument ensured that the surveys would not overlook important issues and that the survey questions would be understandable to respondents. Before implementation, the survey instrument also took into account the cumulative body of research on risk perceptions (Fischhoff et al., 1981; Slovic, Fischhoff, and Lichtenstein, 1985; Slovic, 1987), thereby ensuring consideration of fundamental features of risk perception. Finally, two telephone surveys, one in the state of Nevada and one national (excluding Nevada), were implemented. (A more detailed discussion of these procedures can be found in Desvousges, Kunreuther, and Slovic, 1987).

Conceptual Framework
The risk surveys involve complex conceptual and empirical issues. In part, this complexity stems from their defiance of traditional boundaries of scientific inquiry. Sociological, psychological, and economic factors interact in ways that are only vaguely understood. Further confounding those interactions are the influences of social and political institutions. To organize systematically the variety of issues to be included in the complex analysis, we developed a conceptual framework, presented as Figure 7-1.

   The conceptual framework builds upon the cumulative findings in risk perception, decision processes, and policy analysis pertaining to the siting of noxious or risky facilities. It also incorporates key economic, social, and political factors presumed, on the basis of cumulative evidence, to influence risk decisions. Finally, it includes siting-specific variables, such as previous experience with nuclear facilities and the proximity of residence to the repository. The framework, in brief, argues that perceived risk of a high-level nuclear waste repository is a function of: knowledge of repository issues and previous experience with nuclear issues; a variety of attitudes; subjective characteristics of the repository risk; and background and other individual characteristics. Of particular importance among the attitudes is trust in the federal government, the political body responsible for siting and managing the repository. The subjective risk characteristics, derived from the established taxonomy in the psychometric research tradition, include

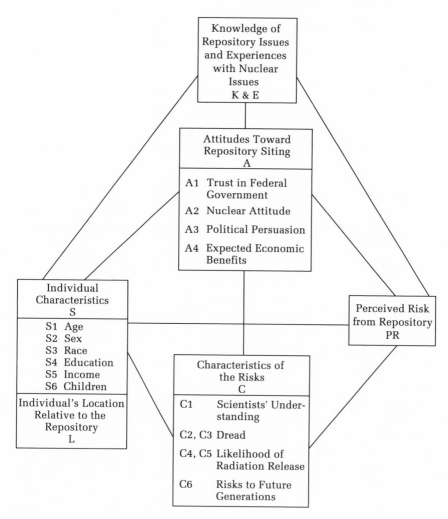

Figure 7-1   Conceptual framework of risk perceptions.

the factors of dread, controllability, scientific understanding, and risks to future generations. The individual characteristics consist mostly of socio-economic background variables. While this outline suffices to introduce the framework, its operational translation into specific variables for modelling purposes is explicated more fully in a later section.

The remainder of this chapter is devoted to the results of the two telephone surveys conducted in 1987. To our knowledge this was the first attempt to assess perceptions of technological or environmental risks among a large sample of respondents across the entire United States (the national survey reported in chapter 3 occurred later). The surveys, therefore, moved risk perception research from the laboratory to "real" risk situations. The results demonstrate the usefulness of survey techniques in obtaining risk perception data. They also permit an assessment of the external validity of previous psychometric findings, almost all of which have been from data collected on small subpopulations, such as students. And, of course, they shed light on the factors shaping perceptions and attitudes toward the siting of a high-level nuclear waste repository.

Data Collection

Two telephone surveys, one of Nevada households with telephones and one of households with telephones within the continental United States but excluding Nevada, provide the data for this study. The samples were drawn from the target populations by using standard random digit dialing techniques.

The questionnaires for the survey evolved as one part of the state of Nevada's socioeconomic impact assessment. As described above and in Kunreuther, Desvousges, and Slovic (1988), the questionnaire development process included focus groups, reviews by various researchers on the impact assessment team, and survey experts who were not associated with the project. After coordinated training sessions, survey groups at the Gordon Black Corporation (national sample) and the University of Nevada-Las Vegas (Nevada sample) conducted the surveys in March and April of 1987.

For the Nevada survey, 5954 telephone numbers were included in the initial sample. Of these, 3887 were residential numbers in service. No contact was made with the target respondent for 1173 (30.2 percent) of these residential numbers, either because of continual busy signals or because, in spite of repeated attempts, the telephone was not answered. Of the 2676 households in which an eligible respondent was reached, 1001 (37.4 percent) provided complete interviews. For the national survey, of the 3419 telephone numbers at which a potential respondent was reached, 1201 (35.1 percent) yielded completed interviews.

Because of the low response rates, our findings must be viewed with caution. Some scholars may object that these response rates are too low to provide useful results. We disagree on various grounds and believe the data do provide insight into the nature of perceptions and attitudes toward the repository. For example, opposition to the repository was so high and widespread, often reaching 80 percent, that populations from which samples are drawn can be assumed to be fairly homogeneous. With homogeneous populations, nearly any sample will be somewhat representative of the parent population. Furthermore, because results from the two separate samples,

Nevada and national, reported in this chapter are so similar, there is lessened concern about nonresponse bias.[1] Despite our position that the results are meaningful, caution is still warranted. In particular, our findings should not be used to guide policy decisions on nuclear waste without considering additional confirmatory evidence, such as that in the other chapters of this volume. Fortunately, as will become apparent, our results are very compatible with those reported in the other chapters, increasing our faith in their validity.

Comparative Perceptions

Seriousness of Pollution Sources

Peoples' perceptions about the risks posed by a HLNWR are related to their attitudes toward nuclear waste. Since there are no absolute standards against which to compare these attitudes, they can best be interpreted in a comparative context. Our surveys asked respondents to rate the pollution problems from a variety of sources, including radioactive wastes from nuclear power plants, on a scale of 1 to 10, with 1 being "not at all serious" and 10 being "very serious." This comparative question preceded any introduction or mention of nuclear wastes in the questionnaires. This procedure was followed to ensure that respondents' ratings were not clouded by a context of nuclear waste information and problems. The results are presented in Table 7-1.

For ease of comparison, mean scores (simple averages) were computed for each pollution source by sample. These results are at the bottom of Table 7-1. For the national sample, water pollution from toxic chemicals, not radioactive wastes from nuclear power plants, had the highest average seriousness rating. National respondents, on average, assigned a seriousness rating of 7.9 to water pollution and 7.4 to radioactive wastes. The average rating for air pollution from cars and factories, the third highest in the sample, was 7.1. Garbage from landfills had the least serious average rating, 6.1.

For the Nevada sample, air pollution received the highest average seriousness rating. Nevadans assigned an average rating of 7.9 to air pollution from cars and factories. Such high ratings for air pollution problems may reflect population clustering in Clark County (Las Vegas) and residents' concerns about air pollution in that area. The rapid increase in population growth, the corresponding increase in motor vehicles, and climatic conditions have all contributed to the air pollution problem in Las Vegas. Nevadans rated water pollution as the second most serious pollution source, with an average rating of 7.8, and garbage from landfills the lowest, with a rating of 5.4. They rated radioactive wastes from power plants an average 6.9—a full

Table 7-1 Frequency Distribution of Seriousness of Various Sources of Pollution: National Sample vs. Nevada Sample (percent)

How Seriously Rated	Sources of Pollution											
	Garbage from Landfills		Air Pollution from Cars and Factories		Radioactive Wastes from Nuclear Power Plants		Water Pollution from Toxic Chemicals		Acid Rain from Power Plants		Radiation from Nuclear Weapons Testing[a]	
	Nation	Nevada	Nation	Nevada	Nation	Nevada	Nation	Nevada	Nation	Nevada	Nation	Nevada
Not at all serious												
1	3.6	6.4	2.1	1.0	5.2	8.4	1.6	1.7	4.0	3.8	NA	11.4
2	3.6	6.7	2.0	1.6	4.5	7.9	1.5	1.7	2.4	4.4	NA	10.3
3	8.0	11.4	3.3	1.6	5.5	5.1	2.2	3.3	4.8	5.3	NA	7.4
4	8.4	9.7	4.1	2.2	4.9	5.1	2.6	4.0	4.8	5.2	NA	7.2
5	23.8	25.4	12.8	8.1	9.2	9.2	6.9	9.8	14.3	14.7	NA	11.3
6	10.8	8.8	10.3	7.2	3.4	4.6	7.9	6.3	10.3	7.6	NA	4.5
7	11.2	9.3	18.8	13.2	6.0	7.1	12.5	10.8	10.9	9.7	NA	4.5
8	12.5	10.8	19.1	23.6	10.3	8.5	17.9	17.2	16.9	15.6	NA	7.6
9	4.9	2.6	9.8	10.9	8.8	5.7	11.5	7.7	10.0	6.9	NA	5.0
10	13.3	9.0	17.8	30.5	42.3	38.5	35.4	37.5	21.5	26.9	NA	30.8
Very serious												
Average rating	6.1	5.4	7.1	7.9	7.4	6.9	7.9	7.8	6.9	6.9	NA	6.1

Actual Survey Question: "I'm going to read a list of several sources of pollution. On a scale from 1 to 10, with 1 meaning 'not at all serious' and 10 'very serious,' please tell me how serious a problem you think each source of pollution is for the United States as a whole."

[a] Respondents in the national sample were not asked to rate the seriousness of radiation from nuclear weapons testing.

scale point below the average for their most serious concern, air pollution. Radiation from nuclear weapons testing was also included as a pollution source on the Nevada questionnaire and received a seriousness rating of 6.1, second lowest of the six pollution sources rated.

Another way of comparing the seriousness of the pollution sources is to examine the percentage of respondents who assign each source the highest rating—a 10 rating—meaning "very serious." For both samples, water pollution and radioactive wastes from nuclear power plants were regarded as the most serious problems. More than 42 percent of national respondents and 38 percent of Nevada respondents assigned the highest seriousness rating to radioactive wastes from nuclear power plants. Water pollution was considered "very serious" by 35 percent of the national and 37 percent of the Nevada respondents.

Nevadans assigned somewhat lower ratings to radioactive wastes than did respondents in the national sample. Increased familiarity among residents of Nevada with radioactive materials may explain this difference. As hosts to the nation's nuclear weapons testing facility, Nevada residents are likely more knowledgeable about radioactive materials and the risks they pose than is the general population. As will be demonstrated in a later section of this chapter, increased knowledge about a risk can result in lower risk perceptions.

Salience and Knowledge of HLNWR Issues

Previous research has shown that risk perceptions and attitudes are related to levels of awareness and knowledge (see, for example, Slovic, 1987). Figure 7-2 provides an assessment of salience by comparing the level of awareness about nuclear wastes between the two samples. As seen in Figure 7-2, twice the numbers in the Nevada sample (35 percent) recalled having read or heard about high-level nuclear wastes more than ten times in the three months before the survey as in the national sample (17 percent). Location of the nuclear weapons testing facility in Nevada and the state's nomination as a possible site for the HLNWR no doubt increased residents' cognizance of nuclear issues. It is not surprising, therefore, that nuclear waste was a more salient issue for Nevada residents than for the nation as a whole.

The surveys also asked how respondents obtained information about high-level nuclear waste issues. Of the respondents who had read or heard about wastes during the previous three months, more than 58 percent in the Nevada sample and almost 49 percent in the national sample had bought a newspaper or magazine or watched a television program specifically to learn about high-level nuclear wastes. Even higher percentages of respondents

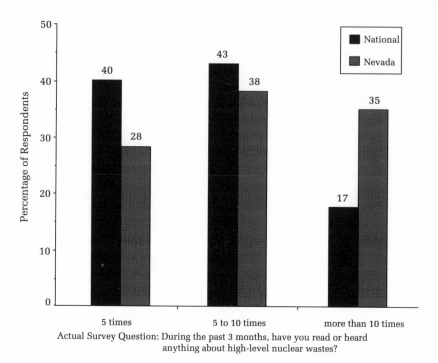

Actual Survey Question: During the past 3 months, have you read or heard
anything about high-level nuclear wastes?

Figure 7-2   Number of times read or heard about high-level nuclear wastes in past three
months.

who indicated awareness about high-level nuclear wastes had discussed the
issues with friends or relatives.

Relatively few respondents attended public meetings to obtain information. Less than 10 percent of the respondents in either the Nevada or the
national sample had attended a public or neighborhood meeting about high-
level nuclear wastes. This finding underscores research by Regan, Desvousges, and Creighton (1990) that indicates that public meetings alone are not
an effective way to communicate information.

Our expectation that salience of HLNWR issues would lead to higher levels
of knowledge was not borne out by the data. We developed three factual
questions that addressed important aspects of high-level nuclear waste disposal. Far fewer respondents answered at least two of these questions correctly than might be expected from the previous results showing high levels
of awareness. The first of the three questions asked: "Do you think most of
the high-level wastes are now stored (a) at the power plants that produced
them, (b) at regional processing centers, (c) at one temporary storage site, or
(d) don't know?" Only about 20 percent of each sample answered correctly

that most high-level wastes are now stored at the power plants that produced them. Far more consistent with expectation was that strong majorities knew that underground disposal is the method being considered seriously in the United States today: 75 percent of the Nevada sample and 64 percent of the national sample. When asked about the length of time for storing high-level nuclear wastes, respondents again demonstrated only meager knowledge; only 27 percent of the Nevada sample and 18 percent of the national sample knew that the repository would store wastes for longer than 1000 years.

General Attitudes Toward the Repository

To capture respondents' overall viewpoints toward the repository, the survey questionnaires asked a wide range of questions assessing attitudes about the repository, its potential benefits and costs, and the equity of developing only a single repository. Answers to these survey questions are summarized in Table 7-2.

Respondents were asked to indicate the extent to which they agreed or disagreed with each of the listed statements. The responses produced an interesting picture of the respondents' views about the repository; highlights of that picture follow:

—Sizable proportions of both the national (48 percent) and Nevada (53 percent) samples agreed or strongly agreed that a repository was the best way to store high-level nuclear wastes. The proportions among only those expressing an opinion was even stronger: 55 percent of the national sample and 68 percent of the Nevada sample strongly agreed or agreed.

—Sizeable proportions, too, strongly agreed or agreed that each region of the country should have a repository: 46 percent of the national sample and 56 percent of the Nevada sample.

—A noticeably small proportion (23 percent) of Nevada residents thought that Nevada was the safest place in the United States for the repository.

—Only about 30 percent of Nevada residents thought that Nevada was the best site for a repository because the weapons testing site was in Nevada.

—The two samples were sharply divided in expectations about the economic growth in nearby communities: only 27 percent of the national sample strongly agreed or agreed that such economic growth would be stimulated, whereas nearly half of the Nevada sample did.

—Few national respondents (25 percent) and few Nevada respondents

Table 7-2 Respondent's Overall Attitudes Toward the Repository: National Sample vs. Nevada Sample (percent)

Attitude	Repository Is Best Storage Method[1]		Each Region Should Have a Repository[2]		Nevada Is Safest Place[3]		Nevada Is Best Place[4]		Repository Stimulates Economic Growth[5]		Economic Benefits Outweigh Costs[6]	
	Nation	Nevada	Nation	Nevada	Nation	Nevada	Nation	Nevada	Nation	Nevada	Nation	Nevada
Strongly agree	8.8	7.4	12.0	16.3	NA	2.9	NA	3.6	3.3	6.2	2.4	3.6
Agree	39.8	45.5	33.9	39.8	NA	20.5	NA	27.0	23.6	38.8	22.7	24.5
Disagree	18.8	19.5	32.1	30.4	NA	40.8	NA	42.3	48.1	40.2	49.0	49.1
Strongly disagree	7.8	4.8	17.2	8.5	NA	22.1	NA	20.0	17.0	6.7	19.2	14.6
Don't know	24.2	22.3	4.7	4.4	NA	12.9	NA	6.5	8.0	7.5	6.5	7.8

Actual Survey Questions:

[1] A repository is the best way to permanently secure high-level nuclear wastes.

[2] Each region of the country should have a repository.

[3] Nevada is the safest place in the United States for a repository. (Respondents in the national sample were not asked this question.)

[4] Nevada is the best place for the repository because the nuclear weapons test site is already here. (Respondents in the national sample were not asked this question.)

[5] A repository would stimulate economic growth in nearby communities.

[6] The economic benefits to nearby communities from a repository would greatly outweigh the risks.

(28 percent) thought that the economic benefits would greatly outweigh repository risks.

Despite majority agreement that a repository was the best storage method, the overall evaluation of the HLNWR was rather negative. National respondents assessed the repository as a bad economic deal; they did not think it would stimulate growth in nearby communities or yield benefits in excess of the risks. Nevada survey respondents seemed unconvinced that their state was the safest place for the HLNWR and seemed somewhat more optimistic about its economic potential, but not to the extent that such potential would outweigh the risks.

Perceived Seriousness of Risks

Previous studies of perceived risks have shown that risk perceptions can be measured using quantitative methods (Slovic, 1987; Fischhoff et al., 1981; and Slovic, Fischhoff, and Lichtenstein, 1985). These studies have produced *cognitive maps* of risk attitudes and perceptions, which show perceptions to be influenced by two main factors: dread risk and unknown risk. Nuclear power and nuclear waste risks rated highly on both the dread and unknown dimensions. Although these studies have not been based on national samples or other general population samples, they suggest high levels of perceived seriousness for the risks from a HLNWR.[2]

Survey respondents were asked to rate the perceived seriousness of seven risks they personally face each year. Included on the list were several nuclear risks as well as such common risks as "an accident at home." The ratings were made on a scale from 1 to 10, with 1 being "not at all serious" and 10 being "very serious". Table 7-3 shows the results for both the Nevada and national samples. Nevada residents rated the risks for a repository that would be located at Yucca Mountain, while national respondents rated risks from a repository that would be located 100 miles from their homes. Generally, Nevada residents ranked all the health and safety risks slightly lower than did the national respondents.

With an average rating of 6.2, national respondents perceived the potential risks from a high-level waste repository as more serious than any of the other risks included in the survey. Perceived risks from exposure to hazardous chemicals from abandoned landfills had the second highest average rating for the national sample, at 5.8, followed by nuclear power risks, at 5.2. Accidents at home (4.3) and at work (3.9) received the lowest average ratings in the national sample.

In addition to rating the same risks posed to national respondents, Ne-

Table 7-3 Frequency Distribution of Seriousness of Various Health and Safety Risks: National Sample vs. Nevada Sample (percent)

How Seriously Rated	Accident at Home		Accident on the Job		Nuclear Power Plant		Exposure to Hazardous Chemicals from Abandoned Landfill		High-Level Nuclear Waste Repository[1,2]		Nuclear Weapons Testing Site[3]		Transport of High-Level Nuclear Wastes[4]	
	Nation	Nevada	Nation	Nevada	Nation	Nevada	Nation	Nevada	Nation	Nevada	Nation	Nevada	Nation	Nevada
Not at all serious														
1	13.6	20.4	16.7	23.9	16.2	33.2	10.8	20.9	7.8	17.1	NA	25.4	NA	14.2
2	16.0	15.8	13.4	12.3	11.2	11.8	8.4	8.8	7.6	9.4	NA	10.9	NA	8.5
3	15.3	14.8	11.9	11.3	10.0	7.0	8.9	7.9	6.9	8.1	NA	8.3	NA	7.3
4	9.6	7.4	8.2	8.9	7.7	5.8	7.0	6.5	7.5	3.7	NA	5.8	NA	4.9
5	21.6	19.9	18.4	18.2	13.5	10.1	13.5	12.5	13.8	13.1	NA	13.2	NA	12.3
6	5.4	6.0	8.5	5.3	5.2	3.6	6.7	5.1	6.4	5.6	NA	4.6	NA	7.8
7	5.5	4.3	7.8	5.6	6.2	4.3	8.7	6.9	10.4	5.9	NA	8.8	NA	5.6
8	5.0	5.5	6.8	6.7	8.3	6.3	10.3	9.3	12.7	9.5	NA	4.7	NA	10.1
9	3.2	1.0	3.2	2.9	3.9	2.3	6.3	4.1	7.5	4.8	NA	2.5	NA	5.2
10	4.9	4.8	5.1	4.9	17.8	15.5	19.4	17.9	19.3	22.8	NA	15.8	NA	24.1
Very serious														
Average rating	4.3	3.9	3.9	3.6	5.2	4.3	5.8	5.2	6.2	5.6	NA	4.7	NA	5.9

Actual Survey Question: "On a scale from 1 to 10, with 1 being 'not at all serious' and 10 being 'very serious,' how serious are the risks you personally face each year from . . ."

[1] Respondents in the national sample were asked about the risks from a high-level repository if it were located 100 miles from their homes.

[2] Respondents in the Nevada sample were asked about the risks from a high-level repository if it were located at Yucca Mountain, which for most residents of Nevada, is approximately 100 miles from their homes.

[3] Respondents in the national sample were not asked about the risks from a nuclear weapons testing site.

[4] Respondents in the national sample were not asked about the risks from the transportation of high-level nuclear wastes.

vadans were asked to rate the seriousness of risks from transporting wastes to Yucca Mountain and from testing nuclear weapons. The average rating of the transportation risks, 5.9, was the highest, followed closely by risks from the repository itself, at 5.6. The other nuclear activities were perceived as somewhat less serious, with the average rating for weapons testing at 4.7 and the rating for nuclear power plants at 4.3. Accidents at home (3.9) and on the job (3.6) also received the lowest average ratings from the Nevada sample.

Overall, on a comparative basis, responses from both the national and Nevada samples perceived a HLNWR as posing fairly high levels of risk. Interestingly, Nevada respondents viewed transporting radioactive wastes to the repository as even more of a risk than the repository itself.

Risk Characteristics

Complex technologies, such as nuclear waste repositories, are multidimensional and comprise a variety of risk characteristics. Assessing the various characteristics associated with repository risks is, therefore, an important step toward understanding risk perceptions and their cognitive mappings. To assess key risk characteristics, the survey included questions on six characteristics that have proven important in previous psychometric studies of risk perception (Slovic, 1987; Slovic, Fischhoff, and Lichtenstein, 1985):

—Accidents at the repository would involve certain death.
—Accidents at the repository would be catastrophic—they would kill many people at one time.
—Scientists understand the risks of repositories.
—People living near the repository could control the risks.
—People would dread living near the repository.
—Repositories pose a serious risk for future generations.

Results from both surveys, national and Nevada, indicate that perceptions of the risks associated with the HLNWR were consistent with previous studies. These results are presented in Table 7-4.

Dread, typically a pivotal factor in risk perceptions, appears to have had a significant role in the formation of perceptions about the repository. Roughly 80 percent of the respondents in both samples either agreed or strongly agreed that people would dread living near the repository. More than 70 percent of both samples thought that an accident at the repository would involve certain death. Moreover, 80 percent of both samples thought that an accident would be catastrophic.

The unknown dimension, frequently emerging as an important factor

Table 7-4 Perceived Risk Characteristics: National Sample vs. Nevada Sample (percent)

Risk

Attitude	Accident Would Involve Certain Death		Accident Would Kill Many People at One Time		Scientists Understand the Risks		People Living Near the Repository Could Control the Risks		People Would Dread Living Near the Repository		Repository Poses a Serious Risk for Future Generations	
	Nation	Nevada	Nation	Nevada	Nation	Nevada	Nation	Nevada	Nation	Nevada	Nation	Nevada
Strongly agree	24.3	24.3	28.7	26.8	13.2	9.0	1.5	1.0	24.9	21.2	32.7	27.6
Agree	49.6	51.6	53.9	54.3	50.3	49.8	8.4	12.0	55.7	56.8	47.9	43.1
Disagree	23.4	21.1	15.7	16.2	29.1	34.0	61.3	65.8	17.4	20.5	16.5	26.0
Strongly disagree	2.7	3.1	1.7	2.8	7.5	7.2	27.8	21.2	2.0	1.6	2.9	3.3

Actual Survey Question: "I am now going to read some statements about the risks from a high-level nuclear waste repository in the United States. Please tell me the extent to which you agree with each."

in previous research, also figured into risk perceptions about the HLNWR. Roughly 37 percent of the national respondents and 41 percent of the Nevada respondents either disagreed or strongly disagreed that scientists understand the risks involved with a repository. Thus, the unknown nature of repository risks also appeared important to respondents but somewhat less strongly than in previous research.

The surveys also clearly indicate that respondents doubted that people who live near the repository could control its risks. Nearly 90 percent of all respondents either disagreed or strongly disagreed with the statement that local people could control the risks. In the focus group sessions that preceded the actual surveys, we found that people had difficulty articulating their concerns about technological control. Since the surveys contained only one question about control, this may not have provided an adequate opportunity for respondents to express their opinions. We suggest that future research efforts explore the dimensions of local control more fully.

Because the wastes in the repository would be stored for very long periods of time, perhaps thousands of years, concern for future generations seems particularly relevant. Such concern was evident in both samples: roughly 70 percent of the Nevada sample and 80 percent of the national sample agreed or strongly agreed that the repository would pose a serious risk for future generations.

The likelihood of accidental large releases of radiation from the repository is another important dimension of perceived risk. Both survey instruments queried respondents about the likelihood of large releases of radiation during the first five or twenty years that the repository would be open (different time periods were randomly assigned to respondents). Respondents were asked to indicate the likelihood of large releases due to each of four causes: an accident at the repository, leakage into ground water, a transportation accident, and terrorist sabotage. Our results were unaffected by the time period, with radiation releases perceived to be just as likely in the first five years as in the first twenty years.

As shown in Table 7-5, a majority of respondents in both samples thought that a large radiation release from any of these sources was somewhat or very likely. The two sources of radiation releases considered to be most likely by both samples were transportation accidents and contamination of underground water. For the national sample, almost 80 percent of respondents thought releases due to transportation accidents were somewhat or very likely, and almost 75 percent thought the same about groundwater leaks. The results for the Nevada sample are identical. Taken together, the results show remarkably high expectations for accidental releases of large amounts

Table 7-5   Perceived Likelihood of Large Accidental Releases of Radiation
from Repository: National Sample vs. Nevada Sample (percent)

	Event							
	Accident at the Repository		Repository Wastes Leaking into Groundwater		Wastes Being Transported to Repository		Terrorist Sabotage	
Attitude	Nation	Nevada	Nation	Nevada	Nation	Nevada	Nation	Nevada
Very likely	21.5	23.7	35.7	39.3	31.3	35.7	25.2	24.5
Somewhat likely	41.1	38.0	39.1	34.8	47.0	44.5	33.3	33.0
Somewhat unlikely	23.9	23.6	16.8	15.6	15.5	14.2	24.6	27.5
Very unlikely	13.6	14.6	8.4	10.3	6.2	5.6	16.8	15.1

Actual Survey Question: "The federal government is planning to make the repository as safe as possible. But there is always some chance that radiation could be released. I'm going to read a list of various ways that a large amount of radiation could be released into the environment from a repository. I'd like you to think about how likely or unlikely each might be. During the first (5 or 20) years a repository would be open, how likely do you think it is that a large amount of radiation could be released from . . ."

of radiation. Indeed, these perceived likelihoods are orders of magnitude greater than estimates provided by technical experts (Peters, 1983).

Risk Perception Models

To evaluate the framework of perceptions and attitudes of risks associated with the HLNWR presented in the introduction, we translated the framework into mathematical models. The aim of the models is to predict perceived risks. The models operationalize the framework's principal concepts into variables that are postulated to be related according to the equation:

$$PR = f(A, K, C, E, L, S)$$

where,

PR = perceived risk from the repository

A = individual's attitudes toward repository siting (and probably nuclear-related issues in general)

K = knowledge about the repository issues

C = characteristics of the risk

E = experiences associated with repository or nuclear issues

L = individual's location relative to the repository
S = individual's socioeconomic characteristics

We used this scheme as the basis for a series of regression models to estimate respondents' perceived seriousness of the risks associated with the HLNWR. This modelling is a first step in beginning to understand the formation of risk perceptions about a repository. Identical models are presented for both the national and the Nevada samples. The models predict the respondent's perceived seriousness of the risks associated with the repository as expressed on a ten-point scale, with 10 as the "most serious" and 1 as the "least serious" rating. The wording of the question eliciting perceived seriousness of the repository and the distribution of ratings, as discussed above, are presented in Table 7-3.

The reported coefficients for the continuous variables are standardized and represent the change in the standard deviation of the scale point rating of perceived risk that results from a one standard deviation change in the independent variable, all other things being equal. Coefficients for qualitative explanatory variables, however, cannot be expressed in this manner and are left unstandardized. These variables either have a value of 0 or 1, which makes it meaningless to standardize their coefficients. Since risk perception is expressed on a scale with the endpoints limited at 1 and 10, two-limit tobit models are the appropriate form for estimation. The tobit procedure, however, produced results very similar to ordinary least squares. We have chosen to present ordinary least squares models because of widespread familiarity with the technique and ease of interpretation.

The variables used in the regression models are defined in Table 7-6. These variables are related to the conceptual framework in Figure 7-1 and Table 7-7, but we need to note a few discrepancies between our conceptual framework and the regression models. In particular, two of the conceptual models' variables are omitted from the regression models. As a consequence, the regression results are tests of a slightly abbreviated model. For the models predicting perceived seriousness of repository risks, there is no variable to correspond with L, the individual's location relative to the repository. This was due to the fact that all respondents were asked to consider that the repository would be located at either Yucca Mountain (Nevada survey) or, equivalently, 100 miles from their homes (national survey). In a subsequent model of voting behavior, we did consider residence in either of two counties closest to the proposed repository site. None of the regression variables corresponds to E, experiences associated with repository or nuclear issues. While it would have been difficult to include experiences with a high-level repository, since no such repositories currently exist, our analysis could

Table 7-6   Description of Variables

Variable	Description
Perceived risk from HLNW repository	A variable that indicates the seriousness of respondent's perceived risk from the location of a HLNW repository near his or her home. (In the national survey, respondents were told that the repository would be located 100 miles from their home. In Nevada, the repository would be located at Yucca Mountain, which is approximately 100 miles from the most populous center of the state.) Response is expressed on a scale of 1 to 10, with 1 being "not at all serious" and 10 being "very serious." In the risk perception models, this is the dependent variable.
Knowledge of HLNW	A scalar variable with values ranging from 0 to 3 that indicates how many of the following questions the respondent answered correctly:
	—"Do you think most of the high-level wastes are now stored . . . a. at the power plants that produced them, b. at regional processing centers, c. at one temporary storage site, or d. don't know?"
	—"Which method for disposing of high-level nuclear wastes is the option being considered most seriously in the United States today? a. putting the wastes on the ocean floor b. burying them deep underground c. shooting them into space, or d. don't know?"
	—"Do you think the high-level nuclear waste repository will be designed to store wastes for . . . a. 1 to 10 years, b. 10 to 100 years, c. 100 to 1000 years, d. longer than 1000 years, or e. don't know?"
Trust in federal government	A variable that the level of trust that the respondent places in federal government officials to make the HLNW as safe as possible. Response is expressed on a scale from 1 to 10, with 1 meaning "no trust" and 10 meaning "complete trust."
Scientists understand risks	A dummy variable that indicates respondent's agreement with the following statement: "Scientists adequately understand the risks from a repository." 1 = strongly agree 0 = all other responses

Table 7-6   *Continued*

Variable	Description
Moderate amount of dread	A dummy variable created to measure respondent's dread of living near a HLNW repository. 1 = strongly agreed with 1 or 2 of the following statements 0 = did not strongly agree with any of the following statements "An accident at a repository usually would involve certain death." "An accident at a repository would kill many people." "People would dread living near a repository."
High amount of dread	A dummy variable created to measure respondent's dread of living near a HLNW repository. 1 = strongly agreed with 3 of the following statements 0 = did not strongly agree with 3 of the following statements "An accident at a repository usually would involve certain death." "An accident at a repository would kill many people." "People would dread living near a repository."
Moderate likelihood	A dummy variable created to measure how likely respondent considered large accidental releases of radiation from certain sources associated with a repository. 1 = respondent thought that 1 or 2 of the following radiation sources would very likely release large amounts of radiation 0 = respondent did not think that any of the following radiation sources would very likely release large amounts of radiation "An accident happening at a repository." "The wastes leaking into underground water." "The wastes being transported to a repository." "Terrorist sabotage at a repository."
High likelihood	A dummy variable created to measure how likely respondent considered large accidental releases of radiation from certain sources associated with a repository. 1 = respondent thought that 3 or 4 of the following radiation sources would very likely release large amounts of radiation 0 = respondent did not think that 3 or 4 of the following radiation sources would very likely release large amounts of radiation "An accident happening at a repository." "The wastes leaking into underground water." "The wastes being transported to a repository." "Terrorist sabotage at a repository."
Future risk	A dummy variable that indicates whether or not respondent strongly agrees with the statement: "A repository would pose serious risks for future generations in Nevada."

Table 7-6   *Continued*

Variable	Description
	1 = strongly agree
	0 = all other responses
Nuclear attitude	A dummy variable that indicates whether or not respondent is in favor of nuclear power.
	1 = does not favor nuclear power
	0 = favors nuclear power
Liberal	A dummy variable that reports respondent's self-described political persuasion.
	1 = very liberal or somewhat liberal
	0 = all other responses
Economic benefits	A scalar variable that indicates how many of the following statements the respondent strongly agreed with:
	"A repository would stimulate economic growth in nearby communities."
	"The economic benefits to nearby communities from a repository would greatly outweigh the risks."
Age	A variable that reports the midpoint of respondent's self-reported age grouping.
Sex	A dummy variable that indicates respondent's sex.
	1 = male
	0 = female
Race	A dummy variable that indicates respondent's race.
	1 = nonwhite
	0 = white
Children	A variable that reports the number of children in the respondent's household under age 12.
Income	A variable that reports midpoint of respondent's self-reported income grouping.
Education	A variable that reports the approximate number of years of education completed by respondent.
Development view	A dummy variable that indicates that respondent either strongly agreed or agreed with both of the following statements:
	"People have the right to change the environment to meet their needs."
	"There are no limits to growth for advanced countries like the United States."
	1 = strongly agreed or agreed with both statements
	0 = all other responses

Table 7-6   *Continued*

Variable	Description
Lincoln	A dummy variable that indicates whether or not Nevada respondent is a resident of Lincoln County. 1 = resident 0 = nonresident
Nye	A dummy variable that indicates whether or not Nevada respondent is a resident of Nye County. 1 = resident 0 = nonresident
Vote	A dummy variable that indicates whether or not respondent would vote for a repository to be located at Yucca Mountain. In one voting model, "vote" is the dependent variable.
Vote with grant	A dummy variable that indicates whether or not respondent would vote for a repository to be located at Yucca Mountain if his or her community would receive a large grant for improved public services as compensation for the repository's location. In one voting model, "vote with grant" is the dependent variable.

have included a measure of familiarity with other nuclear issues. This is an area that merits further consideration in subsequent studies. The focus group results suggest that experience is likely to be an important influence on risk perceptions (Desvousges and Frey, 1989), as do the results reported by Mushkatel, Nigg, and Pijawka in chapter 9.

Results from the models are presented in Table 7-8 (national sample) and Table 7-9 (Nevada sample). Both the F values for the models and the adjusted R^2 values indicate that the models are reasonably good predictors of perceived seriousness of risks associated with the HLNWR. The adjusted R^2 of 40 percent in the final Nevada model is very encouraging.

Model 1 demonstrates the contribution of the variables that represent knowledge of nuclear waste and repository issues and trust in the federal government. In both the national and Nevada models, these variables are significant and negative; increased knowledge of nuclear and repository issues and higher levels of trust in the federal government to operate the repository safely decrease the perceived risk of the repository. As more variables are added in subsequent models, knowledge and trust continue to be significant and negative, but their relative influences on risk perceptions decrease. Both trust and knowledge are potentially affected by risk communication activities related to the siting of the repository. Our results suggest that helping respondents become more knowledgeable about nuclear wastes and

Table 7-7   Relationship between Conceptual and Actual Risk Perception Models

Conceptual Risk Perception Model	Variables from Regression Models
PR Perceived risks from repository	Risk
A Attitudes toward repository siting	A_1 Trust federal government
	A_2 Nuclear attitude
	A_3 Liberal
	A_4 Economic benefits
K Knowledge about the repository issues	Knowledge of HLNW
C Characteristics of the risk	C_1 Scientists understand risks
	C_2 Moderate amount of dread
	C_3 High amount of dread
	C_4 Moderate likelihood
	C_5 High likelihood
	C_6 Future risk
E Experiences associated with repository or nuclear issues	
L Individual's location relative to repository	Lincoln
	Nye
	All respondents in the national sample are assumed to live within 100 miles of repository
S Socioeconomic characteristics	S_1 Age
	S_2 Sex
	S_3 Race
	S_4 Education
	S_5 Income
	S_6 Children

increasing the trust in the federal government to handle wastes effectively would lead to somewhat lower perceived risks. Developing higher levels of trust would, however, require a markedly different process for siting the HLNWR, encompassing much higher levels of public involvement (Regan, Desvousges, and Creighton, 1990).

Perceived characteristics of repository risk are added to the regression in Model 2. Two dichotomous variables, *Moderate Amount of Dread* and *High Amount of Dread*, indicate the degree of dread the respondent expressed of the repository. Construction of these variables is defined in Table 7-6.[3] Their coefficients indicate how much the intercept of the regression changes if the respondent is in either the moderate dread or high dread category in-

Table 7-8 Regression Models on National Data

Dependent Variable: Perceived Risk from HLNW Repository	Model 1 Standardized Coefficients (t-values)	Model 2 Standardized Coefficients (t-values)	Model 3 Standardized Coefficients (t-values)	Model 4 Standardized Coefficients (t-values)	Model 5 Standardized Coefficients (t-values)
F-value	41.314	37.492	32.542	24.029	35.017
Adjusted R^2	.066	.215	.243	.284	.274
Intercept	0.000	0.000	0.000	0.000	0.000
	(39.940)	(22.713)	(20.535)	(12.491)	(13.549)
Knowledge of HLNW	***−0.133	***−0.101	***−0.110	***−0.075	***−0.075
	(−4.658)	(−3.716)	(−4.117)	(−2.687)	(−2.815)
Trust in federal government	***−0.222	***−0.085	**−0.061	**−0.060	**−0.071
	(−7.767)	(−2.958)	(−2.167)	(−2.070)	(−2.553)
Scientists understand risks[1]		**−0.551	*−0.429	−0.329	
		(−2.251)	(−1.783)	(−1.333)	
Moderate amount of dread[1]		***0.864	***0.778	***0.610	***0.636
		(4.593)	(4.195)	(3.220)	(3.528)
High amount of dread[1]		***1.283	***1.086	**0.819	**0.794
		(4.157)	(3.563)	(2.583)	(2.678)
Moderate likelihood[1]		***1.076	***1.040	***1.00	***0.946
		(5.953)	(5.863)	(5.492)	(5.381)
High likelihood[1]		***1.856	***1.771	***1.651	***1.561
		(7.652)	(7.433)	(6.554)	(6.481)
Future risk[1]		***0.946	***0.806	***0.827	***0.838
		(4.884)	(4.211)	(4.247)	(4.461)
Nuclear attitude[1]			***1.046	***0.919	***0.907
			(6.238)	(5.303)	(5.480)
Liberal[1]			0.161	0.257	
			(0.910)	(1.431)	
Economic benefits			**−0.061	**−0.058	**−0.055
			(−2.283)	(−2.129)	(−2.117)
Age				−0.032	
				(−1.196)	
Sex[1]				***−1.018	***−1.003
				(−6.227)	(−6.412)
Race[1]				***0.817	**0.726
				(2.633)	(2.461)

Table 7-8 *Continued*

Dependent Variable: Perceived Risk from HLNW Repository	Model 1 Standardized Coefficients (t-values)	Model 2 Standardized Coefficients (t-values)	Model 3 Standardized Coefficients (t-values)	Model 4 Standardized Coefficients (t-values)	Model 5 Standardized Coefficients (t-values)
Children				*0.046 (1.682)	
Income				0.009 (0.345)	
.ducation				**−0.062 (−2.124)	**−0.057 (−2.085)

Significance levels for *t*-values using two-tailed tests: ***p ≤ .01, **p ≤ .05, *p ≤ .10.
[1]Unstandardized coefficients reported.

stead of the low dread category. The coefficients for both dread variables are significant and positive for national and Nevada data in Model 2 and all subsequent models. Thus, respondents who indicated moderate or high dread of the repository had higher risk perceptions for the repository than did those who had low dread. The coefficient for the high dread group is larger than for the moderate dread group, which also conforms to expectations.

Similarly, *Moderate Likelihood* and *High Likelihood* indicate the respondents' expressions of the likelihood of large accidental releases of radiation from the repository. Both likelihood variables are also significant and positive. The strength of their contributions indicates a definite link between the perceived likelihood of radiation releases and the perceived risk of the repository.[4] The perception that the repository presents risks for future generations is also a positive and significant influence on perceived risk. We have not included future generations into the dread composite because that variable attempts to tap a different dimension of risk—the possible effects on subsequent generations. The different time dimension implied by this variable also influenced our decision to leave it as a separate variable.

In the national sample, the variable that measures respondents' belief that scientists understand the risks associated with a HLNWR (*Scientists Understand the Risks*) is negative and significant in Model 2 but becomes insignificant in other models when additional variables are introduced. This variable is negative but insignificant in all the models run on the Nevada data. The variable was a rough attempt to measure the "known" dimension of perceived risk that Slovic, Fischhoff, and Lichtenstein (1985) found to be important. The lack of significance in the model may reflect our inability

Table 7-9   Regression Models on Nevada Data

Dependent Variable: Perceived Risk from HLNW Repository	Model 1 Standardized Coefficients (t-values)	Model 2 Standardized Coefficients (t-values)	Model 3 Standardized Coefficients (t-values)	Model 4 Standardized Coefficients (t-values)	Model 5 Standardized Coefficients (t-values)
F-value	75.638	67.339	54.745	29.097	61.329
Adjusted R^2	.138	.374	.399	.393	.404
Intercept	0.000	0.000	0.000	0.000	0.000
	(36.077)	(17.385)	(16.395)	(8.403)	(16.451)
Knowledge of HLNW	***−0.244	***−0.169	***−0.162	***−0.124	***−0.150
	(−7.972)	(−6.217)	(−6.057)	(−4.187)	(−5.556)
Trust in federal government	***−0.256	**−0.074	*−0.051	*−0.057	**−0.056
	(−8.360)	(−2.556)	(−1.797)	(−1.897)	(−1.978)
Scientists understand risks[1]		−0.348	−0.195	−0.280	
		(−1.077)	(−0.610)	(−0.840)	
Moderate amount of dread[1]		***1.197	***1.072	***1.096	***0.0147
		(5.703)	(4.179)	(5.018)	(5.127)
High amount of dread[1]		***2.334	***2.072	***2.152	***0.163
		(6.358)	(5.705)	(5.539)	(5.521)
Moderate likelihood[1]		***2.603	***1.392	***1.260	***1.378
		(6.935)	(6.391)	(5.567)	(6.355)
High likelihood[1]		***1.198	***2.394	***2.151	***2.331
		(9.691)	(9.017)	(7.670)	(8.790)
Future risk[1]		***0.142	***0.885	***0.845	***1.053
		(4.652)	(3.910)	(3.539)	(3.699)
Nuclear attitude[1]			***1.138	***1.033	***1.986
			(5.422)	(4.642)	(5.233)
Liberal[1]			−0.022	−0.081	
			(−0.106)	(−0.377)	
Economic benefits			***−0.081	**−0.070	***−0.077
			(−3.061)	(−2.500)	(−2.905)
Age				−0.040	
				(−1.282)	
Sex[1]				***−0.591	**−0.473
				(−3.018)	(−2.562)
Race[1]				−0.350	
				(−1.095)	

Table 7-9 *Continued*

Dependent Variable: Perceived Risk from HLNW Repository	Model 1 Standardized Coefficients (t-values)	Model 2 Standardized Coefficients (t-values)	Model 3 Standardized Coefficients (t-values)	Model 4 Standardized Coefficients (t-values)	Model 5 Standardized Coefficients (t-values)
Children				0.015	
				(0.508)	
Income				−0.032	
				(−1.072)	
Education				−0.041	
				(−1.374)	
Lincoln[1]				−0.167	
				(−0.535)	
Nye[1]				0.077	
				(0.264)	

Significance levels for t-values using two-tailed tests: ***$p \leq .01$, **$p \leq .05$, *$p \leq .10$.
[1]Unstandardized coefficients reported for these variables.

to develop questions that adequately measure this risk characteristic rather than the importance of the characteristic on risk perception. Because of the time limitations in a telephone interview, it was not possible to pursue all risk characteristics with equal thoroughness.

Various attitudinal indicators are introduced to the regression in Model 3. Not surprisingly, the variable that indicates the respondent's position on nuclear power (*Nuclear Attitude*) makes a large, positive, and significant contribution to risk perceptions in the national and Nevada samples in Model 3 and all subsequent models. Opponents of nuclear power have higher risk perceptions for the repository than supporters. *Liberal*, which indicates the respondent's political persuasion, does not appear to be statistically significant in either the national or Nevada samples. *Economic Benefits*, the variable that measures the respondent's perception of the economic benefits associated with the repository, tends to decrease risk perceptions significantly for all models where it was entered. The economic variable has a stronger influence in the Nevada sample, which may reflect the greater overall optimism of at least some Nevadans about the possible economic benefits associated with the repository.

Model 4 includes standard socioeconomic variables in the regression. The respondent's sex is a significant influence on risk perceptions in both the national and Nevada samples. Females tend to perceive the repository risks to be more serious than do males, all other things being equal. This finding is consistent with previous research about attitudes toward nuclear power in particular (see chapter 2 of this volume by Rosa and Freudenburg), and risks in general (Mitchell, 1984 and Desvousges et al., 1990). As for the other socioeconomic variables, they play a moderate role in the formation of risk perceptions for the nation as a whole but barely any role for respondents from Nevada. In the national sample, respondent's race and level of education are significant at the .05 level. Nonwhites tend to view the repository risks as more serious than do whites, and each year of education lowers the respondent's risk perceptions. Neither of these variables, however, is significant for the Nevada sample. The respondent's income, a commonly considered socioeconomic variable, does not appear to be statistically significant for either the national or Nevada sample. This was also true when income was included in other model specifications, implying that risk perceptions for these respondents are not influenced by income levels.

For the Nevada sample, Model 4 also includes two dichotomous variables, *Lincoln* and *Nye*, which indicate whether the respondent was a resident of either of the two counties closest to the proposed Yucca Mountain site. Both of these variables are insignificant, which suggests that the risk perceptions of respondents in the two counties nearest the site were no different from those of other Nevada residents in our sample.

Model 5 includes only the significant variables from Model 4 for each sample. The major differences in Model 5 between the samples lie with the socioeconomic variables. The national model includes the two significant socioeconomic variables, race and education, while the Nevada model does not. Otherwise, the data indicate that respondents from the national sample and from the Nevada sample form risk perceptions very similarly. As implied by our conceptual framework, attitudes about the repository and nuclear waste issues (*Trust in the Federal Government, Nuclear Attitude,* and *Economic Benefits*), knowledge about nuclear waste issues (*Knowledge of* HLNWR), and risk characteristics (*Moderate Amount of Dread, High Amount of Dread, Moderate Likelihood, High Likelihood,* and *Future Risks*) are important influences in the formation of risk perceptions. Proximity and socioeconomic factors appear to be less important.

Position on the Repository

The survey asked respondents: "If a vote were held today on building a permanent repository, would you vote for locating a repository at (a) Hanford in Washington State, (b) Yucca Mountain in Nevada, (c) Deaf Smith County in Texas, or (d) none of the above?" Nevadan's self-projected voting behavior is reported in Table 7-10. Approximately 24 percent indicated that they would vote to locate the repository at Yucca Mountain. More than 40 percent said they would not vote to locate the repository at any of the three proposed sites. Nevada respondents were also asked if they would vote for the location of the repository at Yucca Mountain if their community would receive a large grant for improved public services as compensation. More than half, 59 percent, said they would not vote for the repository even under those conditions.

We have developed a voting-behavior model to explain these results. The model contends that voting behavior is based on risk perception, perceived risk characteristics, attitudes toward the repository, and location. The model predicts the likelihood of an affirmative vote for the repository at Yucca Mountain. Because the dependent variable in the model is dichotomous, that is, can only take either of two values, probit rather than ordinary least squares regression models are appropriate. In these models, the reported coefficient is proportional to the change in probability of voting for the location of the repository at Yucca Mountain that results from one unit of change in the independent variable.[5] In Table 7-11, the dependent variable is the likelihood of voting for the location of the repository at Yucca Mountain.

In Model 1, we use a single explanatory variable, perceived risk, to predict

Table 7-10   Voting Behavior

"If a vote were held today on building a permanent repository, would you vote for locating a repository at . . ."

Proposed Repository Site	Percentage of Nevada Respondents in Favor of Location
Hanford, Washington	4.2
Yucca Mountain, Nevada	24.3
Deaf Smith County, Texas	18.6
None of the above	44.0
Don't know	9.0

"Suppose instead your community were offered a large grant for improved public services like schools, parks, or hospitals to have the repository located at Yucca Mountain. Would you vote to locate the repository under these terms?"

Vote	Percentage of Nevada Respondents
Yes	33.5
No	58.7
Don't know	7.8

Table 7-11   Voting Behavior Model 1

Dependent Variable: Vote for Repository	Model 1 Coefficients (t-values)	Model 1a Coefficients (t-values)	Model 1b Coefficients (t-values)	Model 1c Coefficients (t-values)
Chi-square (χ^2)	230.23	234.81	263.13	272.31
Predicted as percentage of actual	.793	.783	.792	.799
Constant	***0.442	***−0.685	***−0.916	***−1.017
	(5.088)	(−3.428)	(−4.304)	(−4.760)
Perceived risk from HLNW repository	***−0.299			
	(−13.683)			
Predicted risk (from the the risk perception model)		***−0.186	***−0.174	***−0.171
		(−7.678)	(−7.056)	(−6.837)
Trust in federal government		***0.128	***0.125	***0.125
		(6.665)	(6.335)	(6.330)
Knowledge of HLNW		***0.270	***0.277	***0.281
		(4.686)	(4.734)	(4.768)
Economic benefits			***0.721	***0.699
			(4.741)	(4.553)
Prodevelopment view			0.161	0.165
			(1.451)	(1.480)
Lincoln				***0.454
				(2.978)
Nye				0.164
				(1.078)

Significance levels for t-values using two-tailed tests: ***$p \leq .01$, **$p \leq .05$, *$p \leq .10$.

the likelihood of voting for the location of the repository at Yucca Mountain. The significant chi-square value and the number of correctly predicted votes as a percentage of actual votes indicate that the model is fairly successful.

Model 1a adds variables that indicate the respondent's trust in the federal government to operate the repository safely and the respondent's knowledge of nuclear and repository issues to the equation. Using these variables to predict the risk variable presents a problem of simultaneity. To correct for this bias, we use the value for perceived risk predicted by the perceived risk model described above instead of the actual survey response to the 10-point rating scale. We followed this procedure for Model 1c and all subsequent voting models. Both the trust and knowledge variables have positive and

Table 7-12   Voting Behavior Model 2

Dependent Variable: Vote for Repository with Community Grant	Model 2 Coefficients (t-values)	Model 2a Coefficients (t-values)	Model 2b Coefficients (t-values)	Model 2c Coefficients (t-values)
Chi-square (χ^2)	152.70	145.26	171.88	182.53
Predicted as percentage of actual	.705	.703	.704	.716
Constant	***0.551	**0.404	0.126	0.002
	(6.421)	(2.047)	(0.606)	(0.011)
Perceived risk from HLNW repository	***−0.169			
	(−11.867)			
Predicted risk (from the risk perception model)		***−0.197	***−0.194	***−0.187
		(−8.517)	(−8.183)	(−7.850)
Trust in federal government		***0.064	***0.053	***0.052
		(3.954)	(2.899)	(2.875)
Knowledge of HLNW		−0.029	−0.027	−0.023
		(−0.516)	(−0.482)	(−0.401)
Economic benefits			***0.481	***0.394
			(2.929)	(2.741)
Prodevelopment view			***0.402	***0.410
			(3.904)	(3.964)
Lincoln				***0.447
				(3.035)
Nye				0.229
				(1.609)

Significance levels for t-values using two-tailed tests: ***$p \leq .01$, **$p \leq .05$, *$p \leq .10$.

significant influences on predicted voting behavior. Respondents who indicate trust in the federal government or exhibit a high level of knowledge about nuclear waste issues are more likely to vote for the repository than those who do not. The predicted risk value is negative and significant but has a smaller influence than the survey risk variable in Model 1.

Model 1b incorporates variables that indicate attitudes toward economic benefits and development. Not surprisingly, the variable that measures the respondent's expectations of the economic benefits resulting from the repository is positive and significant. Respondents with more optimistic expectations of economic benefits are more likely to vote for the repository. The variable that indicates whether or not the respondent has a prodevelopment view is positive but not statistically significant. As defined in Table 7-6, these are respondents who think that there is no limit to growth for industrialized nations and that people have the right to change the environment to meet their needs. All previously introduced variables retain a significant and fairly constant influence.

Model 1c considers the respondent's proximity to Yucca Mountain in the prediction equation. Dummy variables were created for residents of Lincoln and Nye counties, the counties closest to the proposed repository site. County of residence exerts a positive influence on voting behavior for both counties, meaning that respondents who live in either county are more likely to vote for the repository than those who live elsewhere in Nevada. The variable is statistically significant for Lincoln County but not for Nye County.

With the dependent variable as the likelihood of voting for the location of the repository at Yucca Mountain if the respondent's community will receive a large grant to improve community services, Model 2 is estimated in Table 7-12. The structure of each variation in Model 2 is identical to those of Model 1 discussed above. The chi-square values and correct prediction percentages indicate that, when a large grant for improved public services is linked to the repository's location, all models provide fairly good predictions of the likelihood of an affirmative vote. The contributions of some individual variables, however, differ from their roles in the simpler voting question. In particular, knowledge drops out as a significant predictor while prodevelopment variable (*Prodevelopment View*) exerts a stronger and significant influence on voting behavior in Model 2. Respondents who have a prodevelopment outlook are 40 percent more likely to vote for a repository with a grant program than are those who do not.

Discussion

Overall, the survey results produce an unmistakably negative image of the HLNWR. Nevada and national respondents view the repository as a very undesirable facility due to the seriousness and unacceptability of the risks associated with it. Both samples rated the perceived risks of a HLNWR more seriously than the other six risks included in the survey. Clear majorities believe that repository risks are beyond the control of nearby residents who would dread living near it, that a repository accident would involve certain death to many people, and that a repository poses a serious risk for future generations. Somewhat surprisingly, concern for future generations was the most seriously perceived risk characteristic. A majority of respondents, often sizable, believes in the likelihood that the repository would release a large amount of radiation into the environment as a result of accidents, general deterioration, or human malevolence. Although they show

some recognition of economic benefits associated with the repository, respondents do not believe these outweigh the risks: in effect, the repository
is perceived as a bad deal for local residents.

Both knowledge of nuclear wastes and trust in the federal government
affect risk perceptions, with lower perceived risk existing among the more
knowledgeable and the more trusting respondents. General attitudes toward
nuclear power influence repository risk perceptions, with greater perceived
seriousness among opponents and less seriousness among supporters. Consistent with previous research on other nuclear issues, women in both
samples perceived the risks of the repository as substantially more serious
than men did. None of the other background characteristics had significant
effects in the Nevada sample, though race (nonwhites saw greater risks than
whites did) and education did show some effect in the national sample.

Majorities of Nevadans believe that Nevada is neither the safest (63 percent) nor the best (62 percent) place for the repository. When asked to consider a hypothetical vote, less than a quarter would vote to site the repository
at Yucca Mountain. A plurality, 44 percent, would vote not to locate the
repository at any of the three finalist sites. Even when offered a community
grant, a convincing majority of Nevadans would still vote against siting the
repository at Yucca Mountain.

Conclusions

The data point to a disturbing policy conclusion: it will be extraordinarily
difficult to site a HLNWR, not only in Nevada, but almost anywhere under the
current institutional arrangements. Citizens of Nevada and the nation view
the repository as imposing unacceptably high risks on themselves and on
future generations. Even offers of compensation are insufficient to overcome
the unacceptability of repository risks. The depth of concerns among Nevada
respondents is especially revealing because it shows that the U.S. Department of Energy's (DOE) efforts to reduce concerns about the repository have
generally failed. The failed efforts of DOE may, in part, be due to a mistrust
of the federal government, found in our analysis to be a significant predictor
of the perceived seriousness of repository risk. Whatever the cause, DOE's
risk communication ineffectiveness underscores the importance, as pointed
out by the National Research Council, of two-way communication about
risks; risk information received from the public is as important as information transmitted to the public (National Research Council, 1989; Desvousges
and Smith, 1988). In contrast, DOE's typical risk communication efforts have
typically followed a "top-down" approach, where information is provided
to the public along a one-way channel of communication.

There is little doubt that belief in the safety of the repository is a crucial requisite of a successful siting program. Trust in the federal government's ability to manage the repository is crucial to producing such perceptions of safety. Given the widespread belief in the seriousness of repository risks (or belief in the absence of safety) and given the mistrust of those responsible for the repository, we can expect serious impediments in Nevada and elsewhere to the construction of a HLNWR.

Notes

1   Both samples are biased either in the same way or in very different ways. If biased in the same way, the bias is likely to be reflected in an overrepresentation of middle-class respondents (Dillman, 1978). While this would attenuate the validity of generalizing from the samples, the results would still be useful in understanding the segment of society most active in political matters. If, on the other hand, the bias is quite different, then that reinforces the contention that the populations from which the samples are drawn are homogeneous. Otherwise, one would have to argue the unlikely case that very similar results are due to very dissimilar samples.

2   General population surveys on risk perception have been conducted by Gould et al. (1988) in the states of Connecticut and Arizona. In those studies, nuclear power was perceived to be less risky than predicted from the pyschometric evidence. But, since respondents were not queried about nuclear wastes, it is difficult to extrapolate the findings to predictions about perceived repository risks.

3   Thirty-five percent of national respondents and 32 percent of Nevada respondents fall into the moderate dread category. Nine percent of national respondents and 8 percent of Nevada respondents are in the high dread category. It should be noted that use of these two categories as two separate dichotomous variables yields results analogous to treating high, medium, and low dread as a single, ordinal variable.

4   Forty percent of national respondents and 33 percent of Nevada respondents fall into the moderate likelihood category. Sixteen percent of national respondents and 22 percent of Nevada respondents are in the high likelihood category.

5   This relationship only holds at the mean values of the independent variables (Maddala, 1983).

References

Covello, Vincent T., Paul Slovic, and Detlof von Winterfeldt. 1987. Risk Communication: A Review of the Issues. Washington, D.C.: National Science Foundation.

Desvousges, William H., and James H. Frey. 1989. "Integrating Focus Groups and Surveys: Examples from Environmental Risk Studies." Journal of Official Statistics 5 (4): 1–15.

Desvousges, William H., Howard Kunreuther, and Paul Slovic. 1987. High-Level Nuclear Waste Risk Perception Telephone Surveys: Preliminary Findings. Yucca Mountain Socioeconomic Impact Project, First Year Socioeconomic Progress Report, Appendix A.2.7. Phoenix: Mountain West Research.

Desvousges, William H., and V. Kerry Smith. 1988. "Focus Groups and Risk Communication: The 'Science' of Listening to Data." Risk Analysis 8 (4): 479–84.

Desvousges, William H., V. Kerry Smith, F. Reed Johnson, and Ann Fisher. 1990. "Can

Public Information Programs Affect Risk Perceptions?" *Journal of Policy Analysis and Management* 9 (1): 41–59.

Dillman, Don A. 1978. *Mail and Telephone Surveys: The Total Design Method.* New York: John Wiley & Sons.

Fischhoff, Baruch, Sarah Lichtenstein, Paul Slovic, Stephen L. Derby, and Ralph L. Keeney. 1981. *Acceptable Risk.* Cambridge: Cambridge University Press.

Gould, Leroy C., Gerald T. Gardner, Donald R. DeLuca, Adrian R. Tiemann, Leonard W. Doob, and Jan A. J. Stolwijk. 1988. *Perception of Technological Risks and Benefits.* New York: Russell-Sage Foundation.

Kunreuther, Howard, William H. Desvousges, and Paul Slovic. 1988. "Nevada's Predicament: Public Perceptions of Risk from the Proposed Nuclear Waste Repository." *Environment* 30 (8): 17–20, 30–33.

Kunreuther, Howard, Douglas Easterling, William H. Desvousges, and Paul Slovic. 1990. "Public Attitudes Toward Siting a High-Level Nuclear Waste Repository in Nevada." *Risk Analysis* 10 (4): 469–84.

Kunreuther, Howard, and Douglas Easterling. 1992. "Gaining Acceptance for Noxious Facilities with Economic Incentives." In Daniel W. Bromly and Kathleen Segerson, eds., *The Social Response to Environmental Risk: Policy Formulation in an Age of Uncertainty.* Boston: Kluwer.

Maddala, G. S. 1983. *Limited-Dependent and Qualitative Variables in Econometrics.* Cambridge: Cambridge University Press.

Mitchell, Robert Cameron. 1984. "Rationality and Irrationality in the Public's Perception of Nuclear Power." In William R. Freudenburg and Eugene A. Rosa, eds., *Public Reactions to Nuclear Power: Are There Critical Masses.* Boulder: Westview, American Association for the Advancement of Science. 137–79.

National Research Council. 1989. *Improving Risk Communication.* Washington, D.C.: National Academy Press.

North, D. Warner. 1989. "Risk Analysis for the High-Level Waste Repository: Issues Facing the Nuclear Waste Technical Review Board." In Yacov Y. Haimes and Eugene Z. Stakhiv, eds., *Risk-Based Decision Making in Water Resources.* New York: American Society of Civil Engineers. 135–41.

Peters, Ted F. 1983. "Ethical Considerations Surrounding Nuclear Waste Repository Siting and Mitigation." In Steve H. Murdock, F. Larry Leistritz, and Rita R. Hamm, eds., *Nuclear Waste: Socioeconomic Dimensions of Long-Term Storage.* Boulder: Westview. 41–54.

Regan, Michael J., William H. Desvousges, and James L. Creighton. 1990. *Sites for Our Solid Waste: A Guidebook for Effective Public Involvement.* Washington, D.C.: U.S. Environmental Protection Agency.

Slovic, Paul. 1987. "Perception of Risk." *Science* 236:280–85.

Slovic, Paul, Baruch Fischhoff, and Sarah Lichtenstein. 1985. "Characterizing Perceived Risk." In Robert W. Kates, Christopher Hohenemser, and Jeanne X. Kasperson, eds., *Perilous Progress: Technology as Hazard.* Boulder: Westview. 91–132.

Wald, Matthew L. 1989. "U.S. Will Start Over on Planning for Nevada Nuclear Waste Dump." *New York Times,* November 29, 1, 14.

8 The Vulnerability of the Convention Industry to the Siting of a High-Level Nuclear Waste Repository

Douglas Easterling and Howard Kunreuther

Introduction

One of the major concerns that has been expressed regarding the proposed high-level nuclear waste (HLNW) repository at Yucca Mountain is the potential impact to Las Vegas's convention industry. Conventions and trade shows constitute a major source of revenue for Las Vegas and the state of Nevada. During 1989, approximately 1.5 million individuals attended conventions in Las Vegas, contributing over a billion dollars in gross revenue to the city's economy, according to the Las Vegas Convention and Visitors Authority (1990).[1] The figures have more than doubled over the past decade. With the recent boom in the construction of new hotels, convention attendance can be expected to increase dramatically over the next few years, at least in the absence of any major shocks.

State officials in Nevada have long pointed to the possibility that the repository could adversely impact the Las Vegas visitor industry. For example, ex-governor Richard Bryan contended that a repository could produce losses in convention attendance and tourism, with "catastrophic consequences" for Nevada (Bryan, 1987:36). This claim of economic losses has provided Nevada officials with a rationale for enacting legislation and lawsuits designed to block the Department of Energy (DOE) from characterizing the Yucca Mountain site (Swainston, 1991). For example, the state legislature passed AB222 in 1989, which outlawed high-level waste disposal within Nevada, and the Nevada attorney general filed a suit against DOE calling for the suspension of the repository program [*State of Nevada v. Watkins*, 914 F.2d 1545 (9th Cir. 1990)].[2] These actions have had a tangible impact in delaying DOE's site characterization activities, at least according to DOE officials (Adams, 1990).

The economic argument raised by Nevada officials is endorsed by a majority of the state's residents. In a 1991 survey of 500 Nevadans, 62 percent agreed that "a repository at Yucca Mountain could have a negative impact on the tourist and visitor economy in Nevada" (Flynn, Mertz, and Slovic, 1991). In addition, the Nevada Resort Association acknowledged the potential for visitor losses in approving a resolution that opposes the repository program (Morrison, 1991).

On the other hand, certain key actors in the repository debate argue that a repository would have a negligible impact on the decisions of people who would otherwise visit Nevada. Carl Gertz, DOE's project manager for Yucca Mountain, stated that he believed a tourism impact is "unlikely, but not impossible. You don't see that kind of reaction from the public, in general" (Kerr, 1990). The *Las Vegas Review-Journal* was more adamantly skeptical in a 1988 editorial: "It doesn't really serve Nevada's purposes for the head of the state Nuclear Waste Agency to drag out that old red herring about the proposed high-level nuclear waste dump scaring off tourists. . . . It's not a legitimate issue; it's a bugaboo of strictly political origins."

Thus, while the prospect of economic losses is crucial to the policy question of whether a repository will (or should) be built at Yucca Mountain, there is a good deal of disagreement as to the likelihood that these effects will actually occur. This chapter addresses the issue by reporting the results of two surveys, one of convention planners and one of convention attendees. The first survey examined whether a repository at Yucca Mountain would influence a planner's decision to hold a meeting in Las Vegas, while the second survey considered the question of whether a repository would affect an association member's decision to attend a meeting scheduled for Las Vegas.

Theoretical Rationale for Convention Losses

A repository could impact the Las Vegas convention industry in two distinct ways: by inducing organizations to take their meetings to alternative cities and by decreasing the number of people who attend meetings that are still held in Las Vegas. In this section, we develop theories as to why each of these two effects could occur, by considering first the convention planning decision and then the convention attendee's decision. These theories are then tested in subsequent sections of this chapter.

Declines in the Number of Conventions Held in Las Vegas
The Planning Process. To answer the question of whether a repository might decrease the number of conventions held in Las Vegas, it is important first to

understand the dynamics of the convention planning process. The decision on where to hold a convention is typically made by a professional meeting planner in consultation with the organization that is sponsoring the event. Some planners work directly for the organization, while others are independent consultants. In either case, the meeting planner serves as an agent of the officers of the organization. Given that the officers in turn represent the membership, the planner's selection of a convention site is designed to reflect the preferences and concerns of the association's members.

Planners initiate the selection process anywhere from six months to ten years prior to the meeting date (the larger the meeting, the longer the lead time). In selecting a city for the convention, the planner is guided by two fundamental objectives: (a) to avoid any impediments to a smooth-running meeting and (b) to maximize attendance. As such, the planner takes into account the meeting's physical constraints (e.g., adequate meeting space, a sufficient number of hotel rooms), economic considerations (e.g., travel and hotel costs), and a variety of other factors that make the meeting more desirable to attend (e.g., climate, entertainment).[3]

Introducing a Repository into the Decision. The possibility of a HLNW repository influencing the convention planner's decision is depicted in Figure 8-1. This figure portrays the repository as a new stimulus in the planner's "decision environment." We hypothesize that the repository would influence the planner's perception of Las Vegas in two ways: (a) the city would be perceived as a riskier place to visit and (b) the "image" of Las Vegas (defined as the planner's subjective overall impression of the city) would become more negative. These changes in the perception of Las Vegas, in turn, would lower the chances of the planner selecting the city for the meeting.

The first major assumption of the model is that a repository at Yucca Mountain would influence planners' perceptions of Las Vegas with respect to either risk or image. A number of surveys (e.g., Kunreuther, Desvousges, and Slovic, 1988; chapter 7 by Desvousges, Kunreuther, Slovic, and Rosa; and chapter 3 by Slovic, Layman, and Flynn) have shown that many segments of the population regard the proposed repository as exceedingly harmful and otherwise noxious (e.g., bad, stupid). If the repository became linked with Las Vegas, these same attributions would likely become attached to the city.

Risk Avoidance. Assuming that a repository at Yucca Mountain did cause planners to assign a high risk to Las Vegas, it is likely that the city would be chosen less often as a convention site. Planners have historically paid attention to obvious risks in choosing where to hold conventions. For example, planners in our focus group indicated that they regularly avoid cities on the southern Atlantic coast during hurricane season (Kunreuther, Easter-

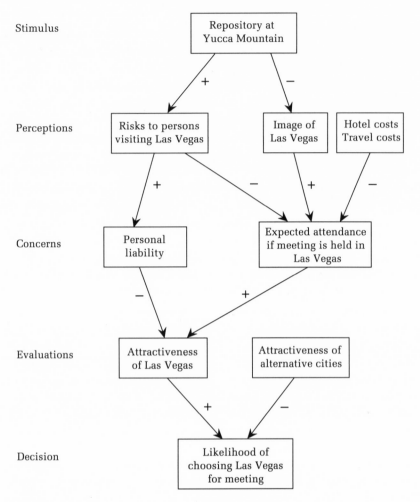

Figure 8-1 Influence of a repository on convention planners' decision to hold a meeting in Las Vegas.

ling, and Kleindorfer, 1988). Risks such as crime, earthquakes, and terrorism also have a deterrent effect.

As Figure 8-1 shows, the planner's attention to risk reflects two underlying concerns. First, if there are highly publicized hazards associated with the host city, attendance at the meeting is likely to decline. For example, international planners were extremely cognizant of the fact that many Americans would not attend a meeting in Europe during the terrorist scare of 1986 (Adams, 1986). The second risk-related concern facing planners has to do with personal liability. In deciding to hold a meeting in a particular city, the planner exposes up to 100,000 delegates to some level of risk. If a tragic accident occurs (e.g., the hotel fire that struck the Dupont Hotel in Puerto Rico), the planner faces the prospect of multimillion-dollar lawsuits. These suits are likely to succeed if it can be shown that the planner was negligent in screening sites on safety criteria (Conlin, 1987). Thus, adding a major new risk to Las Vegas would certainly be of concern to the planner.

Negative Image. There is also considerable evidence for the hypothesis that a repository-induced decline in the image of Las Vegas would reduce the number of conventions held in the city. In a 1983 survey of meeting planners, 59 percent indicated that it was "very important" to select a location with a "glamorous or popular image," at least for meetings that combine business and pleasure (Ziff-Davis, 1984). Las Vegas currently attracts many of these kinds of meetings and thus would suffer losses if its image were tarnished with something like a repository.

The potential for the repository to cause image-mediated losses in the number of Las Vegas conventions is reinforced by a second survey of meeting planners (Survey Research Associates, 1987). In this survey, 63 percent reported that avoiding "unpopular" cities was a "very important" consideration in making a planning decision. The effect of such a label is clear in the case of Philadelphia: a study by the Philadelphia Convention and Visitors Bureau found that black convention planners avoided the city because it has an image of being dirty and hostile to blacks (Sahugan, 1988). This sort of image influences the planner because it leads to less than optimal attendance at a convention.

Moderating Factors. It is important to point out that risk and image are but two of a large number of factors that the planner considers in choosing a city. For some meetings, requirements such as finding a large block of hotel rooms are important, and, in fact, may even rule out many potential cities. For a meeting with especially strong demands, only a few cities may be viable candidates, and some of these may already be booked for the meeting date. If the planner is in a situation where Las Vegas is the only available city that can accommodate the meeting, the repository may be overlooked. In contrast, if many cities are available, the planner will have much less tolerance for risk and negative images.

The question of whether or not a planner will avoid Las Vegas in response to a Yucca Mountain repository also depends on the types of trade-offs that the planner is willing to make. For example, if the officers of the association are particularly concerned with minimizing the costs of attending the meeting, then the less expensive hotel rooms of Las Vegas may be sufficiently attractive to overcome the negative influence of the repository.

The fact that planners must satisfy so many selection criteria suggests that risk may be considered according to a threshold model of decisionmaking. As long as the subjective probability of something going wrong falls below a critical probability, the planner disregards the potential for adverse effects. However, when salient and pronounced risks enter into the decision, the city may be avoided. The question, then, is whether planners will perceive a repository as posing enough of a risk to enter into their decision calculus.

Declines in Attendance at Las Vegas Conventions

A similar model (depicted in Figure 8-2) can be constructed to describe the impact of a repository on convention attendees' behavior. This model is simpler than that for meeting planners, partly because the attendee's decision involves only two alternatives and partly because the attendee is representing only his or her own interests, rather than acting as an agent. However, the prediction is very similar: a repository at Yucca Mountain will alter the perceived risk and image that the attendee associates with Las Vegas, resulting in a decreased probability of attending a meeting that has been booked for Las Vegas.[4]

Risk Avoidance. Although the attendee's decision problem is quite different from the planner's, the attendee obviously has just as much incentive to minimize risk. As such, there are a number of examples of individuals avoiding destinations they perceive to be risky:

> —Fewer tourists visited New York City following a highly publicized subway shooting in 1990 (New York Times, 1990);
> —San Francisco's Fisherman's Wharf was nearly deserted following the 1989 earthquake;
> —The appearance of medical waste on beaches in New Jersey and New York during the summer of 1988 led to visitor losses in the amount of $1.5 billion (Lyall, 1991); and
> —The Bellevue-Stratford Hotel in Philadelphia lost so much business after the 1976 outbreak of legionnaire's disease that it was forced to change its name (Thomas and Morgan-Witts, 1982).

Las Vegas could suffer comparable losses if potential visitors view the repository as a major hazard.

Negative Imagery. The second pathway through which a Yucca Mountain repository could decrease attendance at Las Vegas conventions is via imagery. This process, documented in recent research (Slovic et al., 1991 and chapter 3 of this volume by Slovic, Layman, and Flynn), holds that the images that come to mind when an individual thinks of a place will influence his or her desire to visit there.[5] Thus, if an organization holds a meeting in a city that conjures up all sorts of positive images (e.g., ocean, sunshine,

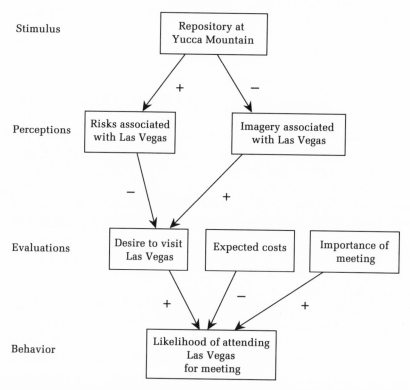

Figure 8-2   Influence of a repository on convention attendees' decision to attend a meeting scheduled for Las Vegas.

fun), members will have a much stronger desire to attend. Conversely, as more and more negative images are elicited in association with the meeting site, the chances of an individual attending the meeting decrease.

As Slovic, Layman, and Flynn demonstrate in chapter 3, a HLNW repository is one of the most noxious images that could conceivably be associated with a city. Thus, if Las Vegas does come to elicit the image of a repository (or related imagery such as danger, death, sickness, or nuclear explosions), one would expect a major drop in attendance at Las Vegas conventions. This linkage could develop simply from the media repeatedly presenting maps indicating that Las Vegas is the closest city to Yucca Mountain.

Moderating Factors. As with convention planners, the attendee's decision is moderated by factors other than risk and image. Individuals considering whether or not to attend a meeting in Las Vegas will take into account the importance of the meeting, the opportunity to meet with colleagues, the chance to gamble, etc. Thus, we hypothesize that the repository will influence the probability that the person attends the meeting, rather than serve as the sole determinant of the decision.

Varying the Severity of the Repository Stimulus

The models presented in Figures 8-1 and 8-2 map out the way in which planners and attendees will respond when a new stimulus—a repository at Yucca Mountain—is introduced into the decision task. However, the degree to which planners and attendees will respond to this stimulus depends in large part on what the repository's track record will be. If the facility operates without incident, and if the current political and scientific controversies abate, then Las Vegas may not suffer any losses to its convention industry. If, however, the repository site is jarred by an earthquake or the facility is plagued by mishaps or mismanagement, then convention losses are much more likely.

More generally, the level of loss to Las Vegas's convention industry should increase monotonically with the severity of repository scenarios. Serious accidents, especially those involving the release of radiation into the environment, will enhance the perceived risk associated with Yucca Mountain and with southern Nevada in general. In addition, heightened media attention will raise the profile of the repository, increasing the chances that it comes to mind when a person thinks of Las Vegas. According to the above models, these sorts of events will increase the rate at which planners and attendees avoid Las Vegas.

For severe repository scenarios, Las Vegas may even become stigmatized, i.e., widely shunned by the rest of the country (Slovic et al., 1991). Stigmatization occurs when a place takes on attributes that indicate contamination to outsiders (Jones et al., 1984; Edelstein, 1988). This process is more virulent than the risk-avoidance and negative-imagery mechanisms described above: if Las Vegas were stigmatized by the repository, planners and attendees would resist any contact with the city for fear of exposure to risk or contamination. Such an effect would undoubtedly require an extreme scenario—something comparable to the accident at Chernobyl.

Analogous Events

One possible way to address the issue of whether a HLNW repository at Yucca Mountain would impact the Las Vegas convention industry is to explore what has happened with analogous facilities, i.e., those that match the repository on key attributes. This approach is valid to the extent that comparable facilities can in fact be identified.

Nevada Test Site

One facility that is often cited by officials at DOE as being informative is the Nevada Test Site (NTS), located just adjacent to Yucca Mountain. This facility, which opened in 1951, has been the site of hundreds of nuclear-weapons explosions. These tests were conducted above ground until the passage of the Limited Test Ban Treaty in 1963, at which point testing moved below ground. The earlier tests released radiation into the atmosphere and, in many cases, exposed downwind residents and observers to substantial radioactive fallout (Fradkin, 1989). Since moving below ground, the explosions still lead to tremors felt as far away as Las Vegas, but radioactive contamination is now confined largely to underground cavities.

DOE officials have argued that the rapid growth that Las Vegas's visitor economy has experienced since 1951 is evidence that NTS has not deterred tourists and conventions. Further, the proposed repository is viewed by DOE as a more benign nuclear facility, so it should have even fewer impacts. "Our feeling is, gee whiz, the underground test program where we blow up bombs hasn't seemed to hurt tourism," according to Carl Gertz, project manager for Yucca Mountain (Kerr, 1990). On the other hand, in a review of NTS's economic effects, Titus (1988) found that the impact to Las Vegas's visitor economy has actually never been studied.

More importantly, even if NTS has had only a minimal effect on the visitor economy of Las Vegas, this does not necessarily mean that every community has been spared. For example, St. George, Utah, which received major doses of radioactive fallout from NTS during the 1950s, suffered a drop in its tourism and convention trade when the increased incidence of leukemia in the area was publicized (Fradkin, 1989). Thus, Las Vegas may have escaped visitor losses only because the tests were deliberately set off at times when the prevailing winds would carry the fallout away from the city (Fradkin, 1989).

Another limitation in using NTS to forecast the impact of the repository is that the two facilities differ in fundamental ways. Whereas NTS has historically (at least until recently) been viewed by the public as integral in maintaining the security of the United States, the repository has a purpose (disposal of extremely unpopular waste) that invites contempt. In fact, the novelty of the proposed HLNW repository makes it extremely difficult to identify strictly analogous facilities.

Radiation Accidents

While there may not be any *facilities* that are truly analogous to the repository, it is informative to consider analogous *events* (i.e., cases where radiation has been released into the environment). These events provide an indication of whether convention losses could occur if things go wrong with the repository.

Three Mile Island. One of the most widely publicized radiation releases to occur in the United States was the March 1979 accident at the Three Mile Island (TMI) nuclear plant near Harrisburg, Pennsylvania. The near-meltdown of the reactor core transfixed the public, although the amount of radiation that actually entered the environment turned out to be extremely small. A study of TMI's economic impacts identified $5 million in visitor losses (including the cancellation of the National Hardware Dealers' spring convention scheduled for Harrisburg) in the thirty days following the accident (Pennsylvania Governor's Office of Policy and Planning, 1980). These losses abated relatively quickly, largely because of the short duration of the radiation threat.

Goiania. A more adverse impact to tourism accompanied a radiation release in Goiania, Brazil, during the fall of 1987 (Petterson, 1988). This release occurred when two men cut into a discarded radiotherapy machine and exposed 100 grams of cesium 137. Children playing in the junkyard were attracted to the glowing material and passed it among themselves and their families. At least 249 people suffered radiation contamination, four of whom died within two months. This event sparked fears throughout Brazil, with severe economic consequences. In terms of the visitor industry, hotel occupancy in the city dropped by about 40 percent for the six weeks following the accident. In addition, scheduled conventions for General Motors, the Corrides Stock Car Association, Comansu Tractors, and 2 regional medical association were all canceled. The effect to the visitor economy largely dissipated over the next four months as it became clear that the threat of contamination had abated. The Three Mile Island and Goiania cases indicate some nuclear-related accidents have potential to disrupt the convention industry, at least in the short run. To determine whether a HLNW repository at Yucca Mountain could incite comparable or worse impacts, we turn to a research approach that considers this facility explicitly.

Overview of Current Research

We undertook two surveys that explored the effect that a repository at Yucca Mountain would have on convention decisions. The first survey addressed the impact of the facility on convention planners' choice of where to hold a convention. A sample of 153 planners who had booked conventions in Las Vegas reconsidered their decision under the assumption that a repository had been constructed at Yucca Mountain. Planners were presented with a number of alternative repository scenarios, ranging from benign (no accidents) to moderately severe (a series of radiation releases). Following each scenario, planners reported the likelihood that they would still choose Las Vegas as the site for their convention. These "intent" data were then input to a forecasting model to generate estimates of the number of meetings that Las Vegas would lose if the proposed scenarios actually occurred.

The second survey looked at the decisions of convention attendees. The sample consisted of approximately 100 members from each of six national organizations (600 total). Each respondent had attended at least one of their association's last four annual meetings. As with the first survey, respondents were asked whether the presence of a HLNW repository would have an impact on their decision. However, the primary focus of this survey was to test the theory that perceived risk and imagery actually influence attendance at conventions.

Convention Planner Survey

Methodology
The convention planner survey, conducted in February 1988, was designed to test whether a repository at Yucca Mountain would have a negative impact on planners who would otherwise choose Las Vegas for a convention. We interviewed a sample of 153 meeting planners who had chosen Las Vegas as the site for conventions or trade shows. Planners were recruited from a list of upcoming meetings scheduled for Las Vegas hotels. Following an initial telephone screening, planners were sent a copy of the questionnaire, and then were called back to report their responses. The response rate was a respectable 66 percent.

The interview first explored the process by which planners came to choose Las Vegas for the target meeting (e.g., the number of cities initially considered, the factors that were taken into account). The remaining questions had the planner reconsider the choice of a convention city under a series of scenarios in which the repository was located at Yucca Mountain.

Repository Scenarios
The scenarios presented to respondents varied with respect to seven experimental factors, the most important of which involved the severity of the events associated with the repository. The event factor had seven levels (see Table 8-1), ranging in severity from no accidents over the first ten years of repository operation (event 2) to an investigative report indicating multiple mishaps and a higher than expected risk (event 7). Events were described in the form of news stories.

In addition to an event description, the scenario consisted of a description of how much media attention had been devoted to the event, either *dampened* (the story appeared only briefly) or *amplified* (extensive coverage by national media). The media attention factor was varied experimentally, so that half the sample saw a given event with dampened attention and the other half saw it with amplified attention.

We were also interested in whether factors such as the price of hotel rooms and meals or providing attendees with gambling chips or free show tickets might ameliorate whatever impact the repository scenarios had on planners' preferences for Las Vegas. We thus modified the scenarios to incorporate a set of amenities, shown in Table 8-2, and varied the level of each of them experimentally.

Each respondent was presented with nine scenarios, consisting of two

Table 8-1 Repository Scenarios Used in Convention Planner Survey

---

I. Events

1. Opening of Repository: Construction has been completed on Yucca Mountain repository. Facility will now begin accepting shipments. (This scenario describes the purpose and physical characteristics of the repository.)

2. Benign History: Repository has been accepting waste for 10 years. Operations have been according to expectations. No releases of radiation or identifiable health effects.

3. Minor Accident at Repository: Accident involving the offloading of a transport canister. Small radiation release on loading dock, but no significant human exposure.

4. Minor Transport Accident: Truck hauling nuclear waste overturns near Las Vegas, but no radiation was released into the environment.

5. Moderate Accident at Repository: Accident involving the transfer of high-level waste from shipment cask to storage container. Three workers were exposed to radiation and required medical treatment, but contamination confined to a small area at repository.

6. Moderate Transport Accident: Truck transporting high-level waste crashed head-on into a gravel truck 40 miles from Las Vegas. Radiation escaped from a defective cask and 4 firefighters were hospitalized for radiation exposure. Traffic detoured for 3 days.

7. Report of Multiple Mishaps: An independent consultant issues a report critical of operations at the repository (e.g., sloppy worker practices and insufficient monitoring by management). The risk of radioactive contamination is higher than previously assumed. Haulers have not abided by transportation regulations. Minor accidents at the facility have led to 15 cases of radiation exposure.

II. Media Attention

1. Dampened: Extensive coverage in Nevada, but only limited mention in national media and only for a single day.

2. Amplified: Extensive coverage by Nevada press and national media; lead story in the *New York Times* and on network news broadcasts; week-long followup.

Table 8-2 Amenity Factors in Convention Planner Study Scenarios

---

III. Price of Hotel Rooms
1. Standard convention rate
2. 10 percent discount from convention rate
3. 20 percent discount from convention rate

IV. Price of Hotel Meals
1. Standard convention rate
2. 10 percent discount from convention rate
3. 20 percent discount from convention rate

V. Gambling Chips
1. No free chips
2. $50 of free chips

VI. Show Tickets
1. No free tickets
2. Free ticket to one big-name show (choice of Wayne Newton or Bill Cosby) for each attendee

VII. Coffee Breaks (including beverage service and snacks)
1. Standard charge for coffee breaks
2. Two coffee breaks per day free

---

replications of event 1 (the opening of the repository), two replications of event 7 (the investigative report), and one occurrence of each of the other five repository events. For events 1 and 7, the two replications differed in terms of the amenities included in the scenario. After reading through a particular scenario, the planner would indicate whether he or she would still choose Las Vegas for the target meeting.

Reductions in the Ranking of Las Vegas
After reading a scenario (defined by the repository event, media attention, and amenity levels), planners indicated whether Las Vegas would be their first choice for the meeting under these conditions. If they said no, they then indicated how Las Vegas would rank among the possible cities. One of the possible responses to this question was "Las Vegas would no longer be considered for this meeting." By comparing Las Vegas's post-scenario ranking to the ranking that the planner initially assigned for the target meeting, we can tell whether the scenario influenced the planner's preference for Las Vegas.[6]

Table 8-3 indicates that for all scenarios, at least 30 percent of the planners lowered their ranking of Las Vegas relative to the initial target meeting. The

Table 8-3 Changes in Planners' Ranking of Las Vegas Under Repository Scenarios

Scenario	Subsample	N	% Who Lower Their Ranking of Las Vegas	% Who No Longer Consider Las Vegas
1A. Opening of repository (base case: dampened media attention & no amenities)	Both	153	32.0	7.8
1B. Opening of repository (replication)				
a. dampened	A	75	32.0	8.0
b. amplified	B	78	37.2	7.7
2. Benign 10-year history				
a. dampened	A	75	30.7	4.0
b. amplified	B	78	30.8	9.0
3. Minor accident on site				
a. dampened	A	75	40.0	6.7
b. amplified	B	78	46.2	14.1
4. Minor transport accident				
a. dampened	B	78	38.5	12.8
b. amplified	A	75	49.3	10.7
5. Moderate accident on site				
a. dampened	B	78	41.0	14.1
b. amplified	A	75	49.3	21.3
6. Moderate transport accident				
a. dampened	A	75	53.3	32.0
b. amplified	B	78	64.1	30.8
7A. Report—recurrent accidents and new risk				
a. dampened	B	78	55.1	32.1
b. amplified	A	75	74.6	42.7
7B. Report—recurrent accidents and new risk				
a. dampened	B	78	55.1	30.8
b. amplified	A	75	74.6	48.0

Note: For each repository event, half the sample was assigned to the dampened media attention condition and half was assigned to the amplified condition. The specific assignment is shown with the subsample column. Each respondent saw event 7 twice. The two replications (7A and 7B) differed with respect to the amenities factors.

greatest shift occurs for event 7 (the release of a report indicating recurrent accidents and safety lapses), combined with amplified media attention. In this case, 75 percent of the sample reduced their ranking of Las Vegas.

Table 8-3 also reports, for each scenario, the percentage of planners who no longer considered Las Vegas to be a viable candidate for the meeting. Under the most benign scenario (event 2 with dampened media coverage), 4 percent of the planners chose to eliminate Las Vegas from their list of possible cities. At the other extreme, the scenario with the report of recurrent accidents and amplified media attention caused 48 percent of the sample to eliminate Las Vegas from consideration.

Likelihood Rating Data

The impact of the repository scenarios on planners' preference for Las Vegas can also be gauged by a second measure: the self-reported likelihood of choosing a city other than Las Vegas for the meeting. This scale ranged from 1 = "definitely would select Las Vegas" to 10 = "definitely would not choose Las Vegas." Summary statistics for this variable are shown in Table 8-4. Higher values reflect a greater intent to avoid Las Vegas in response to the repository scenario; a mean of 1 indicates no avoidance among the sample, while a mean of 10 reflects total avoidance.[7]

The likelihood rating data parallel the ranking data of Table 8-3. Planners are highly resistant to Las Vegas under the amplified version of scenarios 7A and 7B (the report indicating multiple mishaps); the mean rating is greater than 7 for each replication. In addition, even for scenarios where no mishaps have accompanied the repository, the mean likelihood of choosing a city other than Las Vegas is greater than 3.

It should be noted that both Table 8-3 and Table 8-4 classify the scenarios only in terms of event and media attention, pooling over the five amenity factors. The amenity factors were ignored in reporting the data because they had minimal effect on planners' responses. For the likelihood rating data, the five amenity factors together explained less than 1 percent of the within-subject variance. In contrast, the event factor accounted for 29.3 percent and the media attention factor accounted for 3.6 percent of this variance.

Forecasting Actual Losses in Conventions

The data in Tables 8-3 and 8-4 strongly suggest that at least some planners are likely to avoid Las Vegas in response to the repository, especially if the facility is plagued by recurrent accidents. On the other hand, it is difficult to come up with firm forecasts of convention losses based on these data. This is because statements of intended behavior are only imperfect indicators of

Table 8-4   Reported Intent to Choose a City Other Than Las Vegas
Among Those Planners Who Initially Ranked Las Vegas First

| | | | Summary Statistics | | |
Scenario	Subsample	N	Mean	Median	Standard Deviation
1A. Opening of repository (base case: dampened media attention & no amenities)	All	116	3.70	3	2.66
1B. Opening of repository (replication)					
a. dampened	A	58	3.76	3	2.67
b. amplified	B	58	3.62	3	2.68
2. Benign 10-year history					
a. dampened	A	58	3.41	3	2.44
b. amplified	B	58	3.36	3	2.66
3. Minor accident on site					
a. dampened	A	58	3.95	3	2.61
b. amplified	B	58	4.40	4	3.17
4. Minor transport accident					
a. dampened	B	58	3.88	3	3.07
b. amplified	A	58	4.88	5	2.88
5. Moderate accident on site					
a. dampened	B	58	4.12	4	3.12
b. amplified	A	58	5.81	6	3.50
6. Moderate transport accident					
a. dampened	A	58	5.78	6	3.22
b. amplified	B	58	5.81	6	3.50
7A. Report—recurrent accidents and new risk					
a. dampened	B	58	5.45	6	3.58
b. amplified	A	58	7.16	8	3.01
7B. Report—recurrent accidents and new risk (replication)					
a. dampened	B	58	5.42	5.5	3.62
b. amplified	A	58	7.36	8.5	3.03

Note: Intent is measured on a 1-to-10 scale, with 1 = "definitely would choose Las Vegas" and 10 = "definitely would not choose Las Vegas." For each repository event, half the sample was assigned to the dampened media attention condition and half was assigned to the amplified condition. The specific assignment is shown with the subsample column.

how people will actually respond (e.g., Fishbein and Ajzen, 1975; Sheppard, Hartwick, and Warshaw, 1988).

It is, however, possible to model the error surrounding stated intent data. For example, Morrison's (1979) beta-binomial model describes a psychological process whereby a consumer's reported intent to purchase a particular good imperfectly predicts his or her actual propensity to purchase the good. The model explains in mathematical terms the link between stated intent and actual behavior. However, the model's parameters can be estimated only if the researcher has assessed both stated intent and subsequent behavior; it is unable to predict future behavior from current reports of intent.

We have extended Morrison's model to come up with a model that forecasts future behavior in the case where only stated intent has been observed (Easterling, Kunreuther, and Morwitz, 1991). This model translates a set of stated intent data (e.g., the convention planners' ratings of the likelihood of avoiding Las Vegas following a particular repository scenario) into an estimate of the proportion of the sample that will actually engage in the behavior. This forecast is in the form of an interval, $[P^*min, P^*max]$, the width of which reflects the degree of uncertainty in prediction.

In general terms, the model first produces a first-cut estimate of the proportion who will engage in the behavior, P^*. This is the proportion that one would expect if respondents were accurate in predicting their own behavior. P^* is then adjusted to take into account the fact that there is always some degree of bias in predicting one's own behavior. The major step in the forecasting model involves estimating how much bias is present in the intent data that have been observed. This estimation procedure relies heavily on prior studies of intended vs. actual behavior, where one can compute actual bias scores.

In estimating the degree of bias for the convention planner survey, we took special note of the fact that the respondents were predicting a behavior that they had repeatedly performed. This factor tends to attenuate bias. For example, consumers tend to show relatively little bias in predicting the chances of a repeat purchase, such as a new television (Pickering and Isherwood, 1974), but generally overestimate the likelihood of buying novel goods such as a shower radio (Jamieson and Bass, 1989).

After considering a host of factors such as these, we estimated a range of possible bias scores for the convention planners (see Easterling, Kunreuther, and Morwitz, 1991). In particular, we estimated that the actual proportion of planners avoiding Las Vegas following a repository scenario would fall within the following interval:

$$[P^* - s_I, P^* + s_I],$$

Table 8-5 Forecasts of Percentage of Planners Who Would
Choose A City Other Than Las Vegas

Scenario	Estimated Proportion (percent)
1A. Opening of repository (base case: no amenities)	
Dampened	15–39
1B. Opening of repository (replication)	
Dampened	14–42
Amplified	15–37
2. Benign 10-year history	
Dampened	13–35
Amplified	12–36
3. Minor accident on site	
Dampened	17–42
Amplified	18–50
4. Minor transport accident	
Dampened	14–44
Amplified	24–54
5. Moderate accident on site	
Dampened	16–47
Amplified	25–62
6. Moderate transport accident	
Dampened	28–68
Amplified	30–64
7A. Report—recurrent accidents and new risk	
Dampened	25–68
Amplified	43–80
7B. Report—recurrent accidents and new risk	
Dampened	23–66
Amplified	47–80

Note: The range on each forecast reflects uncertainty in the relation between stated intent and actual propensity.

where s_I is the standard deviation of true intent scores for the sample.[8] This unit was adopted because it allows for an easy comparison of bias scores across intention studies. In the case of the planner survey, one standard deviation corresponds to values between 11 and 28 percentage points across the different scenarios.

Table 8-5 presents the model's forecast of the proportion of planners who would move their convention from Las Vegas for each repository scenario.

Even for the least averse scenario (benign ten-year history with dampened media attention), we predict that between 13 percent and 35 percent of the sample would hold the meeting somewhere other than Las Vegas. For the most severe scenario (multiple mishaps with amplified media attention), the forecast range is 47 percent to 80 percent. Although these are relatively wide intervals, none of them includes the value 0, which suggests that the repository will have at least some impact on the convention industry.

To translate these figures into dollar values, we note that approximately 60 conventions are booked for Las Vegas each month, and that an average convention brings in over $1.6 million in revenue (excluding gaming revenue) (Las Vegas Convention and Visitor's Authority, 1990).[9] We forecast that Las Vegas would lose between 15 percent and 39 percent of its bookings under scenario 1A (the opening of the repository, with dampened media attention). If an effect of this magnitude occurred over only the first month, the projected loss would be somewhere between 9 and 23 meetings (between $14.4 million and $36.8 million in current dollars). If, however, the effect lasted a year, we would predict losses of between 108 and 281 meetings (corresponding to $173–$450 million in lost revenue). The predicted losses would be even higher for scenarios 3 through 7.

In making these forecasts, it should be noted that we are assuming that planners' current attitudes toward the repository will persist. In other words, these are provisional estimates of convention losses, which would have to be modified over time if attitudes were found to change.[10]

Convention Attendees Survey

The second survey, conducted in December 1989, examined whether a repository at Yucca Mountain would diminish the willingness of convention attendees to attend a meeting scheduled for Las Vegas. This survey asked about attendance patterns among a sample of 600 individuals who belonged to organizations holding annual conventions. Besides assessing whether a repository would lead attendees to avoid a meeting, this survey also tested the theory that convention attendance is influenced by the perceived risk and imagery associated with the host city.

Methodology

The first step in this survey was to recruit a set of six organizations that hold annual conventions. We imposed the following selection criteria: (a) the organization is national in scope, (b) it holds one convention (not a trade show) each year, (c) the previous four conventions (1986 through 1989) were held in four separate cities, (d) one of these meetings was held in Las Vegas,

Table 8-6 Associations Included in Convention Attendees Study

Association	Location of Annual Meeting			
	1989	1988	1987	1986
American Frozen Food Institute [n=105]	Atlanta	Chicago	Dallas	Las Vegas
American Orthodontic Society [n=72]	Montreal	Las Vegas	Orlando	Dallas
Clinical Laboratory Management Association [n=107]	New Orleans	Las Vegas	Dallas	Atlanta
Joint Council on Economic Education [n=111]	Houston	St. Louis	Las Vegas	Hilton Head
National Purchasing Institute [n=100]	Orlando	Las Vegas	Chicago	Kansas City
United Bus Owners of America [n=105]	Atlanta	Las Vegas	Tampa	New Orleans

and (e) the organization is willing to provide us with its membership list. Eleven organizations met all five of these criteria. We then selected the six groups listed in Table 8-6 based on the objective of maximizing the diversity of the sample. While these six groups are not necessarily representative of the population of organizations that hold meetings in Las Vegas, the sample covers a wide range of professions (e.g., lab technicians, economists, bus owners), which allows us to assess the generalizability of the results.

For each of these six associations, we randomly selected members who had at least some history of attending the annual convention (those who never attend are uninformative in terms of the nature of the attendance decision). To be eligible, an association member had to have attended at least one of the past four meetings. This was determined by an initial telephone contact. Eligible individuals were then mailed a questionnaire with a return envelope. Of the 719 members who were contacted and identified as eligible, completed interviews were obtained from 600, which yielded an impressive response rate of 83.4 percent.

Relation of Imagery and Perceived Risk to Attendance
The first section of the questionnaire asked respondents which of the past four conventions they had attended. We then attempted to predict these attendance decisions as a function of the attendee's perception of the host city. According to the theory developed above, the likelihood of an individual attending a meeting should be determined, at least in part, by the imagery and the perceived risk associated with the city in which it is to be held.

Imagery. The first set of analyses tests the effect of imagery. At the beginning of the questionnaire (directly after the questions on past attendance), respondents reported the imagery that came to mind when they thought of the four meeting cities. For each city, a respondent could report up to six images. In the case of Las Vegas, the most frequently cited images were gambling, entertainment, heat, and bright lights; this ordering varied little across the six organizations.

In order to define the valence of a city's imagery, respondents rated each of their images on a scale from -2 ("very negative") to $+2$ ("very positive"). These were then summed together to form the city's imagery score. With six images per city, an individual's imagery score could conceivably range from -12 to $+12$.

The empirical question, then, is whether these imagery scores can account for an individual's pattern of attendance at the past four meetings. This was assessed with a set of six logistic regressions (one per organization) in which attendance at a meeting (yes or no) was the dependent measure, and the city's imagery score was the predictor. Each attendee contributed up to four data points.[11] To account for individual differences in propensity, we entered as a control variable the percentage of meetings that the individual had attended since becoming a member.

The results of these analyses are shown in Table 8-7. For four of the six organizations, imagery was significant at $p < .05$, while in one other case, the imagery effect had a p value of .07. The only organization in which imagery was clearly nonsignificant was the American Frozen Food Institute. In general then, attendance at past meetings seems to vary systematically according to the imagery associated with the host city.

The regression coefficients indicate the degree to which a change in

Table 8-7   Relation Between Host City's Imagery Score and the Decision to Attend Past Meetings: Results from Logistic Regression Analysis

Association	b	Standard Error	G^2
Amer. Frozen Food Institute	.035	.048	0.53
Amer. Orthodontic Soc.	.177	.055	10.14**
Clinical Lab Management Association	.066	.037	3.28+
Joint Council on Econ. Educ.	.132	.041	10.42**
National Purchasing Institute	.079	.040	3.85*
United Bus Owners	.106	.039	7.28**

Note: Imagery score is computed as the sum of the ratings associated with the images elicited by the city. Respondents could report up to six images for each city. Each image was assigned a score from -2 ("very negative") to $+2$ ("very positive"). Thus, a city's imagery score can range from -12 to 12.
$+p = .07$, $*p = .05$, $**p < .01$.

imagery score leads to a change in the logit associated with the probability (q) of attending a meeting (i.e., $\log(q/1-q)$). For a member of the American Orthodontic Society, the logit increases by .177 for every unit change in imagery score. Translated into probabilistic terms, if a member of this organization initially had a .5 chance of attending the convention, but then the city's imagery score dropped by 10 points, the new probability would be only .15.[12] The percentage-point drop would obviously be smaller if the person started out with a relatively small probability of attending.

Perceived Risk. The next set of analyses was designed to test whether the decision to attend a convention depends on the degree of risk associated with the host city. For these analyses, we asked the attendees to rate each of the four meeting cities on three aspects of risk: (a) crime rate, (b) natural hazards, and (c) pollution and environmental hazards. Each of these risks was rated on a 1-to-5 scale, ranging from 1 = "unacceptable" to 5 = "excellent." Thus, higher scores on these risk indicators correspond to a safer perception of the host city.

A series of logistic regression analyses (one per organization) was used to test whether perceived risk could predict attendance at previous meetings. The three risk indicators were tested simultaneously, so that the effect of each one is estimated controlling for the remaining two. We also controlled for other attributes that might be confounded with a city's risk, including hotel cost, travel cost, climate, and recreation opportunities. Thus, the regression coefficients for the risk factors reflect their incremental effect on attendance, over and above the effect that could be explained by more obvious predictors.

Perceived risk was significantly related to past attendance for three of the six organizations (see Table 8-8). For the American Frozen Food Institute (AFFI), two risk factors were significant: natural hazards, at $p=.01$, and crime rate, at $p=.025$. Crime rate was also significant ($p=.01$) in the case of the National Purchasing Institute (NPI), while pollution was significant, at $p=.005$, within the Joint Council on Economic Education (JCEE).

The size of these effects is indicated by the regression coefficients in Table 8-8. These coefficients indicate how much the logit changes with a unit change in the risk indicator. For JCEE, pollution has a coefficient of .605, indicating that a one-point drop in the pollution rating produces a .605 decrease in the logit function among members of this organization. In terms of probabilities, consider a JCEE member who has a .5 chance of attending a convention in a city that he or she rates as "excellent" on pollution. If this rating drops by 1 unit, the likelihood of attending drops to .353; a two-unit drop reduces the probability to .212; a three-unit drop leads to a probability of .128; and the maximal drop of four points leaves the probability at only

Table 8-8 Relation Between Three Risk Factors and the Decision to Attend Past
Meetings: Results from Logistic Regression Analysis

Association	Crime Rate			Natural Hazards			Pollution/ Env. Hazards		
	b	Standard error	G^2	b	Standard error	G^2	b	Standard error	G^2
Amer. Frozen Food Institute	.450	.201	5.02*	.500	.195	6.60**	.138	.215	0.41
Amer. Orthodontic Soc.	.232	.269	0.75	.620	.322	3.70+	−.537	.302	3.16+
Clinical Lab Management Assn.	.134	.200	0.45	−.343	.191	3.23+	.256	.198	1.68
Joint Council on Econ. Ed.	−.089	.197	0.20	−.025	.174	2.09	.605	.217	7.77**
National Purchasing Inst.	.416	.163	6.54*	.006	.161	0.00	.218	.166	1.74
United Bus Owners	.118	.150	0.62	−.157	.162	0.95	.061	.178	0.12

Note: Risk factors were scaled from 1="unacceptable" to 5="excellent." Thus, a positive regression
coefficient, b, denotes a relation where more *safety* corresponds to a higher probability of attending.
The effect of each risk factor was tested controlling for hotel cost, travel cost, recreation, climate, and
the other two risk factors.
$+p<.10$, $*p<.05$, $**p<.01$,

.074. Similar calculations can be performed for the other two organizations
where risk was significant.

Reported Impact of Noxious Facilities

The above analysis indicates that the attendance decision is somewhat sensitive to the perceived risk and image of the host city. We next tested explicitly whether a repository, along with a number of other noxious facilities, would influence the decision to attend a meeting.

In this part of the survey (which appeared toward the end, after all the items reported above had been asked), respondents were told to assume that they had made a tentative decision to attend a convention. They then found out that a particular facility was located 100 miles away from the host city. Five different facilities were specified: a prison, a nuclear reactor, a hazardous waste incinerator, a low-level radioactive waste repository, and a high-level nuclear waste repository. For each facility, respondents reported whether they would "definitely attend," "probably attend," "probably not attend," or "definitely not attend."

Table 8-9   Willingness to Attend a Meeting After Finding Out That a Noxious Facility was Located 100 Miles Away (percent)

| | Distribution of Response | | | |
Facility	Definitely Would Attend	Probably Would Attend	Probably Would Not Attend	Definitely Would Not Attend
Prison	78.3	20.5	0.8	0.5
Nuclear reactor	70.6	26.2	2.2	1.0
Hazardous waste incinerator	65.7	27.9	5.2	1.2
Low-level radiation waste repository	61.3	28.8	7.1	2.9
High-level nuclear waste repository	49.1	28.0	16.1	6.8

Note: Subjects were told to assume that this was a meeting that they had decided to attend prior to learning of the facility.

The results from these questions are shown in Table 8-9. Little effect is reported for the more common facilities of a prison and a nuclear power reactor: only 1 percent of the sample indicated they "probably" or "definitely" would not attend if a prison were within 100 miles of the host city, while 3 percent reported they would not attend with the reactor. However, the three waste disposal facilities provoked much more of a reaction. For the hazardous waste incinerator, 6.4 percent indicated they "definitely" or "probably" would not attend, while 10.0 percent reported one of these two responses for the low-level radioactive waste repository. The HLNW repository provoked the most extreme response, with 23 percent reporting they "probably" or "definitely" would not attend. Responses to each facility varied only slightly across the six organizations. For example, the percentage who reported they would not attend in response to a HLNW repository ranged from a low of 19 percent (AFFI) to a high of 26 percent (JCEE); this difference was not statistically significant.

These intent data indicate that a HLNW repository is viewed much differently from existing noxious facilities. Thus, the effects on convention attendance that have been historically associated with more familiar facilities do not necessarily generalize to a HLNW repository.

Discussion

In both our surveys, decisionmakers who determine the well-being of Las Vegas's convention industry reported that their behavior would be influenced by a repository at Yucca Mountain. In the convention-planner survey, a third of the sample indicated that they would reduce their ranking of Las

Vegas if the repository opened as planned, and they showed even greater aversion under scenarios where radioactive releases occurred. In the most extreme case tested (an investigator discovers that the repository is plagued with problems similar to what has occurred at DOE's nuclear weapons facilities), three-quarters of the planners reduced their ranking of Las Vegas, while nearly half would no longer consider Las Vegas for their meeting.

Similarly, roughly a quarter of those persons who historically attend conventions indicated that they would not attend if the convention were held in a city 100 miles away from a HLNW repository (the distance from Yucca Mountain to Las Vegas). Based on these surveys, we believe that a repository at Yucca Mountain would potentially trigger major dislocations of Las Vegas's convention industry.

Validity of Intent Data
Although these statements of intent are imperfect predictors of actual behavior, we believe that the nature of the samples improves the validity substantially. In each survey, respondents were reporting on behavior that they had engaged in many times in the past (either choosing a city for a convention or deciding whether to attend a convention). As such, respondents had strong insights into the factors that actually influence the decisions they were describing.

The validity of the intent data was also enhanced by the design of the questionnaire. Prior to the questions that introduced the repository, planners were led through the steps that went into planning their target convention (e.g., the organization holding the meeting, the types of meeting facilities that were required, the importance of hotel costs). Likewise, before attendees reported on the impact of a repository, they first indicated their past attendance at conventions and considered the importance of decision factors such as travel costs. Thus, by the time the repository was introduced, respondents in each survey were firmly within the mindset of the relevant convention decision.

On the other hand, one might argue that the observed intent data reflect simply a distaste for the repository, rather than indicating that people actually will change their behavior in response to the facility. However, this interpretation is less compelling in light of the convention attendees' data. Attendees were presented with a set of facilities that each had strongly negative connotations (e.g., prison, nuclear reactor, hazardous waste incinerator). However, it was only for a HLNW repository that reports of intended behavior were affected in a major way. This suggests that respondents were in fact thinking in terms of likely behavior, rather than just expressing their attitudes toward the facility.

The Importance of Perceived Risk and Imagery
The attendees survey also provides theoretical support for convention losses. Namely, individual attendance at conventions appears to be determined, at least partially, by the perceived risk and imagery associated with the host city. In view of that, planners have good reason to base their choice of where to hold the meeting on factors related to risk and image. Thus, to the extent that the imagery and perceived risk associated with Las Vegas are affected by a repository at Yucca Mountain, both planners and attendees will be less likely to make decisions favorable to Las Vegas.

It should be pointed out that the imagery mechanism can operate among convention attendees even if the risk-avoidance mechanism does not (i.e., the two pathways are not redundant). In particular, attendees who believe that the repository's impacts are confined to the immediate vicinity of Yucca Mountain may nonetheless come to include the facility as part of Las Vegas's image set and thus may still be less likely to visit the city. More generally, an image can affect visitor behavior even if it is unrelated to the person's expectations about what will happen during a visit. For example, the thought of Dallas conjured up images of the Kennedy assassination among fifteen members of the sample. This event occurred almost thirty years ago and thus would not directly confront someone during a visit. Nevertheless, the extremely negative connotations of this image might lower a person's desire to visit Dallas. A similar sort of guilt-by-association effect could occur between Las Vegas and the repository.

Uncertainty
In interpreting the data from these surveys, one must acknowledge the tremendous amount of uncertainty that is inherent in predicting the economic impacts of a repository (Erikson, 1991). A primary source of this uncertainty has to do with our inability to predict how the Yucca Mountain repository will operate; indeed, it is not even clear that it will even be built. The scenario approach we employed in the convention-planner survey provides one way to study economic impacts in the face of this uncertainty. Namely, one sets about the task of *conditional* forecasting, i.e., estimating the losses that would follow distinct repository scenarios. One can then define an aggregate expected impact by assigning probabilities to the alternative scenarios. Thus, two policy analysts with different assessments of the likelihood of severe scenarios (e.g., a DOE official and an official from Nevada) could each make use of the data but might end up generating very different estimates of the overall expected impact of the repository.

Another major uncertainty involved in estimating convention losses has to do with the strength of the link that would develop between the repository and Las Vegas. In other words, would the severe risk that the public associates with the repository extend as far as Las Vegas, and would people spontaneously think of the repository when considering whether to go to Las Vegas? Bassett and Hemphill (1991) argue that such a connection is unlikely to occur, given the fact that very few respondents in the Slovic et al. (1991) surveys (also reported in chapter 3 of this volume by Slovic, Layman, and Flynn) mentioned either the repository or the Nevada Test Site in response to Las Vegas. This lack of nuclear imagery seems to reflect, in large part, the fact that Las Vegas is so strongly associated with gambling and entertainment; these images dominate almost every other attribute of the city. However, this could change if Las Vegas continued to be linked with Yucca Mountain in the media, or if transport accidents involving shipments of nuclear waste occurred close to the city.

Conclusion

Whether or not a repository actually would affect Las Vegas's convention industry seems to depend primarily on the degree to which existing public perceptions of the facility persist over the long term. It is possible that, with time, a repository would be viewed in more benign terms. On the other hand, more information about the repository program might lead to even more severe perceptions of risk and might produce even worse images.

To a large extent, the connotations that a repository takes on will be determined by the degree of scientific and institutional integrity that comes to characterize the program (see chapter 3), as well as whether a need for the facility is convincingly demonstrated (Easterling, 1992). In the absence of such developments, one can expect negative effects on the convention business of cities unfortunate enough to be linked with a repository.

Finally, it should be noted that any visitor effects that did occur would be felt not just by the local community but also at the federal level. The impact-assistance provision of the federal Nuclear Waste Policy Act [Section 116(c)(2) of Pub. L. 97-425] calls for DOE to compensate state and local governments for the negative economic impacts that result from the construction and operation of a repository. To the extent that a host state suffers major losses to its visitor industry, this provision portends significant payouts by the federal government. Such a prospect needs to be seriously considered in judging the suitability of the Yucca Mountain site.

Notes

1   The revenue figures reported by the Las Vegas Convention and Visitors Authority do not include receipts from gaming. Thus, the total contribution to the Las Vegas economy is considerably greater than $1 billion.

2   Although the court rejected Nevada's suit and invalidated AB 222, state officials have indicated that they will find alternative approaches to fight the repository (Swainston, 1991). This prospect casts serious doubts on whether the repository can begin operation by the scheduled opening date of 2010 (Rhodes, 1990).

3   Our understanding of the factors that influence planning decisions is based on a series of interviews with professional meeting planners, along with a focus group of nine planners conducted in Philadelphia in October 1987 (Kunreuther, Easterling, and Kleindorfer, 1988).

4   We are thus considering the case where the repository has not caused the planner to avoid Las Vegas.

5   This imagery mechanism presumes that the convention attendee's decision process is at least somewhat impressionistic: the various images that are elicited while the individual is deciding whether or not to attend the meeting have a direct impact on the choice. In contrast, the image mechanism posited for meeting planners is more analytic, in that a city's image is an explicit decision factor that summarizes the many characteristics associated with the city.

6   The vast majority of the planners (116 out of 153) assigned a ranking of 1 to Las Vegas in the no-repository case. The remainder had booked their meeting for Las Vegas even though this was not their top choice. Las Vegas was chosen in these cases either because the preferred city was not available for the meeting dates or because the planner had been overruled by the officers of the organization.

7   These statistics are restricted to the subsample of 116 planners who initially ranked Las Vegas first for the target meeting. The rating data for the other 37 have ambiguous meaning because it is unclear what value would have been assigned to Las Vegas in the absence of a repository (see footnote 6).

8   A person's true-intent score is his or her current best-guess probability of engaging in the behavior. The distribution of true-intent scores can be estimated from the observed stated intent scores if one adopts the assumptions of Morrison's (1979) model.

9   The upward trend in the Las Vegas convention industry suggests that there will be many more than 60 conventions per month by the year 2010 (the date when the repository is projected to open).

10   One might wish to model this uncertainty regarding the persistence of planners' attitudes by increasing the range of possible bias scores. However, this would have generated an extremely wide range (possibly giving rise to estimates as uninformative as "0 to 100 percent"), due to the very long time lag between 1988 and 2010. Thus, we took the alternative forecasting approach of conditioning the predictions on the assumption that no changes will occur regarding perceptions of the repository.

11   Failure to attend a meeting was ignored if the respondent had not yet become a member of the organization at the time of the meeting.

12   In this example, the logit decreases from 0 (corresponding to $q = .5$) to $0 - (10^{*}.177)$, or -1.77. The probability value corresponding to a logit of -1.77 is .15.

References

Adams, Michael. 1986. "Tourism in Crisis." *Successful Meetings* 35 (4): 220–21.

Adams, Steve. 1990. "Dump Study Control Sought." *Las Vegas Review-Journal*, December 19.

Bassett, Gilbert W., and Ross C. Hemphill. 1991. "Comments on Perceived Risk, Stigma, and Potential Economic Impacts of a High-Level Nuclear Waste Repository in Nevada." *Risk Analysis* 11:697–700.

Bryan, Richard H. 1987. "The Politics and Promises of Nuclear Waste Disposal: The View from Nevada." *Environment* 29 (8): 14–17; 32–38.

Conlin, Joseph. 1987. "Hotel Fire Safety: How You Can Prevent Meeting Tragedy," *Successful Meetings* 36 (3): 22–26.

Easterling, Douglas. 1992. "Fair Rules for Siting a High-Level Nuclear Waste Repository." *Journal of Policy Analysis and Management* 11 (3): 442–75.

Easterling, Douglas, Howard Kunreuther, and Vicki Morwitz. 1991. "Forecasting Behavioral Response to a Repository from Stated Intent Data." In *High Level Radioactive Waste Management: Proceedings of the Second International Conference.* New York: American Society of Civil Engineers. 1540–47.

Edelstein, Michael R. 1988. *Contaminated Communities: The Social and Psychological Impacts of Residential Toxic Exposure.* Boulder: Westview Press.

Erikson, Kai T. 1991. "The View From Yucca Mountain." Mimeo.

Fishbein, Martin, and Icek Ajzen. 1975. *Belief, Attitude, Intention, and Behavior: An Introduction to Theory and Research.* Reading, MA: Addison-Wesley.

Flynn, James H., C. K. Mertz, and Paul Slovic. 1991. *The 1991 Nevada State Telephone Survey: Key Findings.* Report 91-2. Eugene: Decision Research.

Fradkin, Philip L. 1989. *Fallout: An American Nuclear Tragedy.* Tucson: University of Arizona Press.

Jamieson, Linda F., and Frank M. Bass. 1989. "Adjusting Stated Intention Measures to Predict Trial Purchase of New Products: A Comparison of Models and Methods." *Journal of Marketing Research* 26:336–45.

Jones, Edward, Amerigo Farina, Albert Hasstorf, Hazel Markus, Dale Miller, and Robert Scott. 1984. *Social Stigma: The Psychology of Marked Relationships.* New York: W. H. Freeman.

Kerr, John. 1990. "Repository's Effects on Vegas's Economy, Image Raise Concerns." *Las Vegas Review-Journal*, October 26.

Kunreuther, Howard, William H. Desvousges, and Paul Slovic. 1988. "Nevada's Predicament: Public Perceptions of Risk from the Proposed Nuclear Waste Repository." *Environment* 30 (8): 16–20; 30–33.

Kunreuther, Howard, Douglas Easterling, and Paul Kleindorfer. 1988. *The Convention Planning Process: Potential Impact of a High-Level Nuclear Waste Repository in Nevada.* Report prepared for Nevada Nuclear Waste Project Office. Philadelphia: Wharton Risk and Decision Processes Center.

Las Vegas Convention and Visitors Authority. 1990. *Las Vegas Marketing Bulletin—1989 Annual Summary,* Las Vegas: LVCVA.

Las Vegas Review-Journal. 1988. "Red Herring Issue in Nuke Dump Fight" (Editorial), *Las Vegas Review-Journal*, July 15.

Lyall, Sarah. 1991. "Beach Medical Waste: Debris but No Panic." *New York Times*, September 11.

Morrison, Donald G. 1979. "Purchase Intentions and Purchase Behavior." *Journal of Marketing* 43 (Spring): 65–74.

Morrison, Jane Ann. 1991. "Gamers Weren't Approached to Fund Anti-Nuke Dump Ads." *Las Vegas Review-Journal*, September 25.

New York Times. 1990. "Dinkins Assails Bashing of New York." *New York Times*, September 14.

Pennsylvania Governor's Office of Policy and Planning. 1980. *The Socio-Economic Impacts of Three Mile Island Accident*, Harrisburg: Pennsylvania Governor's Office.

Petterson, John S. 1988. *Report on Follow-Up Study of Goiania Incident*, Report prepared for Nevada Nuclear Waste Project Office. La Jolla: Impact Assessment.

Pickering, J. F., and B. C. Isherwood. 1974. "Purchasing Probabilities and Consumer Buying Behavior." *Journal of the Market Research Society* 16 (July): 203–26.

Rhodes Jr., Joseph. 1990. "Nuclear Power: Waste Disposal, New Reactor Technology, Pyramids Underground." Paper presented at the 102nd Annual Meeting of National Association of Regulatory Utility Commissioners, Orlando, FL, November 13.

Sahugan, Miguel Cervantes. 1988. "Report: Minority Visitors Shun City." *Philadelphia Inquirer*, September 14.

Sheppard, Blair H., Jon Hartwick, and Paul R. Warshaw. 1988. "The Theory of Reasoned Action: A Meta-Analysis of Past Research with Recommendations for Modifications and Future Research." *Journal of Consumer Research* 15:325–43.

Slovic, Paul, Mark Layman, Nancy Kraus, James Flynn, James Chalmers, and Gail Gesell. 1991. "Perceived Risk, Stigma, and Potential Economic Impacts of a High-Level Nuclear Waste Repository in Nevada." *Risk Analysis* 11:683–96.

Survey Research Associates. 1987. *MetroPoll II.* Report prepared for *Meetings & Conventions Magazine.* Sausolito: Survey Research Associates.

Swainston, Harry W. 1991. "The Characterization of Yucca Mountain: The Status of the Controversy." *Federal Facilities Environmental Journal* (Summer): 151–60.

Thomas, G., and M. Morgan-Witts. 1982. *Anatomy of an Epidemic.* Garden City, NY: Doubleday.

Titus, A. Constandina. 1988. *Phase IIIA Yucca Mountain Socioeconomic Report: NTS Study.* Report prepared for Nevada Nuclear Waste Project Office, Carson City, NV.

Ziff-Davis Publishing Company. 1984. *The Meetings Market Study—1983.* Report prepared for *Meetings & Conventions Magazine.* New York: Ziff-Davis.

9 Nevada Urban Residents' Attitudes Toward a Nuclear Waste Repository

Alvin H. Mushkatel, Joanne M. Nigg, and K. David Pijawka

As discussed in the introductory chapter of this volume, the Nuclear Waste Policy Act of 1982 and the subsequent amendments to the act in 1987 have had a profound effect on the state of Nevada. In December 1987, Congress amended the original act to force all characterization activities on one site— Yucca Mountain, Nevada—located ninety miles northwest of Las Vegas. Only if the Nevada site were found to be geologically infeasible would other sites again be considered.

Within the state of Nevada, one of the consequences of these congressional actions was to highlight the importance of assessing the impacts that the repository would have on the social and economic well-being of residents. Of particular importance was the effect that the proposed repository would have on the state's most rapidly growing metropolitan area, Las Vegas, given its relatively close proximity to the proposed site. How much concern or anxiety among residents would be produced by the repository siting? What were the attitudes of residents toward the proposed project? Could the repository affect residents' well-being and quality of life if it were constructed at Yucca Mountain?

With these questions in mind, a study was designed to assess the possible social impacts of the proposed repository at Yucca Mountain on residents in the Las Vegas area. One element of the study design was a survey of Las Vegas residents regarding their attitudes toward and beliefs about the repository. The survey results reported in this chapter are among the first efforts to analyze these attitudes. These results complement those from a statewide survey of Nevadans reported by Desvousges, Kunreuther, Slovic, and Rosa in chapter 7 and those from surveys of rural respondents reported by Krannich, Little, and Cramer in chapter 10 of this volume.

In this discussion, five measures of attitudes toward the repository are presented. Next, the impact of four sets of independent variables on residents' beliefs about the repository are considered and placed in the context of previous research that has utilized these variables. Finally, residents' views about the repository are regressed on the significant variables from these four sets of items to determine the relative impact of these variables. Before discussing these analyses, it is necessary to first address the methodology employed in the survey.

Methodology

The purpose of the urban risk survey was to assess attitudes toward the proposed high-level nuclear waste repository at Yucca Mountain for the residents in the Las Vegas metropolitan area. The *Las Vegas metropolitan area* was defined to include the cities of Henderson, Las Vegas, North Las Vegas, and the contiguous area of Clark County, which is urban.

Data collection was conducted through the Survey Research Center at the University of Nevada-Las Vegas. Administration and supervision of the data collection and data reduction were a cooperative effort between UNLV and the authors. Survey interviews were conducted over a ten-week period from March to June, 1988.

A random-digit-dialing (RDD) procedure was used to select 750 households for inclusion in the study. The sample was based on the proportionate distribution of residential households within the sixty-two telephone prefixes that serve the geographic study area.

Once a household was determined to be in-sample,[1] a short telephone interview was conducted with an adult informant. A modified Kish (1949) selection procedure was then used to identify the appropriate random adult respondent in the household and a face-to-face interview was scheduled with that household member. The telephone portion of the interview took approximately ten minutes to complete; the face-to-face interview took between fifty minutes and two hours.

Intensive attention was given to overcoming initial refusals and reluctance on the part of the random respondents to consent to face-to-face interviews. However, by allowing the interviews to take place in public settings (e.g., restaurants), at work places, and at odd hours (several interviews were conducted after midnight when casino and restaurant workers were getting off work), a final response rate of 73.5 percent (n=549) was achieved.

Attitudes Toward Repository Risks

Five items were used to assess perceptions and attitudes that Las Vegas area residents have about the proposed high-level nuclear waste repository: (1) the respondents' concern over the extent to which the repository might produce harmful effects in their community; (2) the relationship between potential benefits and harms for the Las Vegas area if the repository were built; (3) the perceived seriousness of risks of the repository; (4) the risks of transporting wastes to the repository; and (5) the respondents' opinions on whether the repository should be built at all. The results obtained with these questions are reported in Tables 9-1 and 9-2.

To determine how concerned Las Vegas respondents were about the repository, they were first asked:

—"With regard to all possible effects, how concerned are you that the nuclear waste repository could produce harmful effects here in the Las Vegas area?"

—"Are you very concerned, somewhat concerned, not very concerned, or not concerned at all?"

Over two out of five said they were "very concerned," and nearly four out of five were at least "somewhat" concerned about potential harmful effects from the repository (Table 9-1). This high level of concern among Las Vegas residents is consistent with expressions of concern found in other studies of perceptions of nuclear waste risks among various publics, several of which are reported elsewhere in this volume.

Previous risk assessment studies have discovered that the perceived benefits deriving from a potentially hazardous facility may reduce the perceived harms one associates with that facility; that is, the degree to which harms become acceptable is related to the perceived personal and/or community gains that residents believe may occur. The benefits-to-harms dimension has been used in several studies to measure impacts of hazardous technologies (see chapter 5 by Brody and Fleishman in this volume; Kasperson et al., 1980; Flynn, 1979; and Pijawka, 1984).

Respondents were asked a set of five questions about specific benefits or specific harms that they thought would accrue either to themselves or to the communities in which they lived if the repository were built. A summary question was then asked to investigate the trade-off between anticipated benefits and harms that might occur:

—"For the greater Las Vegas area, would you say that the possible benefits of the nuclear waste repository outweigh the possible harmful

Table 9-1   Perceptions of Repository Risks

Perception		Percent	Frequency
General concern about potential health harms:			
Very concerned		42	201
Somewhat concerned		36	176
Not very concerned		14	68
Not concerned at all		8	38
Total		100	483
Seriousness of possible risks at site:			
Very serious	1	23	111
	2	13	64
	3	18	85
	4	15	73
	5	12	59
	6	11	54
Not serious at all	7	8	38
Total		100	484
Seriousness of possible risks in transit:			
Very serious	1	34	166
	2	19	91
	3	17	82
	4	10	51
	5	9	44
	6	7	34
Not serious at all	7	4	17
Total		100	485
Evaluation of perceived benefits to harms:			
Benefits outweigh harms		17	81
Benefits balance harms		30	145
Harms outweigh benefits		53	250
Total		100	476

effects, that the possible harmful effects outweigh the possible benefits, or do they balance each other?"

As can be seen in Table 9-1, over half of the residents expressed the belief that the potential harms from a high-level nuclear waste repository would outweigh the possible benefits from such a facility; nearly a third believed that the benefits and harms would be about equal; while less than a fifth felt that the benefits would be greater than the harms. Hence, while a majority of Las Vegas respondents believed that the repository would result in greater harm than benefit to themselves or their communities, a substantial minority of residents believed that the benefits would be greater than any harms created.

Two questions were then asked to assess the respondents' attitudes toward the potential seriousness of risks both at the repository site and during the transportation of wastes to the site:

> —"People have different ideas about how much risk Las Vegas residents would face if the nuclear waste repository were built at Yucca Mountain.
>
> 1. "How serious a risk would the activities at the Yucca Mountain repository be to the health and safety of residents in the Las Vegas area?"
> 2. "Keeping health concerns in mind, how serious a risk would the transportation of nuclear wastes be to residents in the Las Vegas area?"

Respondents were asked to rate the seriousness of these risks on a seven-point scale from 1="very serious" to 7="not serious at all."

Almost a fourth of the respondents felt that the risks at the repository site —nearly ninety miles away—could result in very serious risks (response 1) and another third believed that serious risks (responses 2 or 3) were a possibility (Table 9-1). Transportation risks were seen as potentially even more harmful—over a third believed them to be very serious (1) and another third serious (2 or 3).

In addition to measuring residents' perceptions of the risks associated with the repository, we also included a question dealing specifically with an opinion toward the construction of the proposed facility. Respondents were asked:

> —"If you were able to make the final decision regarding the location of the nuclear waste repository at Yucca Mountain, would you build it there? Would you say definitely yes, probably yes, probably no, definitely no, or are you uncertain?"

Over two-thirds said they would "definitely not" or "probably not" favor building the repository (Table 9-2). Less than 25 percent of the residents in Las Vegas favored building the repository, and only 9 percent definitely favored its construction. Interestingly, despite all of the publicity over the issue and its importance in the recent U.S. Senate and gubernatorial campaigns, 7 percent of the respondents remained undecided or uncertain about the repository's construction.

In summary, we find that among Las Vegas area residents there was substantial concern about the possible harms from a nuclear waste repository

Table 9-2 Opinion about Building the Repository

	Percent	Frequency
Definitely yes	9	41
Probably yes	15	72
Uncertain	7	36
Probably no	16	79
Definitely no	53	258
Total	100	486

Item: "If you were able to make the final decision regarding the location of the nuclear waste repository at Yucca Mountain, would you build it there? Would you say definitely yes, probably yes, probably no, definitely no, or are you uncertain?"

ninety miles from the metropolitan area and a great deal of opposition to the construction of the facility. Let us now turn our attention to the variables that might influence these attitudes toward the repository and opinion about its construction.

Explanatory Variables

Four sets of variables were examined to explain these different attitudes toward the repository: attitudes toward other types of hazardous facilities; personal, work-related experiences in similar types of facilities; trust in the government's handling of accidents at other facilities; and demographic characteristics. These factors have been found to be important in previous research on public attitudes toward nuclear power plants and other types of hazardous technologies. The remainder of the chapter will examine the importance of each of these four sets of variables in explaining residents' perceptions of and attitudes toward the repository.

Attitudes Toward the Nevada Test Site (NTS)
Several studies have noted the importance of respondents' attitudes toward nuclear power and nuclear testing, as well as other hazardous projects, in explaining peoples' attitudes toward various facilities using or storing radioactive materials. For example, a Battelle study found that attitudes toward nuclear power and nuclear weapons testing were strongly associated with concerns about nuclear waste management and a nuclear waste repository (Lindell and Earle, 1982). More recently, in examining factors that explain perceptions of the superconducting super collider, Stoffel et al. (1988) discovered that *risk perception shadows* may exist. That is, because of previous experience with or exposure to a similar type of hazard or noxious facility,

a predisposition "to distrust projects involving potential adverse health or social impacts" may be present. In addition, the extensiveness of the risk perception shadow will be determined by the range of the potentially feasible adverse impacts, suggesting that the size of a project's potential adverse effects, particularly of projects involving radioactive materials, will cast very large shadows.

The long-term presence of the Nevada Test Site (NTS), a federal facility for the testing of nuclear weapons adjacent to the proposed Yucca Mountain waste repository site, is potentially a "shadow-casting" facility. The NTS provides a similar type of radioactive hazard from which Las Vegas residents can draw conclusions about the potential riskiness of a nuclear waste repository from their own perceptions and experiences.[2]

Several questions concerning respondents' attitudes toward the Nevada Test Site were included in the questionnaire to allow us to examine whether respondents' views about NTS were related to their attitudes toward the proposed repository. The four questions included were:

(1) "How likely do you think it is that above-ground nuclear weapons testing activities at the Nevada Test Site have, in the past, caused harmful health problems for people who live in the Las Vegas area?"

(2) "How likely do you think it is that underground nuclear weapons testing activities at the Nevada Test Site will, in the future, cause harmful health problems for people living in the Las Vegas area?"

(3) "How likely is it that activities at the Nevada Test Site will cause you personally health problems in the future?"[3]

(4) "Generally speaking, would you say that the possible benefits of the Nevada Test Site outweigh the possible harmful effects, that the possible harmful effects outweigh the possible benefits, or do they balance each other?"

Table 9-3 presents the responses to these questions.

The now-discontinued practice of above-ground testing is clearly perceived to have been more harmful to human health than is the current below-ground nuclear testing. However, 39 percent of the Las Vegas respondents believed that the current below-ground testing was likely (responses 5 to 7) to be harmful in general; and 18 percent (responses 5 to 7) believed that they were personally likely to suffer health harms from this activity. Similarly, a fourth of the respondents believed that the harms from the existence of the NTS outweighed the benefits that would be derived by the residents of the Las Vegas area. Thus, for a substantial minority of the respondents, the NTS may have created a risk perception shadow that negatively colors their attitudes toward the proposed high-level nuclear waste repository.

Table 9-3 Attitudes about the Nevada Test Site (NTS)

Attitude		Percent	Mean
Likelihood of harm to health from above-ground tests:			
Not likely at all	1	13	4.5
	2	10	
	3	9	
	4	14	
	5	15	
	6	14	
Extremely likely	7	25	
Total %		100	
Total n		(517)	
Likelihood of harm to health from underground tests:			
Not likely at all	1	15	3.8
	2	18	
	3	14	
	4	13	
	5	14	
	6	11	
Extremely likely	7	14	
Total %		99	
Total n		(520)	
Likelihood of personal health harms from NTS activities:			
Not at all likely	1	28	2.9
	2	22	
	3	13	
	4	19	
	5	10	
	6	4	
Extremely likely	7	4	
Total %		100	
Total n		(520)	
Evaluation of perceived benefits to harms:			
Benefits outweigh harms		32	—
Benefits balance harms		41	
Harms outweigh benefits		27	
Total %		100	
Total n		(510)	

Table 9-4   Relationships between NTS Attitudes and Repository Attitudes
(Kendall's Tau)

| | Perceptions of NTS Activities | | | |
Repository Attitudes	Health Harms in Las Vegas Area from Above-Ground Testing	Health Harms in Las Vegas Area from Below-Ground Testing	Personal Health Harms in Future	Evaluation of NTS Benefits to Harms
General concern about potential health harms from repository for Las Vegas area	tau c = .29 p = .000	tau c = .43 p = .000	tau c = .39 p = .000	tau c = .37 p = .000
Seriousness of possible risks at the site	tau b = .26 p = .000	tau b = .43 p = .000	tau b = .38 p = .000	tau c = .39 p = .000
Seriousness of possible risks in transit	tau b = .26 p = .000	tau b = .40 p = .000	tau b = .34 p = .000	tau c = .36 p = .000
Evaluation of perceived repository benefits to harms	tau c = .28 p = .000	tau c = .44 p = .000	tau c = .37 p = .000	tau b = .43 p = .000
Personal decision whether repository should be built	tau c = .26 p = .000	tau c = .31 p = .000	tau c = .28 p = .000	tau c = .36 p = .000

Relationships between each of these four perceptions of NTS activities and the five primary dependent variables measuring respondents' attitudes toward the repository and opinion concerning its construction were examined. Table 9-4 summarizes these relationships using Kendall's tau, a measure of association appropriate for the ordinal variables of the survey. As can be seen from this table, *all* of the relationships are statistically significant beyond the .001 level. The results consistently show that the more the NTS is perceived as harmful, the more negative the attitude toward the repository.

The weakest relationships are found in perceptions of health harms in the Las Vegas area from *above*-ground testing. Yet, even these relationships are moderately strong (tau ranges from .26 to .29). The strongest relationships are found in perceptions of health harms from *below*-ground testing (tau ranges from .31 to .44), a situation analogous to that of the repository, where the nuclear wastes would be stored in subterranean tunnels. Those respondents who believed that health harms resulted from below-ground testing were also more likely to be concerned about health harms from the repository and to believe that harms would outweigh benefits from the repository;

they were more likely to be more concerned about the seriousness of risks both at the site and in transit to it and would likely not vote to build the repository if it were their decision.

In summary, the findings in Table 9-4 indicate that respondents' perceptions of the various NTS activities were strongly related to negative attitudes toward the repository. These findings suggest the existence of a risk perception shadow emanating from NTS that affects residents' perceptions of potential consequences from the nuclear waste repository.

Experience with Comparable Hazards

Experience with natural hazards has been shown to be a strong explanatory factor in the perception of environmental hazards, as well as in the types of adjustments people make to reduce threat from such hazards (Burton, Kates, and White, 1978; Sorenson et al., 1987; Mitchell, 1974). Kates (1962), in fact, has suggested that people are "prisoners of experience" and tend to perceive hazards based on their past experience. In the area of technological hazards, several studies have suggested that past experience with toxic chemicals or technological accidents is positively related to heightened perceptions of risks associated with technological hazards (Sorenson et al., 1987; Cutter, 1984). In addition, in discussing risk perception shadows and their findings concerning radioactivity risk, Stoffel et al. (1988) examine the widespread belief that experience with radioactive phenomena leads people to become accustomed to such risk and, therefore, to downplay the risk of projects involving radioactive materials. They find, however, that past experience with such projects (i.e., living in close proximity to such facilities) involving radiation *increases* the perceived risks of other proposed nuclear projects. In short, the literature on technological risk perception indicates that past experience with hazards is important in explaining attitudes toward proposed projects, but past experience, as such, does not permit us to predict whether that experience will result in complacency or enhanced concern (Slovic, Fischhoff, and Lichtenstein, 1980; Kates, 1967; Schiff, 1977).

Two sets of questions concerning respondents' past experience with hazards were included in the survey to permit an analysis of the importance of this factor in explaining attitudes toward the repository. The first set, designed to determine whether they had ever worked with or around radioactive materials, were:

—"Have you ever worked at any job that required you to work directly with highly radioactive materials as part of your job?"
—"Have you ever worked at any job that involved being in an area or

a facility where highly radioactive materials were handled and worked with by others?"

The second set, asking about respondents' employment at federal, military, or nuclear facilities within commuting distance of Las Vegas, included:

—"Have you ever worked for the federal government or a government contractor at any of the following places:"[4]
A. "The Nevada Test Site?"
B. "The Tonopah Test Range?"
C. "Nellis Air Force Base?"
D. "Indian Springs Air Force Base?"

The findings regarding past experience are most interesting. As measured by these six questions individually, personal work experience with or around radioactive materials or at federal facilities was *not* significantly related to attitudes toward the repository. However, two composite indices were constructed using these variables—one measuring personal experience with radioactive materials (composed of the first two questions above) and one measuring past experience in federal facilities (composed of the four possibilities in the last question). Personal experience with radioactive materials was still not related to repository attitudes; however, past experience working in Las Vegas area federal facilities was found to be significantly, although modestly, related to three of the risk perception variables. In short, the more experience they had working at these federal facilities, the more likely respondents were to want the repository built (tau b = .16; p = .000), to see its benefits as outweighing its harms (tau c = .07; p = .002), and to be less concerned about any harmful repository effects (tau c = .06; p = .005). This index of experience at federal facilities has been retained for the multivariate analyses discussed below.

Trust in Government and Governmental Agencies
Studies have suggested that citizens' trust in government and governmental agencies responsible for regulating the operation of hazardous technologies is related to their support for similar, new facilities. While governmental trust may be difficult to separate from trust in science and technology, specific questions were developed to measure this concept.

As far back as the earliest election studies conducted by the University of Michigan, trust in government has been found to correlate positively with political efficacy and participation and negatively with cynicism (Campbell et al., 1960). Sonderstrom et al. (1984) discovered that faith in Three Mile Island managers was associated with a desire to restart the facility. Stoffel et al. (1988) found that the level of trust in the agencies involved in previous projects can become a key public risk perception factor, "especially when there is a lack of scientific agreement about the probability of assessed risks."

The trust items examined in this study reflect both a general trust in government as well as specific trust in governmental agencies. The general governmental trust questions were asked with respect to the federal, state, and local levels of government separately. For example, the federal question was posed:

—"Now, let's change topics. I'd like to talk about some of the different ideas people have about government and how you feel about these ideas." First, let's talk about the federal government in Washington. How much of the time do you think we can trust the government in Washington to do what is right. Would you say: just about always, most of the time, some of the time, or almost never?"

The specific trust question, requesting an assessment of the Department of Energy (DOE), the Nuclear Regulatory Commission (NRC), and the Nevada Nuclear Waste Project Office (NWPO), was worded:

—"I'm now going to ask you how you feel about various government agencies and institutions. Please tell me how much trust you have in the ability of each one to make decisions to protect public safety."

Respondents were then read a list of agencies, including the three listed above, and asked to rate them on a seven-point scale from 1= "no trust" to 7= "complete trust."

It was hypothesized that the greater the trust in various levels of government and the energy agencies, the more positive would be one's views on the repository. Table 9-5 contains the summary of the relationships between the trust questions and the five repository attitude variables.

Results for the general trust items, constituting the first three items in Table 9-5, are generally supportive of the above hypothesis with respect to federal and local levels of government. These relationships are modest but significant across all five dependent variables for the trust in federal and local government items. The higher the level of trust in either of these levels of government the lower the concern about the potentially harmful effects of the repository, the less serious site- and transit-related risks were perceived, the more likely benefits were believed to outweigh harms, and the more likely the respondent was to favor building the repository. However, trust in the state of Nevada is only significant for one variable—concern about the harmful health effects of the repository (and then the measure of association is weak).

Table 9-5 Relationships between Attitudes on Government and Repository Attitudes (Kendall's Tau*)

Trust Indicators	Repository Attitudes				
	Concerns About Repository Harms	Seriousness of Risks at Site	Seriousness of Risks in Transit	Harms Outweigh Benefits from Repository	Decision to Build Repository
General trust in government to do what is right:					
Federal	tau b = .21	tau c = .15	tau c = .14	tau c = .17	tau c = .18
	p = .000	p = .000	p = .000	p = .000	p = .000
State	tau b = .07	tau c = .03	tau c = .04	tau c = .05	tau c = .02
	p = .04	p = n.s.	p = n.s.	p = n.s.	p = n.s.
City/county	tau b = .16	tau c = .14	tau c = .09	tau c = .17	tau c = .13
	p = .000	p = .000	p = .004	p = .000	p = .000
Trust in specific agencies to protect public safety if the repository is built:					
DOE	tau c = .23	tau b = .23	tau b = .25	tau c = .30	tau c = .21
(n = 535)	p = .000	p = .000	p = .000	p = .000	p = .000
NRC	tau c = .26	tau b = .28	tau b = .27	tau c = .30	tau c = .23
(n = 533)	p = .000	p = .000	p = .000	p = .000	p = .000
NWPO	tau c = .14	tau b = .15	tau b = .15	tau c = .14	tau c = .09
(n = 503)	p = .000	p = .000	p = .000	p = .001	p = .005

*Negative signs of tau for inverse relationships complicated the table unnecessarily. To simplify the presentation, variables to inverse relationships were recoded so that all taus became positive and were consistent with predicted direction of the relationships.

In summary, the trust people had in the federal government and their own local government affected their attitudes toward the proposed repository; their trust in the state of Nevada "to do what is right" did *not* affect their attitudes about the repository.

Table 9-5 also reports results for the relationships between trust in specific governmental agencies to protect public safety and repository attitudes. Before turning to these relationships, however, attention should be given to the varying level of trust respondents report for these agencies. The state's NWPO received the highest trust rating (mean score of 4.2 on the seven-point scale) to protect public safety; while the two federal agencies with primary responsibility for ensuring public safety—DOE (4.0 mean score) and NRC (3.9 mean score)—received lower trust ratings.

As can be seen from Table 9-5, trust in *all* of the specific governmental agencies was modestly but significantly related to repository risk perceptions. These relationships are all in the hypothesized direction, indicating that greater trust in each one of these agencies is associated with lower levels of concern, higher evaluations of expected benefits to harms, lower assessments of the seriousness of repository-related risks, and personal decisions in favor of building the repository.

An additional trust dimension was also examined in the study. Respondents were asked two questions to indicate the likelihood that the federal government had honestly reported accidents at the NTS and the likelihood that it would report accidents at the repository should it be built. The following wording regarding NTS was replicated in future tense for the proposed repository:

> —"What proportion of accidents at NTS do you believe the government has reported to the public. Have they reported: all accidents, most accidents, some accidents, or very few accidents?"

The frequency distributions for these two items are quite revealing. Only 3 percent of Las Vegas residents in the study believed that the government had reported *all* of the accidents at NTS; while 42 percent believed that few, if any, accidents were reported. Furthermore, only 4 percent of the respondents believed that if the repository were built the government would report all of the accidents, while 40 percent believed that few, if any, repository accidents would be reported.

The relationship between these two items and repository attitudes are moderately strong (taus range from .30 to .42) and all are significant (Table 9-6). Hence, the less likely respondents were to believe that the federal government reported accidents at NTS or would report accidents at the repository, the greater their negative attitudes toward the repository.

The results for the trust relationships (Table 9-5) are very revealing. General trust in government is related to positive attitudes toward the repository, and specific trust in governmental agencies is also related to such support. Yet, trust was not widespread among the respondents, especially trust in DOE and NRC. In addition (Table 9-6), and most importantly, trust that the federal government reported accidents at the NTS or would report accidents at the repository are important predictors of overall attitudes toward the repository.

Table 9-6   Relationships between Repository Attitudes and Beliefs about Governmental Honesty in Reporting Radiological Accidents Involving Radioactive Materials (Kendall's Tau*)

Repository Attitudes	Honesty of Government in Reporting Radiological Accidents at:	
	NTS in the Past	Yucca Mountain Repository in the Future
General concerns about potential health harms from the repository	tau b = .30 p = .000	tau b = .41 p = .000
Seriousness of possible risks at the site	tau c = .33 p = .000	tau c = .42 p = .000
Seriousness of possible risks in transit	tau c = .35 p = .000	tau c = .38 p = .000
Evaluation of perceived repository benefits to harms	tau c = .32 p = .000	tau c = .40 p = .000
Decision to build the repository	tau c = .30 p = .000	tau c = .36 p = .000

*Negative signs of tau for inverse relationships complicated the table unnecessarily. To simplify the presentation, variables to inverse relationships were recoded so that all taus became positive and were consistent with predicted direction of the relationships.

Sociodemographic Characteristics and Repository Attitudes

Research focusing on the effect of various sociodemographic characteristics on environmental concerns, nuclear attitudes, and risk perceptions is quite mixed with regard to its significance and direction (cf., Covello, 1983; Sonderstrom et al., 1987). Despite this lack of consistency in findings, there are several sociodemographic factors which are routinely employed. These factors include: size of family, age of respondent, race or ethnicity of the respondent, socioeconomic status, educational attainment, and gender (also see chapter 10 by Krannich, Little, and Cramer in this volume; Van Liere and Dunlap, 1980; Solomon, Tomaskovic-Devey, and Risman, 1989).

Based on these studies, a number of sociodemographic variables were included in this analysis to determine what impact these variables might have on repository-related attitudes. When these variables were examined to determine their relationships with the five repository attitude measures, only two were found to be significant—gender and race. Relationships for size of household, number of children in household, age, occupation, educational attainment, marital status, and religious preference were not significant.

Table 9-7   Percentage of Respondents Reporting Extreme Negative Attitudes toward the Repository by Gender and Race

Response	Gender		Race	
	Female	Male	White	Nonwhite
Those reporting they are "very concerned" about potential health effects of the repository (%)	47	36	38	62
	(tau $c = -.15$; $p = .001$)		(tau $c = .12$; $p = .001$)	
Those reporting that on-site risks are "very serious"[1] (%)	41	31	35	46
	(tau $c = -.15$; $p = .002$)		(tau $c = .09$; $p = .007$)	
Those reporting that in-transit risks are "very serious"[1] (%)	59	46	50	67
	(tau $c = -.16$; $p = .001$)		(tau $c = .11$; $p = .002$)	
Those reporting that repository "benefits" will outweigh "harms" (%)	13	21	19	7
	(tau $c = .08$; $p = .05$)		(tau $c = -.02$; $p =$ n.s.)	
Those reporting they would "definitely not" build the repository (%)	59	46	50	70
	(tau $c = -.10$; $p = .002$)		(tau $c = .16$; $p = .001$)	

[1]"Very serious" is defined as either a 1 or a 2 on the 7-point scale.

As can be seen in Table 9-7, there are significant gender and race differences in concern about harmful repository effects, assessment of the ratio of costs to benefits from the repository, seriousness of perceived risks associated with the repository, and the decision to build the facility. In the case of gender, females, compared to males, were significantly more concerned about adverse health effects (47 vs. 36 percent), perceived more serious risks would exist both on-site (41 vs. 31 percent) as well as in transit (59 vs. 46 percent), felt that benefits were less likely to outweigh harms (13 vs. 21 percent), and were far more likely to report they would definitely not build the repository (59 vs. 46 percent). In the case of race, nonwhites were significantly more concerned about adverse health effects (62 vs. 38 percent), perceived more serious risks would occur on-site (46 vs. 35 percent) as well as in transit to the site (67 vs. 50 percent), felt that benefits were less likely to outweigh harms from the repository (7 vs. 19 percent), and were far more likely to report they would definitely not build the repository than were whites (70 vs. 50 percent). In short, while most sociodemographic factors do not affect attitudes toward the repository, race and gender do have significant effects.

The question which remains to be answered in the last section of this chapter is the relative importance of these and the previously discussed factors in explaining repository-related attitudes.

Multivariate Analysis of Repository Attitudes

In order to examine further Las Vegas residents' attitudes toward the proposed high-level nuclear waste repository, two multivariate analyses were conducted: one to examine residents' concern about possible harms to the Las Vegas area if the repository were to be built; the other to examine their personal opinion about whether the facility should be built or not.

In both multiple regressions, variables that were found to be significantly related to repository attitudes at the bivariate level were considered for inclusion. In order to make the models more parsimonious and to reduce any problems with multicollinearity, multi-item indicators were constructed, when appropriate, from the individual variables presented in the discussions above.

Table 9-8 presents a summary of the multi-item indicators and the single variables from which they were constructed. Interitem correlation matrices were constructed in order to determine that the items in each proposed com-

Table 9-8   Descriptions of Multi-item Indicators

New Indicator	Individual Items	Cronbach's alpha	Range of Interitem Correlations (r)
Repository attitudes:			
Repository risk index	—Seriousness of risk on site	.85	.55–.74
	—Seriousness of risk in transit		$(p = .000)$[1]
	—Concern about health effects		
NTS attitudes:			
NTS harms index	—Harms from above-ground testing	.67	.45–.63
	—Harms from below-ground testing		$(p = .000)$[1]
	—Likelihood of personal health harms		
Government trust:			
General trust index	—Trust federal level to do what is right	.68	.34–.49
	—Trust state to do what is right		$(p = .000)$[1]
	—Trust local level to do what is right		
Regulatory trust index	—Trust DOE to protect public safety	.78	.38–.61
	—Trust NRC to protect public safety		$(p = .000)$[1]
	—Trust NWPO to protect public safety		
Trust to report accidents involving radioactive materials	—To report NTS accidents	.81	.69
	—To report repository accidents		$(p = .000)$

[1]Probability given is for the lowest value of Pearson's r in the interitem correlation matrix.

posite variable were significantly and strongly related to one another and that their effects were in the same direction. The table presents the minimum and maximum values of Pearson's r, as well as the significance level of the minimum value for each interitem correlation. The correlations appear to be satisfactory, allowing for the aggregation of suggested items.[5] Cronbach's alpha was computed on the indices in order to determine whether the new measure was sufficiently reliable; all appear to be satisfactory (Cronbach, 1951).

There is only one difference between the two regression models. In the second regression, with opinion on building the repository as the dependent variable, the extent of repository-related risks and the assessment of repository-generated benefits to harms were included as explanatory factors. We hypothesized that repository-related concerns about harms to one's community directly affect opinion about whether the facility should be constructed or not; that is, a high level of concern about the repository and a belief that harms outweigh benefits relate to a negative decision about the desirability of constructing the facility.

When the nine variables representing the four constructs are entered all together in the first regression model to predict the repository risk index, a significant amount of variance (46 percent) in respondents' concern about the repository is explained (Table 9-9). Only five of these variables, however, are significant predictors in this explanation. Three of the variables were found to contribute modestly, but significantly, in this analysis. Females were found to be more concerned about repository risks than were males; those with lower general trust in government "to do what is right" were more concerned about repository risks, as were those who lacked trust in the specific governmental agencies responsible for overseeing nuclear facilities.

The two strongest explanatory variables in the model were related to experience with NTS, the NTS harms index, and trust in government to report accidents involving radioactive materials. Those who believed that NTS's activities, both below and above ground, caused health harms for the Las Vegas area in general and for themselves in particular were more likely to be concerned about future repository risks. Similarly, there was concern among those who did not believe that the government was truthful in reporting accidents at NTS or would be truthful in the future with regard to repository-related accidents.

The multiple regression analysis of opinion regarding the construction of the repository—using the same nine variables plus the two repository attitudinal variables—explains 36 percent of the variance in the dependent variable (Table 9-10). Two interesting observations emerge from this analysis. First, as one might predict, the repository risk variables—which were

Table 9-9 Regression of Repository Risk Index on Explanatory Variables

Item	b	beta
Demographics:		
Gender	−.67	−.07*
Race	−1.09	−.08
Past Experience:		
At federal facilities near Las Vegas	.02	.00
With radioactive materials	−.34	−.04
NTS Attitudes:		
NTS harms index	.39	.38**
NTS harms-to-benefits	−.71	−.06
Government Trust:		
General trust index	.28	.10*
Regulatory trust index	−.08	−.09*
Trust to report accidents involving		
radioactive materials	−.99	−.34**
(Constant)	14.01	
$F = 38.97$**		
$R = .68$		
Adj. $R^2 = .46$		

*$p \leq .05$
**$p \leq .001$

newly added in this model—account for almost all of the variance in the model. The more seriously one believes that on-site and in-transit risks will accompany the repository and the greater the amount of harm with respect to benefits one believes will be generated by the repository, the more likely one opposes the construction of the facility. The only other significant predictor in the model is past employment experience with federal facilities near Las Vegas. Such experience increases support for building the facility. Second, the amount of variance explained in the second model—opinion about construction of the repository—is less than that explained by the first model—the perceived risk of the repository—even though the powerful risk perception variables have been used to predict overall attitude toward the repository. This finding seems to indicate that there are other sources of opinion about construction of the repository that have yet to be explored.

Summary

The analysis of the variables underlying the perceived risks of a high-level nuclear waste repository at Yucca Mountain and the respondents' personal

Table 9-10   Regression of Opinion to Build the Repository on Explanatory Variables

Item	b	beta
Demographics:		
Gender	−.08	−.03
Race	−.17	−.05
Past Experience:		
At federal facilities near Las Vegas	−.25	−.13 **
With radioactive materials	.12	.06
NTS Attitudes:		
NTS harms index	−.00	−.00
NTS harms-to-benefits	−.06	−.02
Government Trust:		
General trust index	−.04	−.05
Regulatory trust index	.02	.09
Trust to report accidents involving radioactive materials	−.03	−.04
Repository Risk Attitudes:		
Repository risk index	.09	.30 ***
Harms-to-benefits	−.64	−.26 ***
(Constant)	3.78	
$F = 15.70$ ***		
$R = .62$		
Adj. $R^2 = .36$		

$^*p \leq .05$
$^{**}p \leq .01$
$^{***}p \leq .001$

opinions toward construction of the repository has yielded several interesting findings. The bivariate analyses showed that two sets of variables were strongly and significantly related to the dependent variables. First, respondents' beliefs concerning whether above-ground and below-ground nuclear tests at NTS had health consequences for Las Vegas residents and whether they believed they had personally experienced health harms from these activities are both strongly associated with the dependent variables. The second set of variables that stands out in the bivariate analyses is trust in government. In general, the greater trust in government, the stronger the support for the repository. Unfortunately for supporters of the repository, there is little trust in the agencies that are making decisions concerning construction and later operation of the facility.

The demographic factors which appear significant in the bivariate analyses are gender and race, with nonwhites and women being more concerned about health harms, more likely to perceive serious risks for repository operations, more likely to believe that harms would outweigh benefits, and more likely to oppose building the repository. The final factor, personal experience with comparable hazards, is surprisingly, given previous findings, not a strong predictor of position on the dependent variable. Only one variable is significant from this group—the index of work experience in four federal facilities.

The regression analysis examining the relative importance of these four sets of variables on the first independent variable—concern over repository-related risks and harms—shows that five variables have significant effects. These five variables, which explain 46 percent of the variance, are: gender; the composite index of NTS harms; the general trust in government index; the regulatory trust index for federal agencies; and the composite index measuring trust in government to report accidents involving radioactive materials.

The second regression analysis, with personal opinion on whether to build the repository or not as the dependent variable, explains 36 percent of the variance. The three significant explanatory variables are: work experience in federal facilities, the composite repository risk index, and the belief that harms will outweigh benefits if the repository is built. The fact that attitudes toward NTS, which had a significant effect on perceptions of repository risks, do not have a significant impact on overall opinion of repository construction suggests that the risk shadow may be diluted as one moves from views of repository risks to opinions concerning action options.

The fact that we can predict opinion on repository construction to a lesser degree than we can predict perceived risks associated with the proposed repository, even when risk perceptions are included as predictor variables, is surprising. This suggests that people's opinions on the repository are the result of a complex set of factors—several of which we have apparently not investigated—and not simply a function of their risk perceptions. This indicates the need for additional research aimed at developing a deeper understanding of citizen attitudes on nuclear waste repository siting.

Implications

The implications of this research and some additional analysis recently completed (Mushkatel, Pijawka, and Dantico, 1990) point to a major difficulty confronting the DOE (the federal lead agency) in the siting of a high-level nuclear waste repository. While the federal government may possess the necessary resources and will to site a repository at Yucca Mountain, the level of opposition among Nevada residents to the siting is strong, as our data show, and apparently growing. A 1989 survey of Nevada residents found similar risk perceptions to those in earlier studies but stronger opposition to the proposed repository than reported here (Flynn et al., 1990). In the context of DOE efforts to educate the public about the repository during this period, the risk perceptions of Nevada urban residents and the concomitant opposition to the siting strongly suggest continued political conflict between the state and the federal agencies.

One important implication of the analyses reported here should not be overlooked. None of the key variables explaining either opposition to the construction of the repository or risk perceptions and attitudes toward the facility can be altered by DOE, the federal agency charged with the responsibility for siting the repository. Hence, while the results of these analyses indicated that there may be a complex set of factors underlying risk perceptions, repository attitudes and construction opinions, there is little reason to believe that many of these variables can be readily altered to provide a more conducive opinion climate for the proposed Yucca Mountain repository.

A recent National Research Council (1990) report strongly suggested that one reason the DOE siting was in trouble was the increasing lack of credibility of the agency. The findings discussed in this chapter regarding political trust in DOE strongly reinforce the National Research Council's conclusions and suggest the difficult path which may lie ahead for the agency with regard to urban residents' continuing opposition to the project.

Notes

Research for this chapter was supported by federal funds granted by the U.S. Department of Energy to the Nevada Nuclear Waste Project Office pursuant to the provisions of Public Law 97–425. The views expressed in this chapter are those of the authors and do not necessarily represent the views of the Nevada Nuclear Waste Project Office, the U.S. Department of Energy, or any other individuals or entities associated with the research and studies supported by this research effort.

1   Contacts which identified a "nonsample" telephone number (i.e., one that was nonresidential or not within the geographic boundaries) were replaced immediately. Contacts that did not initially result in an "in-sample" designation were not replaced until after at least ten callbacks had been attempted at different times, on different days, over a four-week period.

2   NTS, like the proposed repository, is operated and overseen by an agency of the federal government. During its history, "downwinders" have claimed to have experienced significant negative health effects from the above-ground nuclear weapons testing at NTS. Even today, with all nuclear testing below-ground, NTS still announces its tests to the population in southern Nevada, since groundshaking due to the detonation of nuclear materials is felt throughout the Las Vegas area, reminding the residents of NTS's continuing operation.

3   For each of these three questions, respondents were asked to provide an answer on a

seven-point scale. A 1 meant that it was "not likely at all," and a 7 that it was "extremely likely."

4   The special location of the "places" are: Yucca Mountain for the NTS; near the mountain to the west for the Tonopah Test Range; bordering the NTS to the north and east for Nellis Air Force Base; and 43 miles northeast of Las Vegas for the small town of Indian Springs. A map of Nevada that highlights these locations is presented in chapter 10 of this volume by Krannich, Little, and Cramer.

5   In two instances, however, suggested items were rejected from inclusion in a multi-item indicator—one in the NTS harms index, and the other in the repository risk index. In each case, the ratio of harms to benefits question was initially considered for inclusion; however, the interitem correlations for these items were low, and their inclusion did not increase the alphas for the composites. However, the relative importance of these items in the bivariate analysis was sufficient argument to include them as single items in the regression analyses.

References

Burton, Ian, Robert W. Kates, and Gilbert F. White. 1978. *The Environment as Hazard.* New York: Oxford University Press.

Campbell, Angus, Phillip Converse, Warren Miller, and Donald Stokes. 1960. *The American Voter.* New York: John Wiley & Sons.

Covello, Vincent T. 1983. "The Perception of Technological Risks: A Literature Review." *Technological Forecasting and Social Change* 23:285–97.

Cronbach, Lee J. 1951, "Coefficient Alpha and the Internal Structure of Tests." *Psychometrika* 16:297–334.

Cutter, Susan. 1984. "Risk Cognition and the Public: The Case of Three Mile Island." *Environmental Management* 8 (1): 15–20.

Flynn, Cynthia. 1979. *Three Mile Island Telephone Survey: Preliminary Report on Procedures and Findings.* Report presented to the U.S. Nuclear Regulatory Commission. Seattle: Social Impact Research.

Flynn, James H., C. K. Mertz, and James Toma. 1989. *Final Report on the Nevada Telephone Survey.* Carson City, NV: Nevada Nuclear Waste Project Office.

———. 1990. "Evaluation of Yucca Mountain Survey Findings About the Attitudes, Opinions, and Evaluations of Nuclear Waste Disposal at Yucca Mountain, Nevada." Eugene: Decision Research.

Kasperson, Roger E., Jerald Berk, K. David Pijawka, Alan B. Sharaf, and James Wood. 1980. "Public Opposition to Nuclear Energy: Retrospect and Prospect." *Science, Technology, and Human Values* 5 (31): 11–23.

Kates, Robert W. 1962. "Hazard and Choice Perception in Flood Plain Management." University of Chicago Department of Geography Research Paper No. 78. Chicago.

———. 1967. "The Perception of Hazard on the Shores of Megalopolis." In David Lowenthal, ed., *Environmental Perception and Behavior.* Chicago: University of Chicago Press. 60–74.

Kish, Leslie. 1949. "A Procedure for Objective Respondent Selection Within the Household." *Journal of the American Statistical Association* 380–87.

Lindell, Michael K., and Timothy C. Earle. 1982. "Public Perception of Industrial Risks: A Free Response Approach." Seattle: Battelle Human Affairs Research Centers.

Mitchell, James K. 1974. "Natural Hazards Research." In Ian R. Manners and Marvin W. Mikesell, eds., *Perspectives on Environment*. Washington, D.C.: Association of American Geographers. 311–41.

Mushkatel, Alvin H., K. David Pijawka, and Marilyn Dantico. 1990. *The Effects of the Proposed Nuclear Waste Repository on the Las Vegas Metropolitan Area*. Carson City, NV: Agency for Nuclear Projects.

National Research Council. 1990. *Rethinking High-Level Radioactive Waste Disposal: A Position Statement of the Board on Radioactive Waste Management*. Washington, D.C.: National Academy Press.

Pijawka, K. David. 1984. "The Pattern of Public Response to Nuclear Facilities: An Analysis of the Diablo Canyon Nuclear Generating Station." In Martin J. Pasqualetti and K. David Pijawka, eds., *Nuclear Power: Assessing and Managing Hazardous Technology*. Boulder: Westview.

Schiff, Myra R. 1977. "Hazard Adjustment, Locus of Control, and Sensation-Seeking: Some Null Findings." *Environment and Behavior* 9:233–43.

Slovic, Paul, Baruch Fischhoff, and Sarah Lichtenstein. 1980. "Facts and Fears: Understanding Perceived Risk." In Richard Schwing and Walter A. Alberts, eds., *Societal Risk Assessment*. New York: Plenum.

Solomon, Lawrence S., Donald Tomaskovic-Devey, and Barbara J. Risman. 1989. "The Gender Gap and Nuclear Power: Attitudes in a Politicized Environment." *Sex Roles* 21:401–14.

Sonderstrom, Jonathan E., John H. Sorenson, Emily D. Copenhaver, and Sam A. Carnes. 1984. "Risk Perception in an Interest Group Context: An Examination of the TMI Restart Issue." *Risk Analysis* 4 (3): 231–44.

———. 1987. *Impacts of Hazardous Technology: The Psycho-Social Effects of Restarting TMI-1*. Albany: State University of New York Press.

Stoffel, Richard W., Michael W. Traugott, Camilla L. Harshbarger, Florence V. Jensen, Michael J. Evans, Paula Drury. 1988. *Perceptions of Risk from Radioactivity: The Superconducting Super Collider in Michigan*. Ann Arbor: Institute for Social Research, University of Michigan.

Van Liere, Kent D., and Riley E. Dunlap. 1980. "Environmental Concern: Does It Make a Difference How It's Measured?" *Environment and Behavior* 13 (6): 651–76.

10   Rural Community Residents' Views

of Nuclear Waste Repository Siting in Nevada

Richard S. Krannich, Ronald L. Little, and Lori A. Cramer

Introduction

The federal government's proposal to site the nation's first high-level nuclear waste repository at Yucca Mountain, Nevada, has created far-reaching political turmoil. Despite its remote and sparsely populated location, Nevada's congressional representatives, as well as state-level political leaders, have been virtually unanimous in their opposition to the project. Public opinion has also revealed widespread dissatisfaction with the proposal. Statewide opinion surveys have consistently indicated that nearly three-quarters of Nevada residents are opposed to having the repository built in Nevada (see chapter 3 of this volume by Slovic, Layman, and Flynn). Opposition to the repository appears to be linked to a variety of factors, including beliefs that it is unfair to force Nevadans to harbor all of the nation's nuclear waste, in addition to concerns about the safety of the storage and transportation of high-level nuclear waste (see chapter 7 of this volume by Desvousges, Kunreuther, Slovic, and Rosa).

Statewide responses to the proposed repository, reflected by both political opposition and public opinion polls, mirror the attitudes and opinions of Nevada's predominantly urban population, which is concentrated in or near Las Vegas and, to a lesser degree, Reno. As a result, such polls provide limited insight into the views of those rural Nevada residents who may be most directly affected by repository construction. While Yucca Mountain is located on federally controlled land about ninety miles northwest of the metropolitan Las Vegas area and is 376 miles southeast of Reno, there are several small rural communities which are relatively close to the proposed repository site. Several others are within reasonable commuting distance of Yucca Mountain. In addition, numerous rural communities in southern Nevada are located along probable waste transportation corridors.

To fully comprehend the consequences of constructing a high-level nuclear waste repository at Yucca Mountain for area residents, it is necessary to focus attention on the small rural communities which are proximate to the project site or major waste transportation corridors. These are the towns and villages which are especially likely to be at risk, that is, communities most likely to suffer any negative environmental and socioeconomic impacts stemming from the repository. The history of western energy developments over the past several decades suggests that, as with other large-scale facilities, repository construction and operation is likely to cause social, economic, and cultural disruptions (Cortese and Jones, 1977; Little, 1977; Davenport and Davenport, 1980; Elkind-Savatsky, 1986).

Typically, rapid population and economic growth associated with large-scale construction projects stimulates a wide array of social changes in nearby rural communities. These communities are vulnerable, primarily because of their small sizes, proximity to the site, and the limited ability of their infrastructures to handle potential inmigration (see Weber and Howell, 1982). Although previous literature generally suggests that rural residents look favorably on efforts that promise local employment and economic benefits, and therefore support most types of large-scale industrial or construction projects (see Little and Lovejoy, 1979), the potential for adverse socioeconomic impacts may result in less positive public views of projects such as the proposed repository.

The special nature of projects involving nuclear materials may also result in less positive responses by community residents (Albrecht, 1983; Williams, 1988). Rural communities, in particular, currently lack the trained emergency response personnel required in the event of a waste-handling accident. Consequently, rural residents living along potential waste transportation corridors may have a heightened perception of the risks associated with a repository. Finally, opposition to the siting of hazardous facilities, and especially nuclear facilities, often reflects a NIMBY ("not in my back yard") response on the part of area residents (Edelstein, 1988; Stoffel et al., 1989; Williams, 1988; Williams and Payne, 1985). For these and other reasons, the views of rural residents whose "back yards" are closest to the proposed repository site provide a particularly important focus for studies attempting to understand human responses to facilities which handle nuclear and other potentially hazardous materials.

The area of southern Nevada encompassing Yucca Mountain provides a unique social and cultural context for siting a high-level nuclear waste facility. Communities in this area have had forty years of experience with nuclear weapons testing programs, as well as with a variety of large-scale military operations and facilities in desert areas north and west of Las Vegas.

Nevada politicians and residents have generally been highly supportive of federal activities ever since the first atomic weapons test occurred at the Nevada Test Site (NTS) in 1951 (Titus, 1986). Despite increased recognition of harmful health effects experienced by military personnel and civilians exposed to radiation during the era of atmospheric testing and despite increasingly strident antinuclear protest activities, there remains widespread recognition of the importance of NTS and other defense programs to the economic vitality of southern Nevada and the state as a whole.

This unique social context, the NIMBY syndrome, the special nature of nuclear projects, and continuing effects of other federal projects and programs all suggest that rural residents' views about a nuclear waste repository may be quite different from those reflected by statewide surveys, such as the one reported in chapter 7 of this volume by Desvousges, Kunreuther, Slovic, and Rosa, or by surveys of urban residents, such as the Las Vegas survey reported in the preceding chapter by Mushkatel, Nigg, and Pijawka. Further, the views of rural southern Nevadans may differ considerably from what might be anticipated on the basis of studies focused on large-scale nonnuclear projects in other rural settings. This chapter addresses this issue by examining perceptions and attitudes toward the repository among residents of six rural communities in southern Nevada. Each of these communities is located relatively close to Yucca Mountain or probable waste transportation routes, and each could experience a variety of impacts from repository construction, operation, and waste transportation. Information on perceptions and attitudes held by residents of these communities concerning the repository provides an important supplement to information on the views held by urban and statewide residents reported in other chapters.

Study Area

The study communities include the towns of Beatty, Amargosa Valley, and Pahrump in Nye County, Indian Springs and Mesquite in Clark County, and Caliente in Lincoln County (see Figure 10-1). A brief description of these communities follows.

The unincorporated town of Beatty is located approximately 115 miles northwest of Las Vegas. The town is bisected by U.S. Highway 95, a potential waste transportation route, and is just eighteen air and forty-five highway miles from Yucca Mountain. Historically, the town has experienced boom-and-bust cycles associated with gold and silver mining. In the past decade or so Beatty has experienced modest growth as a result of employment opportunities linked to the Nevada Test Site (NTS) and other military programs. The town's other economic activities include a low-level nuclear

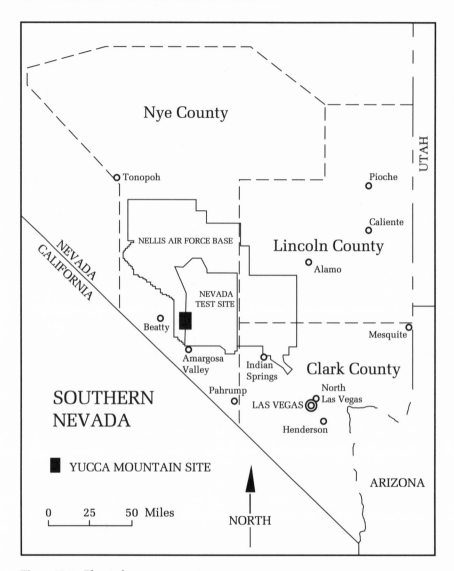

Figure 10-1 The study area.

waste landfill, which is located just a few miles south of town, and service industries oriented toward travelers and tourists visiting nearby Death Valley National Monument. By 1985 the estimated population of the town was approximately 925. However, rapid growth associated with new gold mining ventures began to occur in late 1987, causing the population to increase to approximately 1100 by early 1988.

Amargosa Valley, another unincorporated town, has approximately 600 to 650 residents spread over a 500-square-mile desert area. The northern-most sections of the town are adjacent to U.S. Route 95, and are located within sixteen miles of the proposed Yucca Mountain repository site. Thus Amargosa Valley, like Beatty, could be directly affected by the repository due both to its proximity to the site and its location along a probable waste transportation route. The settlement of Amargosa Valley occurred primarily during the 1950s as a result of homesteading stimulated by the Desert Entry Act. Additional growth occurred in the 1960s and 1970s, first as a result of NTS activity and subsequently due to the development of colemanite and bentonite mining and milling operations. However, NTS employment of local residents declined substantially by the mid-1980s, and the 1986 closure of a large milling operation resulted in the loss of nearly 50 percent of the resident population within a two-year period. The local economy is severely depressed, as evidenced by the closing of several businesses, decreased activity at others, and abandonment of homes and other property.

Located in the southernmost tip of Nye County, approximately fifty-five air miles and sixty-five highway miles from the proposed Yucca Mountain site, lies the unincorporated town of Pahrump. Although not located on a major highway likely to be used for nuclear waste transportation, Pahrump is adjacent to one of the alternative railroad spur routes under consideration for transporting wastes to the repository site. Originally a sparsely popu-lated agricultural valley, in 1970 Pahrump had fewer than 1000 residents scattered over more than 200 square miles. However, rapid growth has oc-curred during the 1970s and especially the 1980s as a result of real estate speculation and housing construction. Agricultural land was subdivided for residential and commercial development, and by early 1988 the town's population had grown to approximately 6500 to 7000. Pahrump is now the largest community in Nye County.

Located forty-three miles northwest of Las Vegas in Clark County, Indian Springs is sixty-one miles southeast of the proposed Yucca Mountain site and, like Beatty, is bisected by U.S. Route 95. Impacts from the repository could result both from the town's proximity to the site and the potential for waste transportation through the community. Prior to World War II there were few people living in Indian Springs. With the war came construction of the Indian Springs Air Force Base and associated support facilities, which caused the community to grow rapidly. Growth continued through the 1950s as a result of weapons testing activities at NTS, the main entrance of which is just 20 miles to the northwest. In late 1987 there was a reduction in Indian Springs's population when responsibility for provision of air base services was shifted from the Air Force to a private contractor whose employees tended to live in Las Vegas. At the same time, the military support services provided by Indian Springs Air Force Base were transferred to other military installations, adding to the loss of residents. However, Indian Springs has recovered from these population losses, largely as a result of NTS activities and in 1988 had approximately 1200 residents.

The city of Mesquite is located in the easternmost section of Clark County near the Utah and Arizona borders, approximately eighty miles northeast of Las Vegas. Although the town's distance from Yucca Mountain (142 air and 184 highway miles) reduces the likelihood of direct impacts associated with construction or on-site operations, Mesquite is bisected by Interstate Route 15, a likely waste transportation route. Mesquite was one of the "downwind"[1] communities most severely affected by radioactive fallout from atmospheric weapons testing during the 1950s and early 1960s (see Fuller, 1984). Consequently, the community's experience with southern Nevada's nuclear heritage has been much different from that of places such as Beatty, Amargosa Valley, Pahrump, and Indian Springs, which have benefitted from NTS employment opportunities and are located west and south of the test site. Originally settled as a Mormon agricultural village, Mesquite remained a quiet, slowly growing, predominantly Mormon rural town until the 1980s, when unprecedented population growth was stimulated by the construction of a large resort and casino complex. The population grew from 922 in 1980 to an estimated 1500 in early 1988. This growth contributed to the establishment of Mesquite as an incorporated city in 1984.

The incorporated city of Caliente is the largest community in Lincoln County and is located approximately 110 air miles and over 250 road miles east of the proposed Yucca Mountain site. The community is bisected by the main line of the Union Pacific railroad, which is likely to be a major route for shipping high-level nuclear waste to Yucca Mountain. Like Mesquite, Caliente was among the "downwind" communities and has not experienced major employment benefits from NTS. Originally developed as a railroad town, Caliente experienced severe economic decline beginning in the late 1940s and early 1950s as a result of the railroad's transition from steam to diesel technology (Cottrell, 1951). Continued deterioration of railroad employment during the past three decades has combined with declining agricultural and mining activities to create a context of persistent economic and demographic stagnation. By early 1988 the community's population was estimated to be 996, just 14 persons more than reported in the 1980 census.

Table 10-1 Sample Size and Response Rates for Rural Nevada Community Surveys

Community	Number Delivered	Number Completed	Response Rate (%)
Beatty	150	111	74.0
Amargosa Valley	123	104	84.6
Pahrump	220	189	85.9
Indian Springs	152	122	80.3
Mesquite	152	110	72.3
Caliente	152	131	86.2
Combined Communities	949	767	80.8

Study Approach

Data Collection

The data for this study have been drawn from surveys administered to representative samples of the adult populations in each of the six study communities. In each community comprehensive sampling frames were developed using water or electric utility records.[2] Following two separate pretests of preliminary versions of the survey instrument, self-completion questionnaires were distributed in March, April, and May, 1988 to randomly selected households in each community. Field workers personally delivered and retrieved the completed survey instruments. Previous research has shown that this technique elicits relatively high response rates (Krannich, Greider, and Little, 1985).

Within each randomly selected household, an individual respondent was selected by identifying the person eighteen years of age or older whose birthday had occurred most recently. This method results in a randomized selection of adult household members, without the complexities or intrusiveness of more traditional respondent selection methods such as those developed by Kish (1949).

Information regarding sample size and the numbers of respondents for each of the study sites is summarized in Table 10-1. Overall, the rates of return for usable questionnaires were excellent, ranging from a low of 72.3 percent in Mesquite to over 86 percent in Caliente. The combined response rate for all six study sites was over 80 percent.

Variables

The questionnaire included a broad array of questions pertaining to respondents' perceptions of the nuclear waste repository program as well as other nuclear and technological programs, perceptions of community characteristics, trust in science and government, personal background characteristics, and a variety of other social, psychological, and cultural dimensions. The analysis presented here focuses on only selected variables that help to identify and clarify the nature of rural residents' responses to several issues surrounding the siting of the repository at Yucca Mountain.

Two measures of respondents' perceptions of and attitudes regarding the repository program are analyzed as dependent variables. The first is an indicator of the degree to which people expressed concern that the repository program might have harmful health and safety effects on community residents.[3] The response scale for this question ranged from 0 to 10, with the extreme values labelled as "not at all concerned" (0) to "extremely concerned" (10). The second measure of attitudes toward the repository was a question asking respondents whether they would build the repository at Yucca Mountain if the decision were theirs.[4] Responses were measured on a five-point scale which ranged between "definitely yes" and "definitely no." Although these two variables were highly correlated ($r = .73$), they appear to address two important and distinct dimensions of respondents' views concerning the repository and its possible effects.

In order to account for the variation in respondents' views of the proposed Yucca Mountain repository, several questions were treated as independent variables in this analysis. First, respondents' community of residence was considered, since preliminary analyses have indicated very substantial differences in attitudes regarding the repository across the study communities (Krannich and Little, 1988).

Another independent variable included in the analysis was a measure of the degree of local economic harm or benefit which respondents anticipated could result from repository development.[5] This variable, measured on a scale ranging from 0 ("entirely harmful effects") to 10 ("entirely beneficial effects"), provides a means of determining the degree to which anticipated economic opportunities may attenuate risk perceptions or repository opposition.

Since individuals' attitudes about other analogous activities may influence perceptions concerning proposed facilities (see Stoffel et al., 1988, 1989), an index measuring perceptions of health and safety risks of nuclear weapons testing was developed by summing responses to two questions. The first addressed the perceived likelihood of past harmful effects from atmospheric testing,[6] and the second involved the perceived likelihood of future harmful effects from current underground testing at the NTS.[7] Both questions were measured on the same scale of 0 to 10, with responses ranging between "not at all likely" (0) and "extremely likely" (10); the correlation between the two items was very high ($r = .76$). The resulting summed index has a potential response range from 0 to 20.

In addition, a measure of trust in science was included as an independent variable, because attitudes about science and technology are likely to influence views about specific technological projects such as a nuclear waste repository. This variable was measured as a summed index comprising responses to five related questionnaire items.[8] Interitem correlations for these questions ranged between .20 and .61, and item-to-corrected total correlations ranged between .38 and .63. The internal consistency of the index is substantial, as reflected by a value of .72 for Cronbach's alpha coefficient of reliability (Cronbach, 1951).

Trust or distrust of organizations viewed as being responsible for management of hazardous projects appears to play a key role in attitudes and opinions about such projects (Edelstein, 1988; Stoffel et al., 1989). Therefore, a question addressing respondents' trust in the federal government to provide accurate information on nuclear programs was also used.[9] Responses to this question were recorded on a 0-to-10 scale ranging from "not at all confident" (0) to "extremely confident" (10).

In addition, several questions pertaining to respondents' sociodemographic characteristics were incorporated as control variables, since prior research has indicated potentially important relationships between sociodemographic characteristics and attitudes toward nuclear facilities (Nealey, Melber, and Rankin, 1983; Freudenburg and Rosa, 1984). Such research suggests that women tend to express greater concern than men about the risks of nuclear facilities as well as the risks of other hazardous events and toxic episodes (Harris and Associates, 1976; Hamilton, 1985; Mitchell, 1984; Mushkatel and Pijawka, 1989; Nealey, 1990; Nealey, Melber, and Rankin, 1983) and are generally more likely to express concern about the environment (McStay and Dunlap, 1983; Van Liere and Dunlap, 1980). Therefore, respondent gender was included in the analysis. Previous research has also suggested that being a parent affects attitudes and perceptions about potential hazards (Hamilton, 1985); thus the number of living children reported by respondents was included. Years of residence in the present community was included in order to reflect differences in the degree to which residents have shared experiences with nuclear weapons testing. That is, the longer respondents have resided in one of these communities, the more they are likely to be sensitive to ways in which the community may have been affected, either positively or negatively, by NTS programs, especially early atmospheric testing activities. Finally, respondent age was included in the analysis, since age has previously been demonstrated to influence general environmental attitudes (Van Liere and Dunlap, 1980).[10]

Results

Response distributions for both of the dependent variables vary substantially across the six study communities. As shown in Table 10-2, levels of concern about harmful health and safety effects stemming from a repository are lowest in Amargosa Valley and Beatty, the two communities located nearest to Yucca Mountain. In Beatty over 66 percent of responses were on the "not concerned" end of the scale (values 0 through 4), while slightly more than 60 percent of Amargosa Valley responses were in this same range. Levels of concern were higher in Indian Springs and Pahrump, both of which are located somewhat farther from the proposed project site. In Indian Springs approximately 53 percent of responses were below the scale midpoint, while only about 36 percent of Pahrump respondents indicated a similar lack of health and safety concern. Concern was highest in Caliente and Mesquite, the two communities which are furthest from Yucca Mountain. Only 24 percent of Mesquite responses and 28 percent of Caliente responses were in the range of scores falling below the scale midpoint that represent low concern about harmful effects.

A similar pattern emerges when the distribution of responses to the

Table 10-2 Distribution of Perceived Repository Health and Safety Risks for Six Study Communities (percent)

Response	Beatty	Amargosa Valley	Pahrump	Indian Springs	Mesquite	Caliente
"Not at all concerned"						
0	25.7[a]	26.2	14.9	18.8	7.5	7.4
1	15.2	14.6	8.6	11.1	2.8	5.7
2	12.4	11.7	6.3	11.1	3.8	2.5
3	10.5	4.9	2.9	6.8	6.6	7.4
4	2.9	2.9	2.9	5.1	3.8	4.9
5	14.3	11.7	14.9	5.1	8.5	10.7
6	1.9	5.8	2.3	4.3	4.7	5.7
7	0.0	2.9	5.1	6.0	12.3	6.6
8	4.8	5.8	9.7	6.8	13.2	8.2
9	1.0	1.9	7.4	3.4	5.7	4.9
10	11.4	11.7	25.1	21.4	31.1	36.1
"Extremely concerned"						
N	105	103	175	117	106	122
Mean	3.3	3.6	5.6	4.7	6.7	6.6

[a]Percentages may not total to 100 due to rounding.

$F = 18.91$, d.f. $= 5, 722$, $p < .0001$

Table 10-3   Distribution of Support/Opposition for Construction of Repository (percent)

Response	Beatty	Amargosa Valley	Pahrump	Indian Springs	Mesquite	Caliente
Definitely yes (1)	45.3[a]	47.1	20.0	25.9	7.8	12.8
Probably yes (2)	28.3	28.4	23.9	28.4	16.7	20.8
Uncertain (3)	13.2	10.8	21.1	18.1	24.5	24.8
Probably no (4)	6.6	4.9	7.2	9.5	16.7	11.2
Definitely no (5)	6.6	8.8	27.8	18.1	34.3	30.4
Combined percent "yes"	73.6	75.5	43.9	54.3	24.5	33.6
Combined percent "no"	13.2	13.7	35.0	27.6	51.0	41.6
N	106	102	180	116	102	125
Mean	2.0	2.0	3.0	2.7	3.5	3.3

[a]Percentages may not total to 100 due to rounding error.

$F = 23.27$, d.f. $= 5, 725$, $p < .0001$

question regarding opposition to and support for a repository is examined. Almost 74 percent of the Beatty respondents would definitely or probably construct the repository at Yucca Mountain if the choice were theirs, while 75.5 percent of Amargosa Valley respondents gave similar responses (see Table 10-3). Analogous to the pattern observed with health and safety issues associated with the proposed repository, Pahrump (43.9 percent) and Indian Springs (54.3 percent) respondents were slightly less supportive, and Mesquite (24.5 percent) and Caliente (33.6 percent) respondents were the least supportive.

These results contrast sharply with findings from the Las Vegas urban area survey reported in the preceding chapter by Mushkatel, Nigg, and Pijawka. Response patterns in all of the rural communities except Mesquite and Caliente reflect much higher levels of repository support than is evident from the Las Vegas data. Even in Mesquite and Caliente, respondents were considerably less likely to express opposition than was the case among urban area residents, tending instead to report higher levels of uncertainty about a repository. These differences may be attributable in part to rural-urban differences in economic development levels and needs and to a related tendency for rural area residents to view a repository as a potentially important source of future economic growth opportunities.

Insofar as all of the study communities face potential impacts from the transportation of nuclear wastes to Yucca Mountain, the obvious differences in repository orientations held by residents of the six study communities must be linked to some other factors. Therefore, an attempt to account for intercommunity variation in the dependent variables requires an examina-

Table 10-4   Means, Standard Deviations, and ANOVA Results Comparing Response Patterns on Measures of Anticipated Economic Effects, Trust in Science, Trust in Government, and Perceptions of NTS Health Effects

Variable	Beatty	Amargosa Valley	Pahrump	Indian Springs	Mesquite	Caliente	F
Effects on economy[a]							
Mean	7.79	8.06	6.91	7.27	4.84	5.91	20.16**
Std. dev.	2.20	2.43	2.91	2.63	2.80	2.88	
Trust in science[b]							
Mean	32.19	31.24	30.92	31.97	29.82	28.83	2.51*
Std. dev.	9.04	7.62	8.80	8.93	7.42	8.51	
Trust in government[c]							
Mean	5.67	5.32	4.03	4.82	3.30	3.22	13.23*
Std. dev.	2.84	3.10	3.22	3.22	2.62	2.94	
Perceived NTS health effects[d]							
Mean	6.85	7.40	10.36	8.37	14.67	13.16	29.38**
Std. dev.	5.73	6.14	6.68	6.38	5.57	5.98	

[a]Values range between 0 (entirely negative effects) and 10 (entirely positive effects)
[b]Values range between 0 (no trust) and 50 (total trust)
[c]Values range between 0 (not at all confident) and 10 (extremely confident)
[d]Values range between 0 (health effects not at all likely) and 20 (health effects extremely likely)
*$p < .05$, **$p < .0001$

tion of additional variables. As noted previously, some of the differences can be attributed to anticipated local employment and other economic benefits, which are more likely to be significant in the communities located nearer to Yucca Mountain. Moreover, the economic decline experienced in Amargosa Valley and the boom-and-bust history of Beatty may help to account for the more supportive orientations observed in those two communities. Residents of Beatty and Amargosa Valley tend to place a high priority on the need for economic growth and a stable employment base (Trend, Little, and Krannich, 1988a, 1988b).

Response distributions to a question concerning anticipated economic effects of the repository lend support to an explanation based on anticipated economic benefits. As reported in Table 10-4, residents of Amargosa Valley and Beatty were on average more likely to expect beneficial economic impacts from the proposed repository than residents of the other four communities. Pahrump and Indian Springs have the next highest mean expectations, while the anticipation of beneficial economic effects is lowest in Caliente and Mesquite. Furthermore, the standard deviations suggest that

Beatty and Amargosa Valley have greater intracommunity consensus on this question than the other study sites.

Table 10-4 also indicates that there are important differences across the study communities for several other variables that may influence repository perceptions. Trust in science, a four-item index which could explain intercommunity differences, yielded mean scores which suggest a moderate amount of trust in science in all six communities. Even though differences in mean responses on this variable were statistically significant (p ≤ .05), the magnitude of intercommunity differences was substantively insignificant. Nevertheless, it should be noted that the downwind communities of Mesquite and Caliente, while generally trusting of science, nonetheless demonstrated the least trust in science of any of the study communities.

Another variable with the potential to explain intercommunity differences was the respondents' trust in the federal government to provide honest information on nuclear program safety. The mean community scores for this variable hovered about the midpoint of the scale (5.0), indicating neither great trust nor great mistrust of the federal government on this issue. The F test revealed statistically significant (p ≤ .0001) differences among the communities. Caliente and Mesquite exhibited the greatest distrust of the federal government, Amargosa Valley and Beatty the greatest trust.

The last attitudinal variable examined to explain community differences was perceptions of the health effects of nuclear testing activities at the NTS. Residents of the downwind communities of Caliente and Mesquite were substantially more likely than residents of other communities to believe that nuclear weapons testing programs result in adverse health effects for area residents. In contrast, responses from Beatty and Amargosa Valley residents reflect very low average concern levels over the consequences of activities at the NTS.

An examination of bivariate correlations between the two dependent variables and the several independent variables reveals a number of potentially important relationships (see Table 10-5). Aggregating the combined responses from the six communities, the relationship between perceived economic effects and concern about health and safety effects was moderate. The correlation (r = −.49) demonstrates that respondents who believe the repository will bring beneficial economic effects have lower levels of concern over health and safety issues. The relationship between the measure of anticipated economic effects and the measure of support/opposition was even stronger (r = −.60). Thus, the greater the anticipation of economic benefits, the less the opposition to repository construction. Statistically and substantively significant correlations were also observed between each of

Table 10-5   Zero-Order Correlations Between Dependent and Independent Variables
for Combined Communities

Independent Variables	Dependent Variables	
	Health/Safety Concern	Repository Support/Opposition
Effects on economy	−.49*	−.60*
Perceived NTS effects	.65*	.61*
Trust in science	−.28*	−.27*
Trust in government	−.61*	−.64*
Sex	.10*	.10*
Number of children	−.01	.05
Age	−.03	.08
Length of residence	.02	.05

*$p < .01$

the dependent variables and the perception of NTS-related health risks, indicating a tendency for levels of repository concern and opposition to be
highest among rural residents who believe that there are health risks associated with nuclear testing. Levels of concern and opposition also tend
to be higher among those who express low trust in science and low trust
in government. While the correlations for the former are somewhat meager (−.28 and −.27 respectively), the trust in government question explains
approximately 36 percent of the variation for each of the two dependent
variables.

There is also a statistically significant but relatively weak correlation between both dependent variables and respondents' gender, which reflects a
tendency for women to express somewhat higher levels of concern over and
opposition to a repository than men. Length of residence in the local community, number of children, and age exhibited virtually no linear relationships
with either of the dependent variables.

Although the bivariate relationships examined to this point indicate some
potentially important interrelationships, a multivariate approach is required
to sift out the interplay between community differences and the influence
that respondents' perceptions and sociodemographic characteristics may
have on attitudes toward repository siting. For each of the dependent variables, a multiple classification analysis (MCA) was undertaken, using community as a six-category independent "factor" and the measure of anticipated
economic effects, NTS risk perceptions, trust in science, trust in government,
respondent gender, number of living children, length of residence, and age
as control variables. The MCA approach allows us to determine the extent
to which observed differences among the study communities in attitudes

toward siting a repository at Yucca Mountain are due to differences in the control variables, e.g., perceptions of the effects on the economy and the health risks, trust in science and government, and selected demographic characteristics (gender, number of children, age, and length of residence).

Considering first the measure of perceived health and safety risks from the repository, the results summarized in Table 10-6 indicate that the cross-community differences noted in Table 10-2 persist but are less substantial after the effects of covariates (control variables) are taken into account. Overall, the value of eta, which reflects the bivariate correlation between the community factor and the risk perception measure, is moderate, at .33. After taking into account the influences of the eight covariates, however, the partial correlation (beta) for the community factor is substantially smaller, at .12.

By examining the coefficients listed in the columns labelled "unadjusted deviation" and "adjusted deviation," it is possible to determine the degree to which community differences are evident both before and after the effects

Table 10-6   Multiple Classification Analysis of Health and Safety Concerns Regarding the Repository

	N	Unadjusted Deviation	Adjusted Deviation	F	Significance
Health and Safety Concern[a] (Grand mean = 5.01)					
Main factor (community)				25.57	.000
Amargosa Valley	83	−1.37	−0.35		
Beatty	84	−1.76	−0.64		
Pahrump	128	0.53	0.60		
Indian Springs	94	−0.42	0.19		
Mesquite	75	1.30	−0.31		
Caliente	92	1.47	0.13		
Eta and beta		.33	.12		
Covariates				66.20	.000
Effects on economy				13.88	.000
Perceived NTS health effects				89.84	.000
Trust in science				0.40	.525
Trust in government				65.91	.000
Sex				1.17	.280
Number of children				8.30	.004
Age				3.21	.074
Length of residence				0.36	.548
$R^2 = .55$					

[a]Values range between 0 ("not at all concerned") and 10 ("extremely concerned")

of covariates are taken into account. The grand mean for the risk perception variable (e.g., the mean obtained when responses from all communities are pooled together) was 5.01. The unadjusted deviations from the grand mean indicate that, ignoring the influence of the control variables, the mean response values on this scale are lowest (reflecting low levels of concern) in Beatty (-1.76), and only slightly higher in Amargosa Valley (-1.37). In contrast, concern levels were highest in Caliente ($+1.47$) and Mesquite ($+1.30$). After controlling for the effects of the covariates, the remaining community differences, reflected by the adjusted deviation values, still indicate that levels of concern were lowest in Beatty (-0.64) and Amargosa Valley (-0.35). However, the highest adjusted deviation value was for Pahrump ($+0.60$), indicating a tendency for levels of concern to be greater in that community after controlling for the covariates. In contrast, the very high concern levels initially observed in Caliente and Mesquite appear to be largely attributable to the influence of the covariates, since the adjusted deviations for these communities are rather small.

Among the covariates, or control variables, the variables of primary importance in accounting for variation in the measure of repository risk perceptions are the measures of anticipated economic effects, perceptions of NTS health effects, trust in government agencies responsible for nuclear program management, and respondents' number of children. When all of the independent variables are considered simultaneously, the results reflect a tendency for perceptions of repository health and safety risks to be higher among those who anticipate few economic benefits for their community, believe that nuclear testing activities at NTS are associated with adverse health effects, believe that federal agencies fail to deal honestly with the public regarding nuclear program safety, and are parents. The relationship between health and safety concerns and the covariates representing trust in science, respondent gender, age, and length of residence are all statistically insignificant. In combination, the community factor and the eight covariates account for a substantial 55 percent of the variation in this dependent variable (multiple $R^2 = .55$).

Turning to the measure of support for or opposition to construction of a repository at Yucca Mountain, results of the multiple classification analysis indicate a similar tendency for community differences to become less pronounced after the effects of the covariates are taken into account (Table 10-7). The bivariate correlation between the community factor and the support/ opposition variable is moderate, as indicated by an eta value of .36. However, the partial association (beta) after inclusion of the covariates drops to .11, reflecting a considerable attenuation of the differences across communities. After adjusting for the influence of the covariates, levels of repository sup-

Table 10-7   Multiple Classification Analysis of Support/Opposition to the Repository

	N	Unadjusted Deviation	Adjusted Deviation	F	Significance
Support/Opposition to Repository[a] (Grand mean = 2.69)					
Main factor (community)				35.05	.000
Amargosa Valley	83	−0.71	−0.27		
Beatty	84	−0.65	−0.20		
Pahrump	129	0.16	0.10		
Indian Springs	93	−0.13	0.10		
Mesquite	73	0.72	0.12		
Caliente	92	0.57	0.09		
Eta and beta		.36	.11		
Covariates				77.00	.000
Effects on economy				78.09	.000
Perceived NTS health effects				28.48	.000
Trust in science				0.02	.901
Trust in government				99.52	.000
Sex				2.44	.119
Number of children				4.32	.038
Age				0.74	.389
Length of residence				0.70	.402
R^2 = .59					

[a]Values range between 1 ("definitely yes") and 5 ("definitely no")

port are highest in Amargosa Valley and Beatty (adjusted deviations of −0.27 and −0.20 from the grand mean, respectively). Although responses from Mesquite yielded a rather high unadjusted score (0.72), the adjusted deviation score (0.12) is virtually indistinguishable from the remaining three communities. This suggests that most of the differences observed initially among these four communities (Pahrump, Indian Springs, Caliente, and Mesquite) are attributable to variations in community distributions for the control variables.

The relationships between support/opposition and the eight covariates examined in Table 10-7 indicate that the variables of primary importance in accounting for variation in this dependent variable are, in order of relative magnitude, trust in government, the measure of anticipated economic effects, perceptions of NTS health effects, and number of children. These are the same variables that exhibited significant partial relationships with the measure of perceived health and safety risks. Once again, the partial relationships involving trust in science, respondent gender, age, and length of residence are statistically insignificant. The multiple R^2 of .59 indicates that,

Table 10-8   Multiple Classification Analysis of Support/Opposition to the Repository, with Risk Perception as a Covariate

	N	Unadjusted Deviation	Adjusted Deviation	F	Significance
Support/Opposition[a] (Grand mean = 2.68)					
Main factor (community)				41.40	.000
Amargosa Valley	83	−0.70	−0.20		
Beatty	84	−0.67	−0.10		
Pahrump	127	0.14	0.05		
Indian Springs	95	−0.11	0.09		
Mesquite	76	0.73	0.11		
Caliente	94	0.55	0.03		
Eta and beta		.36	.07		
Covariates				106.54	.000
Effects on economy				62.74	.000
Perceived NTS health effects				2.85	.092
Trust in science				0.27	.601
Trust in government				44.80	.000
Sex				1.57	.211
Number of children				0.66	.417
Age				2.28	.132
Repository risk perceptions				118.81	.000
$R^2 = .66$					

[a]Values range between 1 ("definitely yes") and 5 ("definitely no")

in combination, the community factor and the covariates are able to account for 59 percent of the variation in levels of repository support/opposition.

As a final step in the analysis we reexamined the possible predictors of repository support or opposition by replacing the length of residence variable, which, as discussed above, provided little explanatory power in predicting levels of support or opposition, with the measure of perceived health and safety risks associated with the repository. This resulted in an even greater attenuation of cross-community differences, as indicated by adjusted deviation values which are relatively small for all communities (see Table 10-8). The partial association (beta) between the community factor and support/opposition was relatively small (.07), indicating that most of the observed bivariate association between these variables is accounted for by variation in the control variables. With the risk perception variable incorporated as a covariate in the analysis, the variables of primary importance in accounting for variation in levels of support/opposition are, in order of relative magnitude, perceptions of repository health and safety risks, anticipated

economic effects, trust in government agencies responsible for nuclear programs, and community of residence. Neither perceived NTS health effects
nor number of children exhibited a significant relationship with support/
opposition when the measure of perceived repository health and safety risks
was included as a covariate. None of the other covariates exhibited statistically significant partial relationships with the dependent variable. Overall,
the community factor and these covariates accounted for a very substantial
amount (66 percent) of the variation in this measure of support/opposition.

Discussion

The results of this analysis demonstrate that several factors influence rural
community residents' views of the proposed Yucca Mountain high-level
nuclear waste repository. First, the substantial differences across communities in perceived health and safety risks and in levels of support/opposition
suggest that attempts to assess local attitudes and perceptions as components of overall social impacts (Albrecht and Thompson, 1988) must take
into account the unique sociocultural contexts of individual community
settings.

The results also clearly indicate that attitudes about a potentially hazardous facility such as a nuclear waste repository are linked to expectations about project-related benefits. Not surprisingly, the extent to which
survey respondents anticipated positive economic effects of the repository
for their communities exerted an important influence on both perceptions
of health and safety risks and overall support/opposition regarding the repository program. Like many rural areas, these six communities have all
experienced some degree of economic instability and uncertainty as a result of fluctuations associated with dependence on a single major economic
activity (Krannich and Luloff, 1991). Such dependence includes Caliente's
reliance on now-obsolete railroad technologies, boom-bust mining cycles in
Amargosa Valley and Beatty, fluctuating levels of defense-related programs
in Indian Springs, and shifts from agricultural enterprise to low-wage service
industries based on tourism and retirement in Mesquite and Pahrump.

Although the extent of economic difficulties has varied widely across the
study sites, they all share a general concern about the need for more economic stability and increased local economic opportunities, a concern that
is common to many other rural areas. The economic context results in a
greater willingness of many rural residents to accept potentially dangerous or noxious facilities than is likely to occur with most urban residents
(Krannich and Luloff, 1991). Under such circumstances, the willingness to
accept potentially harmful facilities increases dramatically when project

proponents promise that local residents will obtain high-paying jobs and that there will be a concomitant increase in area business activity.

Repository perceptions and attitudes are also influenced by experience with and perceptions of other, possibly analogous, projects and programs. To a substantial degree, the different views expressed by residents of these rural communities appear to be linked to their beliefs about the public health effects of past and present nuclear testing programs. The "downwinder" experiences of some southern Nevada residents, especially those living in Caliente, Mesquite, and other communities that are northeast of the Nevada Test Site, contrast sharply with those in Amargosa Valley, Beatty, Indian Springs, and Pahrump. These latter communities are not only upwind of NTS but have also experienced the economic benefits of NTS employment as well as jobs related to the operation of a nearby low-level nuclear waste repository, which, to date, has generally been problem-free. Thus, the "risk perception shadows" (Stoffel et al., 1988, 1989) cast by NTS and other nuclear projects in southern Nevada may be quite different for area residents, depending upon past experiences with things nuclear. Those experiences account for important differences in community views of the potential consequences of storing high-level nuclear waste at Yucca Mountain.

Also linked to these experiences is the extent to which residents believe that the federal government can be trusted to provide honest information about the safety of nuclear programs. Nevadans share a general antigovernment orientation common throughout the rural West. In addition, many residents are convinced that the government has been dishonest in its dealings with the public over such events as the nuclear contamination from atmospheric weapons testing, nuclear contamination from improperly contained underground nuclear tests, and attempts to site MX missiles in rural areas of Nevada and Utah. Moreover, public controversy erupted in early 1988 over the alleged suppression by the Department of Energy of government scientists' reports questioning the geological suitability of Yucca Mountain as a repository site. These circumstances have created a context of increased hostility and distrust of the federal government that appears to have strongly influenced rural Nevadan's views about the repository program.

The prospect of a high-level nuclear waste repository also creates a context in which many residents are confronted with the question of cross-generational risks. Some individuals may be willing to expose themselves to risks or to accept trade-offs between risks and economic opportunities that may benefit them as individuals or that can improve general community economic conditions. Even in the case of a project characterized by the potential for radiation releases, some persons may express fairly low levels of concern because they assume that they are relatively immune to the health risks posed by future accidents or the long-term nature of health threats from low levels of radiation exposure. The perceptions of parents, however, are influenced not only by concerns about personal health consequences but also by concerns about the well-being of their children and grandchildren. Such concerns are likely to be especially important in determining responses to a facility such as the proposed nuclear waste repository, which would not become operational until after the turn of the century and which would be required safely to isolate highly toxic radioactive materials for more than 10,000 years.

Although cross-generational risks appear to have some influence on residents' attitudes and risk perceptions, more general perceptions of health and safety risks are among the strongest predictors of rural residents' attitudes about the proposed Yucca Mountain repository. Respondents who reported high risk-perception levels were much more likely to express opposition to repository construction than were those who reported low risk-perception levels. Clearly, perceptions of health and safety risks are among the important "special" effects of nuclear projects that must be considered when assessing the human impacts of such facilities.

Although the observed community differences in repository attitudes and perceptions were substantially attenuated when other variables were included in the multivariate analysis, significant cross-community differences remained unaccounted for by the control variables. These remaining community differences may involve a variety of factors not considered here. For example, variations in the extent to which local populations are geographically mobile may be a factor, since residents who anticipate moving away from the area may believe that they will not be exposed to any of the long-term risks associated with repository operations. This could help to explain the low levels of concern expressed by respondents from Beatty, where a mining boom has attracted a more transient population. Also, the desperation which forces economically depressed communities to support virtually any growth opportunity (see Gallaher and Padfield, 1980; Krannich and Luloff, 1991) may help to account for the high levels of repository support in Amargosa Valley. Perhaps there is some underlying community or regional ethos which determines, at least in part, the manner in which residents of different communities respond to federal government projects in general or the nature of local views about the acceptability of risk.

In any event, the importance of community differences clearly necessitates a focus on the unique characteristics of individual communities rather than on an undifferentiated rural impact area. Our findings suggest that it is very important to understand how responses to nuclear and other hazardous projects may differ across various settings and local contexts. In sharp contrast with some other studies of response to proposed nuclear facilities (e.g., Stoffel et al., 1989), our results indicate that opposition and concern are strongest in the communities farthest from Yucca Mountain, and lowest among those located nearest to the repository site. These findings fly in the face of the oft-cited NIMBY syndrome and suggest that the relationship between proximity and opposition is not universal and probably far more complex than previously suggested.

In sum, the responses of rural Nevada residents to the proposed high-level nuclear waste repository at Yucca Mountain appear to be influenced by a complex set of factors, ranging from the unique sociocultural settings of specific local communities, to the widely divergent experiences and perceptions which are linked to past and present nuclear testing, to the cross-generational concerns and risk perceptions that appear to be uniquely important when addressing the long-term toxicity of hazardous and radioactive materials. Residents of these rural study communities, particularly those nearest to Yucca Mountain, generally express lower levels of concern over, and greater support for, a repository than has been observed among urban Nevadans. However, such positive views about a repository are far from universal. The differences in views that are evident among communities, and among individuals who exhibit different perceptions and personal characteristics, suggest that the social and psychological costs stemming from repository development will not be borne evenly by all area residents.

Notes

This research was supported in part by a research contract with Coopers and Lybrand, Inc. (formerly Mountain West Research) for the Nevada Nuclear Waste Projects Office. Additional support was provided by the Utah Agricultural Experiment Station.

1 *Downwind* refers to the fact that the prevailing winds move eastward from NTS toward Mesquite, Caliente, and other rural areas of southern Nevada and Utah. *Downwinders* are those people who were in the path of the radioactive fallout from atmospheric testing of thermonuclear weapons at NTS.

2 Employees of the public utility companies went through customer lists with members of the research team to certify that no service connections had been added or deleted since the list was printed. Additionally, these same employees noted instances where multiple families occupied a dwelling with only a single utility hookup. If the number of households was not known with certainty, an on-site inspection by a team member resolved the question. Whenever it was suspected that utility records were inaccurate, team members mapped the locations of all housing units in areas of question. In one instance the entire community was mapped by team members.

3 The question asked, "If the repository is built at Yucca Mountain, how concerned are you that it might have harmful effects on public health and safety in this area?"

4 The question asked, "If you were able to make the final decision regarding the location of the nuclear waste repository at Yucca Mountain, would you build it there?"

5 The question asked, "How likely do you think it is that the repository would affect the economic well-being of residents or businesses in this area?"

6 The question asked, "How likely do you think it is that above-ground nuclear weapons testing activities at the Nevada Test Site have in the past caused harmful health problems for people who live in this area?"

7 The question asked, "How likely do you think it is that underground nuclear weapons testing activities at the Nevada Test Site will in the future cause harmful health problems for people who live in this area?"

8 The items were as follows:

(a) "Scientists generally work for the well-being of the public."
(b) "Scientists often make sensational announcements just to get publicity."
(c) "Science attempts to increase the knowledge we can apply to our everyday lives."
(d) "Science creates more problems than it solves."
(e) "Scientists can almost always be trusted when they say something like a product or procedure is safe."

For purposes of index construction, the responses to items (b) and (d) were reverse coded.

9 The question asked, "How confident are you that federal agencies have provided the public with honest and accurate information about the safety of the government's nuclear programs?"

10 Preliminary analyses also examined respondents' education and employment experience at NTS as possibly important independent variables. However, the SPSS-PC statistical package used to analyze the data restricted the multivariate analysis to a maximum of eight independent variables in addition to the variable representing community of residence. Since neither education nor NTS employment experience exhibited meaningful relationships with the dependent variables, they were not included in the final analysis.

References

Albrecht, Stan L. 1983. "Community Response to Large-Scale Federal Projects: The Case of MX." In Steve H. Murdock, F. Larry Leistritz, and Rita R. Hamm, eds., *Nuclear Waste: Socioeconomic Impacts of Long-Term Storage*. Boulder: Westview. 233–50.

Albrecht, Stan L., and James G. Thompson. 1988. "The Place of Attitudes and Perceptions in Social Impact Assessment." *Society and Natural Resources* 1:69–80.

Center for Survey Research. 1988. *Public Opinion in Nevada: The Future*. Las Vegas: University of Nevada.

Cortese, Charles, and Bernie Jones. 1977. "The Sociological Analysis of Boom Towns." *Western Sociological Review* 8:76–90.

Cottrell, William F. 1951. "Death by Dieselization: A Case Study in the Reaction to Technological Change." *American Sociological Review* 16:358–65.

Cronbach, Lee J. 1951. "Coefficient Alpha and the Internal Structure of Tests." *Psychometrika* 16:297–334.

Davenport, Judith, and Joseph A. Davenport, eds. 1980. *The Boom Town: Problems and Promises in the Energy Vortex.* Laramie: University of Wyoming Press.

Edelstein, Michael R. 1988. *Contaminated Communities: The Social and Psychological Impacts of Residential Toxic Exposure.* Boulder: Westview.

Elkind-Savatsky, Pamela D., ed. 1986. *Differential Social Impacts of Rural Resource Development.* Boulder: Westview.

Freudenburg, William R., and Eugene A. Rosa, eds. 1984. *Public Reactions to Nuclear Power: Are There Critical Masses?* Boulder: Westview/American Association for the Advancement of Science.

Fuller, John G. 1984. *The Day We Bombed Utah.* New York: New American Library.

Gallaher, Art, and Harland Padfield, eds. 1980. *The Dying Community.* Albuquerque: University of New Mexico Press.

Hamilton, Lawrence C. 1985. "Who Cares About Water Pollution: Opinions in a Small-Town Crisis." *Sociological Inquiry* 55:170–81.

Harris, Louis, and Associates, Inc. 1976. *A Second Survey of Public and Leadership Attitudes Toward Nuclear Power Development in the U.S.* New York: Ebasco Services.

Kish, Leslie. 1949. "A Procedure for Objective Respondent Selection Within the Household." *Journal of the American Statistical Association* 44:380–87.

Krannich, Richard S., Thomas Greider, and Ronald L. Little. 1985. "Rapid Growth and Fear of Crime: A Four Community Comparison." *Rural Sociology* 50:193–209.

Krannich, Richard S., and Ronald L. Little. 1988. "Differential Orientations of Rural Community Residents Toward Nuclear Waste Repository Siting in Nevada." Presented at the Annual Meetings of the Rural Sociological Society, Athens, GA, August.

Krannich, Richard S., and Albert E. Luloff. 1991. "Problems of Resource Dependency in U.S. Rural Communities." In Andrew Gilg, David Briggs, Robert Dilley, Owen Furuseth, and Geoff McDonald, eds., *Progress in Rural Policy and Planning*, Vol. 1. London: Bellhaven. 5–18.

Kunreuther, Howard, William H. Desvousges, and Paul Slovic. 1988. "Nevada's Predicament: Public Perceptions of Risk From the Proposed Nuclear Waste Repository." *Environment* 30:16–33.

Little, Ronald L. 1977. "Some Social Consequences of Boom Towns." *North Dakota Law Review* 53:401–25.

Little, Ronald L., and Steven B. Lovejoy. 1979. "Energy Development and Local Employment." *Social Science Journal* 6:27–49.

McStay, Jan R., and Riley E. Dunlap. 1983. "Male-Female Differences in Concern for Environmental Quality." *International Journal of Women's Studies* 6:291–301.

Mitchell, Robert Cameron. 1984. "Rationality and Irrationality in the Public's Perception of Nuclear Power." In William R. Freudenburg and Eugene A. Rosa, eds., *Public Reactions to Nuclear Power: Are There Critical Masses?* Boulder: Westview/American Association for the Advancement of Science. 137–79.

Mushkatel, Alvin, and K. David Pijawka. 1989. *The Analysis of the Las Vegas Urban Survey Data—Final Report.* Prepared for the Nevada Nuclear Waste Projects Office. Las Vegas: Mountain West Research.

Nealey, Stanley M. 1990. *Nuclear Power Development: Prospects in the 1990s.* Columbus, OH: Battelle.

Nealey, Stanley M., Barbara D. Melber, and William L. Rankin. 1983. *Public Opinion and Nuclear Energy.* Lexington, MA: Lexington Books.

Stoffel, Richard W., Michael W. Traugott, Camilla L. Harshbarger, Florence V. Jensen,

Michael J. Evans, and Paula Drury. 1988. *Perceptions of Risk From Radioactivity: The Superconducting Super Collider in Michigan.* Ann Arbor: Institute for Social Research, University of Michigan.

Stoffel, Richard W., Michael Traugott, Carla Davidson, Florence V. Jensen, John Stone, Gail Coover, and Paula Drury. 1989. *Social Assessment of Siting a Low-Level Radioactive Waste Isolation Facility in Michigan: A Summary of Two Studies.* Ann Arbor: Institute for Social Research, University of Michigan.

Titus, A. Constandina. 1986. *Bombs in the Backyard: Atomic Testing and American Politics.* Las Vegas: University of Nevada Press.

Trend, Michael G., Ronald L. Little, and Richard S. Krannich. 1988a. *Summary Ethnographic Report: Amargosa Valley.* Prepared for the Nevada Nuclear Waste Projects Office. Las Vegas: Mountain West Research.

———. 1988b. *Summary Ethnographic Report: Beatty.* Prepared for the Nevada Nuclear Waste Projects Office. Las Vegas: Mountain West Research.

Weber, Bruce A., and Robert E. Howell. 1982. *Coping with Rapid Growth in Rural Communities.* Boulder: Westview.

Williams, R. Gary. 1988. "Perceived Knowledge and Perceived Risk." Presented at the Second Symposium on Social Science in Natural Resource Management, Urbana, IL, June.

Williams, R. Gary, and Barbára A. Payne. 1985. "Emergence of Collective Action and Environmental Networking in Relation to Radioactive Waste Management." Presented at the Annual Meetings of the Rural Sociological Society, Blacksburg, VA, August.

Van Liere, Kent D., and Riley E. Dunlap. 1980. "The Social Bases of Environmental Concern: A Review of Hypotheses, Explanations and Empirical Evidence." *Public Opinion Quarterly* 44:181–97.

Part IV

Summary and

Policy Implications

11 Prospects for Public Acceptance of a High-Level Nuclear Waste Repository in the United States: Summary and Implications

Eugene A. Rosa, Riley E. Dunlap, and Michael E. Kraft

Nature is still there, however. She contrasts her calm skies and her reasons with the madness of men. Until the atom too catches fire and history ends in the triumph of reason and the agony of the species. But the Greeks never said that the limit could not be overstepped. They said it existed and that whoever dared to exceed it was mercilessly struck down. Nothing in present history can contradict them. (Camus, 1955:137)

Technology and Democracy?

Politics and technology, science and the state, democratic governance of risks, technology as the force of social change—these are the central topics of this volume. Central topics, too, on the political agendas of postmodern societies, they are drawn together in this book around a central policy challenge of our age: the disposal of very long-lived, highly toxic nuclear wastes, the unavoidable by-products of nuclear technology. The governance problem has been magnified in the past several decades because the social fabric has increasingly become technologically textured (Short, 1984). The increasing reliance of societies on complex and risky technologies has accelerated a process and exacerbated a dilemma, both of whose origins are synonymous with the development of nuclear science itself. The process it accelerates is the indispensable and growing need for the contribution of scientific expertise to public policy, thereby strengthening the mutual dependency of science and the state. It also eases the way for technical experts to exercise privileged influence on policy outcomes. This, in turn, leads to the dilemma of how to weigh the specialized expertise against the views of citizens who are not knowledgeable about the scientific details of technological issues. After all, very few of us, including experts, can understand technologies in all their details.

Indeed, an unavoidable partner to the growth of science and technological complexity is personal ignorance. A homey case of this pervasive feature of modern life is the video cassette recorder which, according to a recent survey by a popular magazine, defeats not only most of us ordinary folk, but also Nobel laureates as well (Stebben, 1992). More consequential is the fact that managers of technological systems often acknowledge that they cannot comprehend the systems for which they are responsible in all of their complexity (LaPorte, 1988).

The issue of how to solve the nuclear waste problem through democratic process draws the dilemma into sharp focus. Indispensable, on the one hand, is specialized knowledge to perform rigorous risk assessments and to inform technological choices. Yet because, in the words of Justice Louis Brandeis, "the highest office in the land is the citizen," it is citizens who must ultimately judge the acceptability of risks and proposed solutions, especially citizens most clearly affected by them. But citizens must rely on technical expertise in order to make informed choices. That reliance further magnifies the dilemma because the lion's share of expertise is held by agencies of the national state (Dietz and Rycroft, 1987), such as the Department of Energy (DOE), whose veracity and competency in the minds of citizens—as the evidence in this volume amply shows—is very much in question.

Bricolage

A key step in unraveling the dynamics of siting a nuclear repository through the democratic process is to see where, in fact, the citizenry stands on the waste issue. The question of citizen stance was first addressed by Rankin and Nealey (1978) and Nealey and Hebert (1983), who, drawing on a compilation of all nuclear survey data then available, attempted to trace the emergence and level of citizen awareness about the nuclear waste issue. They reported that some, albeit small, awareness was evident by the early 1970s, but that citizen's concerns with waste were far overshadowed by a variety of nuclear safety issues. That pattern of a low but consistent level of waste awareness coupled with a high concern for nuclear safety persisted for decades.

With the passage of the 1982 Nuclear Waste Policy Act and its 1987 amendments, an urgency arose to assess directly public awareness of the nuclear waste issue and the acceptability of proposed solutions. The preceding eight chapters, comprising the original empirical evidence of this volume, are aimed directly at these issues. It is the most comprehensive collection of such evidence in any one source. Each chapter is a rigorous, self-contained empirical study. Too rich in detail, too finely nuanced, too carefully crafted to be reduced to brief synthesis[1]—they can largely stand on their own.

Our role in this, the final chapter, then, is like that of the *bricoleur*, one who snaps up "found objects," in order to reconfigure them in ways that create a new vision or that uncover fresh insights. It is a form of creative recycling with a functional dependence on the efforts of others. The found objects are the empirical evidence from the preceding chapters—the proximate empirical factors shaping the acceptance of siting a high-level nuclear waste repository (HLNWR). Our specific goal here is to delineate which of these factors have most markedly shaped the current impasse over siting a HLNWR and will, therefore, likely shape the future of the nuclear waste disposal. We proceed toward our goal by first developing an overall profile of public acceptability.

Empirical Profile of Acceptability

The previous eight chapters of detailed, original evidence address empirically the pivotal questions raised by the Nuclear Waste Policy Act and its amendments: What is the psychological, social, and political acceptability of a HLNWR in the United States? Will the national state realize its policy goal of a single, permanent geologic repository? These questions, though simply stated, are studded with some of the most challenging issues facing all nuclear societies. Because they are complex and multifaceted, understanding these issues requires systematic analysis of each of their components. This was the objective of the empirical chapters of this volume.

The extensive coverage and considerable breadth of those chapters tends to overwhelm the senses with such a surfeit of details that the overall picture recedes from view. Our goal here is to refocus the overall picture of the empirical analyses with the development of a profile of key findings. As a tool for this refocus we begin by presenting two summary tables of the results. The first of these recounts the location, samples, techniques, and measurements of public acceptability employed in each study. As can be seen in Table 11-1, the geographical coverage is extensive, ranging from the nation as a whole to small counties and local communities in the impact zones of proposed repositories. Samples were drawn according to accepted procedures and nearly all are of sufficient size to be within conventionally acceptable error tolerances. The most common data collection technique was the telephone interview, but other techniques were used to suit particular needs and circumstances. Most survey questions were asked in widely used formats, and composite indexes, where used, were constructed in accordance with standard statistical procedures. Deserving of emphasis is the fact that three studies (see chapters 3, 5, and 6) elicited open-ended answers

Table 11-1 Summary of Methods and Samples for Each Chapter Presenting Original Data

Chapter Number	Geographical location of sample	Sample Size		Survey Technique	Survey formats and variables measured
3 (Slovic, Layman, Flynn)	3-1 National 3-2 State of Nevada 3-3 Phoenix, AZ 3-4 Southern California Nevada counties: 3-5 Nye 3-6 Lincoln 3-7 Esmeralda	825 500 802 801 204 101 101		Telephone	(1) Free association via method of "continued association," followed by (2) Rating scales of affect (3) Trust ratings of state and federal government (4) Ratings of distance acceptable for each of ten facilities (5) "Referendum" vote by Nevadans (6) Demographics
4 (Kraft, Clary)	4-1 Wisconsin 4-2 Maine 4-3 North Carolina 4-4 Georgia	243 481 209 115		Content analysis of public hearings	(1) Extensive coding scheme of public comments by citizens attending hearings
5 (Brody, Fleishman)	In proposed Texas repository impact zone: 5-1 Deaf Smith, Oldham, and Swisher counties Outside impact zone: 5-2 Two counties 5-3 Amarillo	1984 605 236 —	1986 340 253 609	Telephone	(1) Likert-type attitude questions (2) Open-ended items (3) Ratings of likelihood of potential impacts (4) Demographics
6 (Dunlap et al.)	Washington state counties:[1] 6-1 Benton 6-2 Franklin	 426 232		Telephone	(1) Open-ended questions (2) Knowledge about HLNWR (3) Likert-type attitude items (4) Ratings of likelihood of potential impacts (5) Attitude indexes (6) "Referendum" vote on repository (7) Demographics
7 (Desvousges	7-1 National 7-2 State of	1201		Telephone	(1) Rating scale of environmental problems

Table 11-1 *Continued*

Chapter Number	Geographical location of sample	Sample Size	Survey Technique	Survey formats and variables measured
et al.)	Nevada	1001		(2) Knowledge questions (3) Rating scales of risks of HLNWR (4) Likert-type attitude scales (5) "Referendum" vote by Nevadans (6) Demographics
8 (Easterling, Kunreuther)	Specialized national samples: 8-1 Convention planners 8-2 Attendees at past conventions Las Vegas	153 600	Mail following telephone contact	(1) Choice among repository scenarios (2) Likelihood of locating conventions in Las Vegas (3) Imagery stimulus (4) Reported past convention attendance (5) Likert-type rating of various noxious facilities
9 (Mushkatel, Nigg, Pijawka)	9-2 Las Vegas metropolitan area	549	Short telephone interview followed by face-to-face interview	(1) Likert-type attitude questions (2) Likelihood ratings of effects of Nevada test site (3) Reports of nuclear work experience (4) Trust ratings of government agencies (5) Demographics
10 (Krannich, Little, Cramer)	Rural towns in Nevada: 10-1 Beatty[2] 10-2 Amargosa Valley[2] 10-3 Pahrump[3] 10-4 Indian Springs[3] 10-5 Mesquite[3] 10-6 Caliente[3]	111 104 189 122 110 131 — 767	Self-completed question-naires delivered and collected by field workers	(1) Rating scales of perceptions and attitudes (2) Index of safety risks (3) Rating scales of trust in federal government (4) Demographics

[1]Also presented were results of 5 statewide polls (average sample size=580), 4 Tri-City area polls (average sample size=373), and the results of a November 1986 statewide referendum.

[2]Nye County

[3]Clark County

about nuclear waste from respondents. This all but assured that people's first impressions of nuclear waste were not being influenced by question wording or by other items in the questionnaire.

Another measurement feature worth emphasizing was the hypothetical repository referendum included in several samples. As pointed out by Kraft, Rosa, and Dunlap in chapter 1, the origination of public opinion polling stemmed from a populist movement aimed toward "enlightened democratic governance" through the proxy vote of people's opinions. Asking how they would actually vote substitutes a direct proxy for an indirect one (opinion), thereby more closely aligning perceptions with behavioral intentions.

Table 11-2 presents a chapter-by-chapter summary of the key variables and major findings in the analyses. The principle conclusions to emerge from the findings, taken together, are:

—Given an opportunity to vote on having a HLNWR in their state, an overwhelming number of citizens *would vote against it.*

—Compared to other environmental hazards, high-level nuclear waste is believed to be one of the *most dangerous.*

—Compared to any other industrial or municipal facility, a HLNWR is the *least desirable,* often by wide margins.

—Images of a HLNWR are so negative that they reveal *deep dread,* revulsion, and anger.

—Perceptions of the risks of a HLNWR are closely tied to *trust* in government and trust in agencies responsible for repository management, especially the DOE.

—There is widespread belief in the *likelihood of accidents* at a repository or in the transportation of waste to it and great concern over such accidents.

—Economic concerns appear in some analyses but are often overwhelmed by concerns over *safety and health,* including that of future generations.

—*Knowledge* is not a consistent factor in citizen opposition to a repository; it is correlated sometimes positively, more often negatively, and sometimes not at all with repository attitudes.

—If a repository were to be sited at Yucca Mountain, Nevada, the city of Las Vegas and the state would likely *suffer substantial economic losses* due to a significant convention decline.

—*Past experience* with other hazardous facilities or with other federal facilities is of only minor significance.

—That *opposition* to a HLNWR *is deep and widespread* is revealed in the

Table 11-2   Summary of Key Findings in Each Chapter Presenting Original Data

Chapter Number	Key Dependent (or response) Variables	Key Independent (or control) Variables	Major Findings
3 (Slovic, Layman, Flynn)	(1) Willingness to live near various facilities (2) Perceptions of transportation accidents (3) Whether would vote for repository	(1) Images of a nuclear waste repository (2) Trust of federal and state officials	(1) Nuclear repository was the most undesirable facility to live near (2) Solid majorities believe transportation accidents will occur (3) Widespread mistrust of DOE (4) Sizable majorities (70–80%) would vote against repository (5) Images of repository are overwhelmingly negative (6) Strong inverse relationships between trust and perception of risk of HLNWR
4 (Kraft, Clary)	(1) Support/opposition toward a repository (2) Locating repository within state (3) Acceptability of radioactive waste risks on 5 dimensions	(1) Levels of knowledge (2) Localism (3) Trust and confidence in government (4) Political and social concerns (5) Environmental and economic impacts (6) Group affiliation: citizen, environmental, government, public integrity, industry/utilities, industries (7) State location (8) Demographics	(1) A solid majority (70%) opposed DOE siting plans (2) A sizable majority (58%) opposed a repository in their state (3) A majority of people (57–83%) were knowledgeable of the issues (4) Environmental impacts (water resources and public health) and economic impacts (local economy and tourism) are important (5) Strongest predictor of opposition to repository siting is mistrust of DOE (6) Knowledgeability is *positively* correlated with opposition (7) Results barely affected by demographics, group affiliation, or state location of hearings

Table 11-2 *Continued*

Chapter Number	Key Dependent (or response) Variables	Key Independent (or control) Variables	Major Findings
5 (Brody, Fleishman)	(1) Attitude toward construction of repository in Deaf Smith county and whether repository would benefit county—combined into an index (2) Environmental risks	(1) Levels of knowledge (2) Environmental risks (3) Socioeconomic impacts (4) Attitudes toward industrial development (5) Site location (6) Demographics	(1) High level of opposition to repository (over 80%) at all sites (2) Harm to agriculture followed by water contamination most important reasons for mentioning opposition (3) Solid majorities (average of 70%) rated variety of risks as likely: transportation, water contamination, food contamination, health problems, etc. (4) Among 9 types of industrial development, preference for low-level nuclear waste disposal was a distant last along with nuclear power plants (5) In a risk/benefit model predicting attitudes toward a repository, environmental threats exceeded economic benefits while knowledge had no significant effect (6) Results are barely affected by site location or by demographics
6 (Dunlap et al.)	(1) Attitudes toward repository in statewide and site location polls (2) Three survey attitudes—combined into an index	(1) Perceptions of various types of accidents (2) Political ideology (3) Faith in science and technology (4) Attitude toward nuclear power (5) Perceptions of nuclear waste	(1) In statewide polls strong majorities (average of 70%) oppose repository while a slim plurality at site location (Tri-Cities) favor repository (2) Survey results show that: (1) a slim plurality favors repository; (2) a vast majority favor a proposed testing program; and (3) a sizeable majority would vote for repository if were proven safe (3) A solid majority (over 65%)

Table 11-2 *Continued*

Chapter Number	Key Dependent (or response) Variables	Key Independent (or control) Variables	Major Findings
		(6) Confidence in DOE (7) Levels of knowledge (8) Environmental and health impacts (9) Economic benefits (10) Nuclear stigma (11) Nuclear radiation (12) Likelihood of accidents (13) Demographics	of state residents saw a variety of accidents as likely while about half of Tri-Cities residents felt that way (4) Majorities—from overwhelming (94%) to slim (51%)—saw most economic benefits and costs and environmental costs as somewhat likely (5) A majority of Tri-City respondents (56%) were confident in DOE, but this was below expectation, given ties to DOE (6) In model predicting attitudes toward repository the most significant effects, in order, were: environmental and health impacts, confidence in DOE, economic benefits, attitude toward nuclear power, perception of nuclear waste, accident likelihood (7) Virtually no demographic effects in model
7 (Desvousges et al.)	(1) Perception of risks from repository (2) Seriousness of various forms of pollution (3) "Referendum" vote on repository	(1) Levels of knowledge (2) Variety of attitudes toward repository (3) Ratings of seriousness of various health and safety risks (4) Trust in federal government (5) Expected economic benefits	(1) High level of knowledge about method of disposing of waste but low level about length of isolation (2) Among pollution problems, water pollution and radioactive wastes were seen as most serious (3) Respondents in the national sample perceive HLNW repository risks as most serious while Nevadans were most concerned about transportation (4) Solid majorities of all respon-

Table 11-2 *Continued*

Chapter Number	Key Dependent (or response) Variables	Key Independent (or control) Variables	Major Findings
		(6) Character-istics of risks: knowledge, dread, accidents, future generations (7) Demographics	dents believed that: accidents would involve certain death; and kill many people; people living near the repository couldn't control the risks and would dread living there; and a repository poses serious risks for future generations (5) A majority of respondents also thought the accidents presented to them were likely (6) Majorities would vote against repository (7) In a model predicting risk perceptions with very similar results for national and Nevada samples the most significant effects are: trust in the federal government, nuclear attitude, economic benefits, knowledge, and risk characteristics (dread, accident likelihood, and threats to future generations) (8) In a model predicting voting behavior—that nests the perception model—the most significant effects (above risk perceptions) are trust in the federal government, knowledge, economic benefits, and proximity to Yucca mountain
8 (Easterling, Kunreuther)	(1) Scenario-by-scenario likelihood of choosing Las Vegas convention size	(1) Repository scenarios presented to convention planners (2) Media coverage of scenarios (3) Convention amenities	(1) Between ⅓ and ¾ of convention planners lowered their rating of Las Vegas for all repository scenarios (2) For most scenarios, planners shift convention preferences from Las Vegas to other cities (3) With least aversive scenario,

Table 11-2 *Continued*

Chapter Number	Key Dependent (or response) Variables	Key Independent (or control) Variables	Major Findings
			estimated losses of 12–36% of conventions; 47–80% with most severe scenario (4) Estimated dollar losses for least aversive scenarios of \$14–\$32 million; \$173–\$450 million for most severe scenarios
	(2) Patterns of past attendance by convention attendees	(1) Images of Las Vegas (2) Perceptions of risk: crime, natural hazard, pollution, and environmental hazards (3) Scenarios of various noxious facilities within 100 miles of convention (4) Membership associations of convention attendees	(1) Image of Las Vegas—gambling, entertainment, etc.—was predictive of post convention attendance (2) Perceived risk predicted past attendance for some associations (3) Of the 5 noxious facilities (prison, nuclear reactor, hazardous waste incinerator, low-level radioactive waste repository, and HLNWR) the HLNWR elicited more than twice the response on attendance than the next closest facility (a low-level waste facility)
9 (Mushkatel, Nigg, Pijawka)	(1) Attitude toward building repository (2) Perception of harmful effects on community (3) Perception of risks (4) Transportation risks (5) Perceived benefits versus harms	(1) Attitudes toward other hazardous facilities (2) Experience with past hazards (3) Trust in government (federal, state, county) and government agencies (DOE, NRC, NWPO) (4) Past accident reporting practices	(1) Solid majority (69%) says no to repository (2) Majorities are concerned with harmful effects of repository and said harms (3) Majorities felt that repository and transportation risks were serious (4) Majorities said above-ground testing at NTS was harmful but not other NTS activities (5) Past experience is unrelated to repository attitude (6) The NWPO received the high-

Table 11-2   *Continued*

Chapter Number	Key Dependent (or response) Variables	Key Independent (or control) Variables	Major Findings
		(5) Demographics	est trust rating while the DOE and NRC received the lowest (7) In a model predicting repository risks (an index of 3 items) the most significant effects are: beliefs in harms associated with the NTS and trust in government (8) In a model that predicts repository attitude the most significant effects are: perceived risks, ratio of benefits to harms, and past experience at nearby federal facilities (9) Virtually no demographic effects
10 (Krannich, Little, Cramer)	(1) Attitudes toward public health and safety (2) Locating repository at Yucca mountain	(1) Perceptions of local economic harm or benefit (2) Perception of harms in past from NTS tests (3) Trust in science (4) Demographics	(1) Concerns about health and safety are *inversely* related to distance from Yucca mountain (2) Opposition to the repository is likewise inversely related to distance (3) Unlike the rest of the state and nation the four communities closest to Yucca mountain *favor* the repository (4) In a model predicting risk perception the most significant effects are perceived NTS health effects, trust in government, economic impacts and community residence. (5) In a model predicting attitude toward repository—that nests the perception model—the most significant effects are: economic impacts, trust in government, and community of residence

fact that, almost without exception, perceptions and attitudes toward a repository are unaffected by standard demographic variables.

Two of these conclusions merit brief elaboration. Exceptions to the nearly unanimous disapproval of a repository are the four communities adjacent to Yucca Mountain: Beatty, Amargosa Valley, Pahrump, and Indian Springs. Why a majority of these residents favor a repository is not entirely clear to us. In the case of Beatty, support could be due to the mobility of the population as suggested by Krannich, Little, and Cramer in chapter 10. Or, for all four communities more generally, it could be due to the desperation of attracting any kind of economic growth to an area depressed economically. Whatever the reasons, the fact remains that this is an area of very low population (the total population for the four towns is barely 4000), hardly representative of Nevada as a whole. Consequently, the state's position on the repository is likely unaffected by the support of these local communities.

Perceptions of risk are markedly influenced by the level of trust in institutions responsible for managing the risks. The close inverse relationship between trust in government and risk perceptions and attitudes again underscores and reaffirms a consistent finding in the sizable literature on risk perception: contextual features, usually qualitative ones, exercise significant influence on perceptions. What this means is that formal risk analysis, a key tool of the nuclear subgovernment, with its emphasis on quantitative measures of risk, abstracts away some of the features of risk most important to people (Gould et al., 1988; Dietz, Frey, and Rosa, 1993). This neglect likely reinforces the mistrust of agencies, such as DOE, which partly shaped risk perceptions in the first place.

The empirical findings, showing the connection between trust and risk perceptions, further illustrate the importance of social and historical context to public acceptability. The perceptions of risk associated with nuclear waste, narrow in focus, are embedded in the more diffused focus of perceptions of government agencies charged with disposing of nuclear wastes. The latter set the broader context for the former. But perceptions of agencies are, in turn, embedded in a still broader context: trust in institutions more generally. What is the level of public trust in American institutions?

In . . . We Trust

One of the most striking social trends of the past thirty-odd years to capture the sociological imagination has been the decline in public trust and confidence (Barber, 1983). From the late 1950s and early 1960s, when polls on the subject first started, until the early 1980s, with the election of Ronald

Reagan, there was a nearly uninterrupted downward trend in confidence expressed by citizens toward nearly every major American institution (Lipset and Schneider, 1987). For Americans, a gap steadily widened between expectation and performance of these institutions. When queried over time about their confidence or trust in the government, business, labor, the media, medicine, education, the military, organized religion, or nearly any other institution, people expressed increasingly negative feelings about the performance of these structures of society. Nearly half the electorate had a great deal of confidence in these institutions in the mid-1960s; by the early 1980s, the percent expressing such confidence had been cut in half.

This downward spiraling trend bottomed out in late 1983, on the verge of the Reagan reelection, and made a noticeable partial recovery in the first two years of his second term. Even so, these early 1980 confidence levels significantly lagged behind those from the first polls of two decades earlier. Then, with the Iran-contra scandal in December 1986, nearly all that had been regained was lost. The scandal reversed the incipient rising trend (Lipset and Schneider, 1987). Public confidence again declined to near-bottom levels and has, available evidence suggests, remained there since.

Accompanying—and perhaps contributing to—the declining institutional trust was a steady erosion in political efficacy and a steady increase in feelings of alienation among the public. Political efficacy is a measure of how much people believe their views can affect politics or what the government does. From the first University of Michigan biennial poll on the subject, in 1952, until the 1980s, people expressed, with minor exceptions, a steady decrease in their feelings of efficacy: their vote didn't count much, politics was getting more complicated, and they didn't have much say in what government does (Lipset and Schneider, 1987).

Alienation, a sense of detachment or estrangement from the mainstream of social and political life, showed a similar pattern. From the mid-1960s, when first polled, until the late 1980s, people increasingly answered in the affirmative to statements such as: "What I think doesn't count anymore." "The rich get richer, and the poor get poorer." "The people running this country don't really care what happens to me." "Most people with power try to take advantage of people like myself." "I feel left out of things going on around me" (Lipset and Schneider, 1987).[2]

Taken together, the trend evidence coalesces around a picture of Americans hardly ebullient about their social and political lives. Rather, they appear disaffected by their institutions, frustrated in their ability to direct them, and resigned to personal disengagement. Americans are especially critical, the evidence also shows, of the leaders of major institutions: of their competency, their trustworthiness, and their integrity. Revealed, too, in these results may be a more fundamental and more disturbing message, stated eloquently by anthropologist Roy Rappaport (1988:190): "Much more serious than discrediting of particular authorities is the risk to the social system generally of loss of confidence in its own information processes—its basic values, meanings, and understandings of the world, its trust in the probity not only of particular authorities but authority generally." This, then, was the social environment into which the topic of nuclear waste would enter in the early 1980s with the passage of the 1982 Nuclear Waste Policy Act.[3] Under a canopy of general mistrust of institutions, and of the government in particular, the way was paved for the mistrust of government agencies responsible for the execution of waste policy. Thus, the DOE and other responsible agencies stepped into an arena of profound skepticism. Had they entered the arena forewarned and prepared for this skepticism, they might have had the opportunity to establish their own public trust, or at least forestall the effects of the negative halo already present. Instead, as the evidence of this volume clearly shows, mistrust of the DOE and other agencies—due, no doubt, to the general mistrust as well of specific agency actions—has deepened.

Fix Fixe

In view of the long-term erosion in institutional trust, and in view of the proximate factors revealed in the empirical evidence assembled here, what is the likely response of the national state to the challenge of disposing of nuclear waste? Because of considerable institutional inertia, we suspect the response will be deeply rooted in past practices. New programs may appear with new names and new vigor, but it seems unlikely that they will be able to overthrow the cake of institutional custom.

Problems encountered in the development and implementation of technology by the national state have attracted a singular strategy: the application of a variety of fixes. Nuclear technologies are an exemplary case of this general strategy, as evident in the policy styles of lead federal agencies such as the U.S. Atomic Energy Commission (AEC) and its successors, the DOE and the U.S. Nuclear Regulatory Commission (NRC). The development and promotions of the technology, as well as problems arising from the regulation and management of nuclear power, have typically produced one of several types of policy fixes from these agencies. Three, in particular, can be identified, each with a long history in nuclear development and regulation: technological fix, judicial/legislative fix, and knowledge fix.

Versions of all three fixes are currently either under consideration or in some stage of development or implementation. Together they have held steady aim toward a nuclear target with shifting bull's-eyes; at certain times emphasizing development and promotion of the nuclear option, while at other times seeking solutions to management problems, such as the disposal of nuclear wastes. It is especially apt to discuss the three fixes against the background of empirical evidence presented in the previous chapters regarding the public acceptability of a geological solution to the high-level nuclear waste problem.

Technological Fix

A long standing theme of Western culture has been to praise the virtues of technology, even to exalt technology to utopian heights.[4] In the past, such a deep belief in technology as ultimate savior have often reached absurd proportions. For example, as philosopher of technology Don Ihde reminds us:

> Science-technology, rightly applied and developed, it was believed, would eventually solve most, if not all human social and personal problems. Certainly the major ones like poverty, crime, disease, pests, and the like would once and for all be eliminated. Today one seldom finds such globalistic utopianism. But what might be called specific or single system utopianism still abounds in various beliefs in the *technological fix*. (1990:6–7)

While somewhat tempered, the beliefs have retained their core optimism; within well defined scopes of applicability, it is still believed that certain problems can be solved solely with technological know-how. One source of this "techno-fix" ethos is, of course, technologists themselves: scientists, engineers, technicians, and other professionals committed to the creation or improvement of technologies.

Nuclear power, perhaps more than any other modern technology, has, throughout its brief history, been underpinned by the optimism of technological fix. The decision to develop the technology in the first place was driven by this optimism. While knowingly exaggerated, the phrase "too cheap to meter" nonetheless revealed the deep belief that nuclear power would "fix" energy problems, inevitably making the world a much better place to live. Such grand utopianism no longer shrouds the technology. A long series of setbacks has forced a more sober reassessment of the benefits of nuclear power. Nevertheless, the optimism of "specific system utopianism" lives on in the belief that whatever ails nuclear power is ultimately curable—curable via technological fix. Safety problems, for example, can be addressed with new reactor designs, so-called "passively safe" reactors.

Despite the lengthy history of the theme, the actual term "technological fix" is of recent vintage. It was coined by Alvin Weinberg (Teich, 1990) in the mid-1960s, before nuclear power's setbacks and while it was still widely supported by Americans. So popular was the technology then that countless applications were envisioned for it, among them the desalination of sea water to cultivate the poor farming regions of the world. In explicating the new term, Weinberg's key point cannot be overemphasized because it cemented into place the fundamental proposition of the technological fix: that social problems are more quickly and more efficiently solved through the application of technology rather than by relying on a multitude of people to act rationally (1966). This notion it has been the bedrock of the technological fix ever since and is evident in the other fixes as well.

Better Containers as Techno-Fix. In the context of nuclear waste issues, there are two proposals of technological fix for moving beyond the current impasse. The first of these, favored by Weinberg (1992a), stems from a recognition that the safe, long-term storage of high-level nuclear wastes in any geological formation is problematic at best. The solution to this problem is one technological fix: shift scientific effort away from the geology toward the canisters that will house the waste itself. In particular, the United States, following the lead of Sweden, should develop copper canisters that will be resistant to leakage and, in Weinberg's words, "catastrophe proof." What makes the case for copper compelling for Weinberg and others is the fact that native copper has been in the earth and stable for millions of years. Presumably, then, canisters made of copper would be more than adequate to meet the needs of nuclear decay taking tens of thousands of years.

Alchemy as Techno-Fix. A second technological fix would be to burn nuclear waste via transmutation. Several national governments, principally through their national laboratories, have begun to investigate this possibility. The physics of this type of transmutation has been known since the 1960s, at least, but the concept failed to attract the concentrated research effort needed to make it practical. In the United States, growing accumulation of nuclear wastes, and stalled efforts to establish a permanent geological repository, have renewed interest in the concept. Feasibility projects are underway at Los Alamos and Brookhaven National Laboratories, as well as at DOE's Hanford Nuclear Reservation, and experimental work is being done at Argonne National Laboratory in Idaho.

Transmutation, the transformation of one element into another, was the medieval alchemist's dream. The dream was partly realized in this century when scientists discovered how to create nuclear reactions: in fission and fusion processes fuel elements are transformed into entirely new elements. Indeed, it is transmutation that produces one of nuclear power's major problems—nuclear waste, the very subject of this volume. For example, enriched uranium fuel is transmuted during the fission process into a whole host of highly toxic radioactive elements, such as strontium 90 and cesium 137, and a small amount of plutonium (an element that is both extremely toxic and a raw material for the fashioning of nuclear bombs). Were the irradiated fuel to be reprocessed, in order to squeeze out more of its fissionable potential, this chemical separation process would produce even vastly greater quantities of plutonium. So, while nuclear power produced neither energy "too cheap to meter" nor alchemist's gold, it did produce transmuted byproducts: highly toxic nuclear wastes.

Burning nuclear waste would, in essence, be a second-order transmutation process. The undesirable products of the first-stage transmutation, the nuclear wastes, would themselves be transmuted into other products that decay more quickly into stable elements. Alternative approaches to waste transmutation, often involving fundamentally different processes, are being pursued at the four research sites mentioned above. Rather than describing each one, we will examine the proposal viewed by some experts to be the most feasible (Gibson, 1991; Hinman, 1991a)—the one by Los Alamos National Laboratory. Called the accelerator transmutation of waste (ATW) project, its principal goal is to transform dangerous radionuclides into less harmful substances. If successful, the procedure could reduce the lethality of high-level nuclear wastes from the current tens of thousands of years to a few centuries or even a few decades.

Transmutation takes place at a subatomic level, typically with the action of neutrons on the atomic nucleus. For example, reactions in a typical nuclear power plant begin with the bombardment of a fissionable material, such as enriched uranium, with a neutron. The neutron splits the nucleus, releasing huge amounts of heat energy and other neutrons, setting off a chain reaction that is self-generating. Similarly, the Los Alamos ATW project would use neutrons for waste transmutation. Long-lived radioactive wastes—both the lighter atoms, such as technetium, and heavier atoms called actinides, such as plutonium—would be extracted and showered with neutrons. Some of the atoms would absorb the neutrons to create short-lived elements that soon decay into nonradioactive isotopes. Thus, technetium 99, with a half-life of 200,000 years, would be transmuted into ruthenium 100, a substance that is not radioactive. Similarly, iodine 129, with a half-life of 16 million years, would be transmuted into stable xenon 130. The nuclear structure of the actinides, such as plutonium, would be destabilized by the bombarding neutron resulting in a much faster decay of their radioactivity, from thousands to hundreds of years. Thus, much of the transmuted waste would still be dangerous and require repository-type isolation, but for centuries not millennia.

The flow of neutrons needed for transmutation would come from high-energy linear accelerators. It was the recent improvement in these accelerators—due ironically, to another technological fix project, the Strategic Defense Initiative—that drew the transmutation concept into the realm of feasibility. The accelerators would be used to direct intense beams of protons onto heavy metal targets of lead, bismuth, or tungsten, resulting in the release of a sizable number of neutrons. This neutron flux would then shower the nuclear wastes.

Even if it proves practical, waste transmutation is, both supporters and opponents agree, a painstakingly slow process. For example, it is estimated that it will take thirty to forty years to transmute the nation's military wastes (Lawrence, 1991), even longer for the substantially larger quantity of civilian wastes (Gibson, 1991). By raising other issues about transmutation techniques, opponents have made the technology controversial within scientific circles. Among the questions raised by skeptical observers and opponents is whether, in fact, the amount of waste may be increased, rather than reduced. One of the technology's most informed and most persistent critics points out that transmutation will also activate materials, introducing radiotoxicity into materials that were previously free of it. Furthermore, transmutation involves very sophisticated reprocessing, raising once again the troubling issues of nuclear proliferation and the safeguarding of dangerous nuclear materials (Pigford, 1991).

Despite the continuing controversy over its feasibility, the DOE is supporting the efforts of several researchers to assess the technology's feasibility (Hinman, 1991b). These efforts are largely due to the intractability of siting a permanent geological repository. If successful, transmutation technology would reduce markedly one of the major anxieties associated with a repository—the length of time needed to isolate dangerous wastes. Would this technological fix, by reducing a key anxiety, also result in a social and political fix? Would communities now uniformly opposed to a local nuclear waste site be willing to reconsider the acceptance of a transmuted waste site? Would the reduced risks be socially and politically acceptable?

The evidence presented in this volume has assessed the acceptance of a permanent nuclear repository within the context of technologies believed already to be feasible. Thus, to address the public acceptability of transmuted wastes takes us beyond the scope of the empirical evidence presented. A direct and thorough assessment of that acceptability, therefore, awaits future research. In the meantime we can, nevertheless, inform the issue with some reasonable speculations based upon the data that are available and assembled here. Chapter 7, by Desvousges et al., and chapter 10, by Krannich, Little, and Cramer, discuss the concerns about cross-generational risks stemming from the need for long-term isolation (at least 10,000 years) of a HLNWR. Cross-generational problems should be solved in part if the new transmutation technique proves successful.

In our view, successful transmutation might mollify public concerns somewhat but not nearly enough to ensure the easy acceptance of a waste facility. All the available evidence points to deep public concern with almost any feature of nuclear power, but there are especially deep-seated concerns with radioactive wastes. These concerns are due only in part to the risks associated with the wastes. Just as important to public concern is the lack of trust in the institutions and agencies responsible for the disposal of wastes. And public trust will be difficult to regain. Thus, not only are proponents of transmutation technology faced with the challenge of demonstrating the feasibility of the technology, they are also faced with a much more difficult challenge: how to convince a skeptical public that their evidence of feasibility can be trusted.

Perhaps the issue was nowhere better summarized than by *Science* magazine (1991:1613), in its brief coverage of the ATW project: "But [the] technological challenge might be simpler than the political and environmental challenges posed by any scheme for coping with nuclear waste." This conclusion is consistent with both the available evidence and our considered judgment. For the lesson to be learned (and, unfortunately, to be relearned) from the evidence in this volume is that technological fixes, on the one hand, are seldom a problem solution in themselves and, on the other hand, they often attract as many problems as they solve.[5]

Judicial/Legislative Fix

For much of the 1980s, nuclear waste was the bull's-eye aimed at by the judicial/legislative fix. With the passage of the Nuclear Waste Policy Act (NWPA) in 1982 and the Nuclear Waste Policy Act Amendment (NWPAA) in 1987, this fix had put into place, it was assumed, an effective program for disposing of wastes. Once the program was in place, the focus of waste disposal activity shifted away from legal and legislative issues and toward the implementation of the NWPA fix itself. Implementing the siting of a repository, the orienting perspective of this volume, continues to be the central action of the national state in dealing with nuclear wastes.

Having "successfully" hit the waste bull's-eye, the aim of the judicial/ legislative fix shifted back at the close of the 1980s to issues associated with the revitalization and growth in nuclear power itself. In 1989, as pointed out by Rosa and Freudenburg in chapter 2, the NRC had approved procedures for streamlining the licensing of new nuclear power plants. A key provision of the procedures was struck down in November 1991 by a three-judge panel in the United States Court of Appeals for the District of Columbia.

The ruling was appealed by the NRC, who requested a rehearing en banc (before the entire court), rather than before only the three-member panel. Oral arguments for the rehearing were presented in late November 1991, with a decision still pending as of mid-1992 (Dunkle, 1992). The expectation is that whatever the outcome of this round of judicial action, the struck-down provision will be subject to litigation that will finally be settled by the U.S. Supreme Court—realistically, not before mid-1994 (U.S. Senate, 1991a).

With the appeal in process, the Bush administration moved to different ground in order to pave the way for the ambitious revitalization of nuclear power contained in its national energy strategy. Because it was estimated that over 25 percent of that strategy requires legislative action (Rogers, 1991), it was to the Congress that the administration turned for support. While Congress, since the early 1980s, has made numerous attempts to reform nuclear licensing, none has been successful. In the late 1980s and early 1990s, despite these earlier failures, Congress renewed its efforts to reform nuclear licensing procedures with a variety of new bills.[6] Key among these was the proposed National Security Act of 1991 (S. 341), a comprehensive energy bill introduced by Senator Bennett Johnston,[7] Democrat, of Louisiana, and Malcolm Wallop, Republican, of Wyoming. Consistent with Bush administration policy, key provisions of the bill (titles XII and XIII) were designed to aid the nuclear industry—explicitly included was a provision for combining the construction and operating licenses of nuclear power plants.

Due to other controversial provisions of the bill, S. 341 began losing momentum on the floor of the Senate in the summer of 1991. Fearing its defeat, Johnston and Wallop submitted replacement bill S. 1220, with the most controversial provisions of S. 341 deleted. But even the watered-down version of the massive energy bill was derailed on November 1, 1991, when a 50 to 44 cloture vote in the Senate fell short of the required 60 votes (Dunkle, 1992). Despite its defeat, Johnston was still hoping to push ahead with S. 1220 by offering a version trimmed further with the elimination of still another remaining controversial provision (Greenwald, 1991).

In February 1992, the Senate passed a comprehensive energy bill (S. 2166) by an overwhelming majority, 94 to 4 (Idelson, 1992a). It was the first effort to overhaul the nation's energy laws in a decade. A key provision of the bill was the adoption of the Nuclear Regulatory Commission's combined licensing rule, the one struck down but still under review in the federal courts. The streamlining of the licensing process has long been viewed as the key incentive needed to spur renewed growth in nuclear power.

Then, in May 1992, the House of Representatives passed its own version of an energy bill (H.R. 776), likewise by an overwhelming majority, 381 to 37 (Idelson, 1992b). The House version went even further in paving the way for the rejuvenation of nuclear power. Like the Senate bill, it included a provision for streamlined licensing with a one-step licensing procedure. But, it also included a provision to speed up efforts to site the nation's first HLNWR at Yucca Mountain. Specifically, the House bill would preempt state laws requiring the issuance of permits to perform the necessary geological studies at Yucca Mountain. Because of the preemption provision, Nevada's senators, Democrats Richard H. Bryan and Harry M. Reid, have vowed to fight the passage of the entire energy package (Idelson, 1992c).

The final version of the energy bill easily passed the House by a 363 to 60 margin and the Senate by voice vote on October 8, 1992. It was signed by President Bush on October 24 to become law (PL 102–486). Upheld in the final version were the two provisions most germane to the construction of nuclear facilities: the one-step licensing proposal of the NRC for building new nuclear power plants and provisions to ease construction of a HLNW repository at Yucca Mountain, Nevada. (For a history of the law's enactment see Kraft, 1993). Whether this last round of legislative fix will markedly alter the fortunes of nuclear power or nuclear waste is problematic given the disfavor of nuclear power shown by the Clinton Administration (Anderson, 1993).

Knowledge Cum Public Relations Fix
Massive efforts to promote nuclear power, not only in the early days of commercialization but also for decades afterwards, are so well-known and so well-documented that the issue requires only brief treatment here (Weart, 1988; Rhodes, 1986; Boyer, 1985). The nuclear subgovernment, then as now, was guided by the unshakable belief that increased public understanding— the knowledge fix—would translate into support for nuclear technologies. All that was required was thoughtful public relations to convert the dull, scientific knowledge into interesting, convincing public knowledge. Belief in a knowledge fix was given credence by early successes of the AEC's public relations efforts. Among its many activities were traveling exhibits, stationary exhibits and demonstrations at AEC facilities or those of its contractors, and what might best be described as "nuclear fairs" that featured such attractions as "Theater of the Atoms" and "atomic pinball machines" (Boyer, 1985). Visitors to the General Electric exhibit at one of these "fairs," the 1948 month-long "Man and the Atom" exhibit in New York's Central Park, "received free copies of *Dagwood Splits the Atom*, a colorful comic book produce by King Features Syndicate in consultation with the AEC. . . . Over 250,000 copies were distributed, leading GE to order a further printing of several million" (Boyer, 1985: 296).

The intensity and pace of nuclear promotion continued nearly unabated for decades. As noted in chapter 2 by Rosa and Freudenburg, more than 40 million people attended screenings of AEC films in the 1960s, and another 158 million watched the films on television. These early successes, it appears, became an embedded feature of the culture of nuclear promoters and government agencies. For, despite a complete turnaround in the image of the technology (Weart, 1988) and despite a similar turnaround in public support for it (Rosa and Freudenburg, chapter 2), those early successes have led to a persistent belief in the efficacy of the knowledge fix. One need only travel to any national laboratory, the facilities of many nuclear contractors, or the reception area of many utilities to see that public support via "knowledge through exhibit" is still a guiding belief of the nuclear culture.

Siting Solutions Through Public Relations? Evidence of the persistent belief in the knowledge fix is illustrated even more dramatically with nuclear wastes. During the final stages of this volume's preparation, the nuclear industry, in direct response to the overwhelming opposition by Nevadans to the siting of a nuclear repository in their state, prepared to launch a concerted public relations and advertising campaign. The industry was apparently convinced that "[m]ajor shifts in public perceptions and attitudes are achievable only through a sustained advertising program aimed at Nevadans. . . ." (Associated Press, 1991:B1–B3). By late 1991, radio and television advertisements supporting industry goals had already begun to air in Nevada. And, by then, $800,000 had already been spent to "neutralize the political resistance and public opposition to the site," and, subject to final approval by utility executives in early 1992, an additional expenditure of $9–$10 million over the three-year period of the campaign was anticipated. The monies would come from the fifty utilities around the country that own or operate nuclear reactors, each contributing between $33,000 and $100,000 (Schneider, 1991; Associated Press, 1991).

Key objectives of the campaign were contained in a blueprint prepared by the American Nuclear Energy Council, an industry trade association. "According to the confidential plan, the nuclear industry was out to win the hearts and minds of Nevada residents . . . to establish the nation's only high-level nuclear waste repository at Yucca Mountain. . . ." (Schneider, 1991:A11). Furthermore, the plan predicted that the media and public relations effort would convince a majority of the state's citizens to support construction of the site in their state. The plan's operational goal was to counter the "misinformation, misstatements, and false statements" made by Nevada politicians, state officials, and organized opponents to the repository siting.

The principle means for accomplishing this goal would be media blitzes consisting mostly of extensive television, radio, and newspaper ads. But the campaign also called for the training of teams of DOE scientists—sometimes referred to by industry advisors as "truth squads" or "attack-response teams"—in public relations. The job of these public relations teams would be to respond to media reports, whether in newspapers or magazines or on radio or television, appearing in Nevada and thought by prorepository forces to be inaccurate or misleading. The DOE would, by eliminating inaccuracies and by correcting misleading impressions, educate the public on the scientific facts associated with the site. Presumably, then, the public—now armed with scientific facts—would recognize that opposition to the repository was unfounded and shift to a favorable attitude, perhaps even enthusiastic support. This shift, according to the plan, would occur remarkably quickly—within two years.

That the idea for the campaign originated with a leading advertising executive in Las Vegas is hardly surprising. That it is apparently modeled after the type of campaign for which advertising and public relations firms have enjoyed great success—the selling of products or candidates—is, too, hardly surprising. What is surprising is the willingness to ignore decades of scientific social research, including the focused research of this volume, in order to embark on an expensive campaign whose success is, at best, dubious. Nearly all available empirical evidence gathered systematically by social scientists points to the conclusion that public concerns over nuclear power cannot be made to disappear with public relations efforts (Freudenburg and Rosa, 1984). Indeed, the results of previous industry efforts to increase the public acceptability of nuclear technology might best be described as a Chinese finger trap: the harder they tried to pull opinion in that direction, the more elusive became success.

That conclusion is not only reinforced by the evidence presented in this volume; it is also deepened by revealing its direct connection to repository siting issues. A nuclear waste repository is, simply put, too deeply etched with dreaded, stigmatized images in the public mind for the public to be easily swayed by advertisement. Advertising may be effective in getting consumers to make inconsequential choices at the surface of consciousness, such as purchasing laundry detergent A over detergent B. However, deeply held concerns with inordinately greater consequences, such as those associated with nuclear waste, are another matter entirely. Furthermore, DOE officials—the very members of the scientific public relations teams who are expected to change public attitudes—are consistently mistrusted by citizens. There is little reason to believe that, once these teams appear in the media with the "scientific truth," they will be any more trusted than in the past. Indeed, a reasonable prediction, based upon the cumulative evidence, is that the campaign will, contrary to the industry's keen optimism, not only fail but exacerbate matters further. Quantity of media coverage has been systematically shown to *increase*, not decrease, opposition to nuclear technologies (Mazur, 1990). Taken in total, the evidence leads us to predict that the campaign would likely increase opposition to the repository and mistrust in DOE officials.

These predictions are partly confirmed by results from a subsequent Nevada statewide survey conducted from October 25 to November 5, 1991 (Flynn, Mertz, and Slovic, 1991; Flynn, 1992). A vast majority of respondents had seen the ads (72 percent), most of them via television (86 percent). At this time about 75 percent of Nevadans were opposed to the repository. Respondents were asked whether the advertisement had made them more supportive of the Yucca Mountain repository program, less supportive, or about the same as before the ads. A majority (53 percent), reported no change in their position on the repository program. Of those who changed, twice as many (32 percent) shifted toward *less support* than those (15 percent) who shifted toward more support.

While the survey questions do not permit a direct test of our second prediction, increased mistrust, some of the evidence is nonetheless germane to it. In particular, when respondents who had heard about the ads were asked," Would you tell me, what is your single most important opinion or feeling about these advertisements?" the most common response (49 percent) was "disbelief." Another 15 percent either disagreed with the ads or felt insulted by them. The negative cast of these responses, especially the widespread disbelief, suggests that the ad campaign will simply strengthen the pervasive distrust of Nevadans toward repository siting.

Democracy as Imbroglio

Camus's epigram that opened this chapter sends us back to classical Greece, to the origins of the very governing procedures that challenge us here. For the democracies of that era, the *idiot* was defined as a totally private person, the citizen disengaged from political life (Mills, 1967). It was also classical Greece where plays were used, not solely for entertainment, but principally to make people think and to give them moral guidance. Greek tragedy provided guidance by warning of unalterable forces acting on mortal lives. Here it provides a vivid and useful metaphor to understanding a complex web of self-perpetuating forces that act as constraints to democratic process and as obstacles to solving the nuclear waste impasse.

The nuclear waste problem has been defined as a scientific and technical problem. Defining an inherently political problem as scientific or technical creates a process of self-fulfilling defeat of the tenets by which democratic governance are presumed to operate (also see Jacob, 1990). Citizens are placed into an inextricable double bind. They may wish to take an active role in the politics of technology (thereby avoiding the epithet *idiot* in the classical Greek sense). They may wish to challenge, for example, the basis upon which judgments about acceptable risks are made or the acceptability of the risks. Yet, because their role has been distanced from central decision making, the only avenues of critical expression are obstructionist: to object, to naysay, or even to protest decisions by the state. With such a narrowly defined role as this, those citizens who do express concern or objection are viewed as driven by narrow, self-serving, and uninformed interests.

Defining a problem as scientific or technical sets the stage, arranges the props, and casts the actors who are "qualified" to take part in its solution.[8] Because citizens lack an understanding of the scientific details of complex technological problems, they are pushed off the stage of key decisionmaking—viewed by some insiders, if not publicly as idiots, at least as unqualified to make informed judgments. Science is thus used to justify insider decisionmaking; the effect is to disempower the electorate. Protest, via a questioning of the credibility of the scientific data and of the means for incorporating it into policy, prompts a familiar response from government agencies discomforted with outside criticism: application of a fix. Deeply lodged in a culture of fixes, such agencies see this as a problem of knowledge fix: the reason citizens are objecting is that they lack all of the pertinent evidence; the masses are technologically uneducated (another meaning of the term *idiot*).[9] Once provided with the "true facts" of the matter, no thinking, reasonable person could fail to see that the solution of the experts is also the solution of the people. As an article of faith, this belief lives on in the culture of the policy establishment. As a circumvention of democratic process it is barely discussed.

Will Fixes Fix?

For all their apparent differences, the separate categories of fix and accompanying policy styles are bound together with a common emphasis and singular effect: minimization of direct public participation. By erecting barriers to the equal access to technological decision making, citizens are displaced from the center of policymaking. This persistent practice likely raises suspicion about the openness and commitment to democratic process by the agencies responsible for managing nuclear power.[10] Indeed, such suspicion can only reinforce and exacerbate the widespread public mistrust of the siting process and of the agencies responsible for it.

As currently constituted the technological fix is doomed to failure so long as the agencies responsible for solving the waste problem continue to be mistrusted. A regaining of trust will require a fundamental modification of their cultures, cultures inherited from a military past of secrecy and insulation from public accountability. This lag in the culture of responsible agencies, recognized by not only their critics, but also the secretary of energy himself (as pointed out in chapter 1 by Kraft, Rosa, and Dunlap), must be addressed—regardless of what fix is applied—if the waste problem is to be solved within current institutional arrangements.

Other problems court any fix strategy. The continued ownership of the nuclear waste problem by the DOE all but guarantees their definition of the problem, and of solutions that will be congruent with the expertise of the agency. The persistent practice of defining the problem in terms of available fixes, or those believed the easiest to achieve, not only reduces the available alternatives, but also reduces the range of solution strategies. Agency orientation, deeply rooted in its own history, makes a recasting of the problem extremely difficult. The sanctification and legitimation of the problem as one requiring available fixes reaffirms agency authority, on the one hand, while simultaneously delegitimating alternative approaches. The three conventional fixes discussed above overlook the most important fix of all: a fix through participatory democracy. In view of the body of evidence presented in this volume, the application of any of the three conventional fixes, devoid of a focused effort of public involvement, will prove to be a step in the wrong direction.

Retrospect and Prospect

What is the level of public support and political acceptability of a HLNWR in the United States? In this volume citizens from every region of the nation, from every social strata, from every political persuasion have, through the instruments of social science, expressed their views on this question. Against the rigorous evidence provided by them, an unequivocal conclusion emerges: A HLNWR is unacceptable to a majority of Americans for the foreseeable future. Majorities of citizens from Nevada, Texas, Washington state, Arizona, California, and the nation as a whole, while expressing a wide variety of views on the siting of a HLNWR, are unanimous about their rejection of such a facility. For people everywhere and in every walk of life a repository is simply, and unequivocally, unacceptable; it is far too risky and those responsible for its management are not to be trusted.

What is extraordinary about this conclusion is not its overawing surprise but a remarkable consistency across the various studies leading up to it. The assembled empirical results cover the map: the geographical map (nation, repository sites, rural versus urban, etc.), the social map (sex, race, social status, etc.), and the cognitive map (raw impressions, perceptions, ponderous assessments, and voting intentions). Yet, despite the breadth of coverage and despite the intricacies of this complex issue, the results converge to form a reasonably orderly picture. Rare are instances (even rarer for complex issues) when social scientific investigations produce such consistent results as reported in this volume. That the results stem from such a diversity of methods and locales makes the case all the more compelling.

So widespread is the disapproval of this especially dreaded facility that it does not conform neatly to the NIMBY syndrome. Rather, the message citizens are providing the national state and policymakers is both NIMBY and NIABY ("not in my back yard," "not in anyone's back yard"). The results from the states of Nevada and Washington are especially instructive, for they are both states with a long history of experience with and support for nuclear technologies.

Noteworthy, too, is the profile of the citizenry to emerge from the data: that of individuals willing and able to weigh the importance of social issues and political concerns. It is not the profile of *homo oeconomicus*, the cold, rationally calculating citizen who, devoid of values or politics, acts to further immediate self-interest. It is not with economics that citizens are most concerned but with issues of health, safety, and political trust and with risks imposed not only on themselves but on future generations as well.[11]

People's concerns are neither superficial nor confined to the surface level of thought but rather reach the inner depths of consciousness with submerged images of dread. Because of their depth, these images, and their expression in attitude and action, are likely to be durable—extremely difficult to change over a short period of time. This problem is confounded further, as noted above, by the pervasive and profound mistrust of the management of nuclear technology: mistrust of science, mistrust of industry, and mistrust of agencies of the national state. The credibility problem of the DOE and other lead agencies is magnified, there is little doubt, by this broader milieu of skepticism. The dread and mistrust is compounded further by perceptions of inequity—the inequity of siting a repository in Nevada, a state that generates none of the waste to be housed in the repository, and the inequity of burdening future generations with the externalities of past and present generations.

As pointed out in chapter 2, by Rosa, and Freudenburg and in chapter 3, by Slovic, Layman, and Flynn, regaining public trust is decisive to solving the nuclear waste problem. But, as also pointed out in those chapters, the establishment and loss of trust is an asymmetrical process: trust is far more easily lost than regained. Thus, the difficulty of reestablishing trust cannot be exaggerated. The difficulty may be exacerbated further by the very nature of the problem itself. The risks of nuclear waste are so dreaded that it may be difficult for *any* organization to establish the trust and credibility needed to attract public support necessary for effective management.

As for Nevada itself, home of the only candidate site for a repository (at Yucca Mountain), opposition is high, persistent, and widespread. Furthermore, the state of Nevada, as well as its congressional delegation, remains staunchly opposed to the repository. In addition to concerns about risk, Nevadans are especially concerned about the fairness of the siting process—a process, in their eyes, biased to saddle them with the repository whatever their objections. Nevadans can also be justifiably concerned that the siting of a repository may result in stigmatization of their state and lead to a variety of unwanted consequences. Among these, as documented by Easterling and Kunreuther in chapter 8, are sizable impacts to the $1 billion per year contribution to the Las Vegas economy from conventions and trade shows. The impacts on Las Vegas's gambling revenues, estimated by some sources as $5½ billion annually, though not yet specifically assessed, would surely be great, as well.

The evidence compiled and analyzed in this volume builds to a conclusion about the social feasibility of siting a HLNWR at Yucca Mountain or anywhere else in the United States. Arguing on grounds of excess scientific uncertainty and of inattention to the ethics issue of protecting future generations, Shrader-Frechette (1993) argues that we *ought* not site a HLNWR at this time. Rather, we should postpone permanent disposal until there is a greater scientific base from which to predict consequences and until there is protection for future generations.

The problem of siting a HLNWR can be best summed up, not in the words of distinguished scientists, nor in the pronouncements of leading statesmen, nor in the prognostications of social pundits, nor in the eloquent disquisitions of poets or philosophers, but in the words of Jerry Scoville, president of U.S. Ecology, a company specializing in the disposal of hazardous wastes, when he said, in connection with the siting of a low-level nuclear waste dump, "Involving the public does not guarantee success, but not involving the public just about guarantees failure" (1989).

Succinctly stated, this is the lesson learned in attempting to site a permanent HLNWR in the United States; it is the lesson crafted by the rigorous empirical work of this volume; it is the lesson that citizens of the nation have been providing the national state and its agencies; but it is a lesson that is yet to penetrate the culture of decision makers charged with the task of solving the nuclear waste problem. Until that lesson is learned, the biggest risk may not be from a waste storage facility but from a siting process that chooses to ignore the views and preferences of citizens. Until that lesson is learned, the siting of a permanent HLNWR in the United States through democratic process is all but an assured failure.[12]

Notes

For this the final chapter we had the great fortune to receive critical comments and suggestions from many scholars, but we are especially thankful for the comments of nuclear physicists George Hinman and Alvin Weinberg, chemical engineer John Sheppard, philosopher Kristin Shrader-Frechette, historian Jerry Gough, and sociologists Chip Clark, Chris Cluett, James Flynn, Bill Grigsby, Greg Hooks, Loren Lutzenhiser, Allan Mazur, Angela Mertig, and James F. Short, Jr.

1   We consciously avoid the word *synthesis* because that would be tantamount to the claim that we are also able to synthesize the various disparate perspectives and methods of the social science disciplines that produce the empirical results.

2   These five items, forming an index, are used by the Harris surveys to measure alienation.

3   Another key factor was the growing strength of public concern about environmental quality (Dunlap and Scarce, 1991).

4   On this well-known point, distinguished historian of technology, Thomas P. Hughes, writes: "Technological enthusiasts throughout history have tended to be utopian, assuming that available technology will be used to fulfill their particular visions of the future to establish a postmodern technology" (1989:458).

5   The point was made over a decade ago by Metlay (1978:1) in an NRC report: "The persistent faith in a technological fix has produced a myopic vision of the waste management problem. . . . Those who believe in technological fix strive to eliminate the human factor—an element which, it is generally held, can only produce noise."

6   Bupp and Derian, in their historical analysis of nuclear power which became a near-instant classic, are unguardedly blunt in their assessment of such efforts: "Licensing reform is not merely administrative rationalization. It is modification of *basic* rights of political participation" (1981:xx).

7   Johnston is chairman of the Senate Energy and Natural Resources Committee, a powerful body in the setting of national energy policy.

8   Consistent with this argument is the excellent analysis of change in nuclear policy by Baumgartner and Jones (1991). Political actors within any policy subsystem, such as nuclear policy, manipulate the degree of outside influence on policymaking by, among other things, establishing the venues for policy dialogue. Indeed, this practice is a basis for a common criticism of pluralistic governance: "One characteristic of pluralist government often and accurately noted by political scientists is the ability of single-industry economic interests to insulate themselves from the influence of large-scale democratic forces through the creation of relatively independent depoliticized subgovernments" (1991:1045).

9   The "knowledge fix" solution is further embarrassed by an indiscretion in logic, the failure to define how much knowledge a layperson must have to be considered "knowledgeable" on the issues. On the one hand, this reasoning violates the tenets of

any experiment, whether an actual or thought experiment, that requires a baseline or standard of comparison in order to yield interpretable results. What is the standard against which to compare lay knowledge? On the other hand, it is neither practical nor logical to expect laypersons to have knowledge equal to experts. Since expertise is defined as *specialized* knowledge, to evenly distribute specialized knowledge among everyone would deny anyone the status of *expert*.

10 Again eschewing euphemism in discussing the control of nuclear power more generally, Bupp and Derian observe that there is "another unhappy image linked to nuclear power: that [of] a small, elite group of scientists and technicians making decisions about what is good for society" (1981:12).

11 Citizen concern for future generations introduces an historical irony. The older generation of nuclear scientists favored a permanent storage facility in order to avoid burdening future generations with the problem (Weinberg, 1992b).

12 The energy policy establishment, especially the DOE, has apparently begun to recognize the compelling message of findings such as those of this volume. In late 1991 both DOE's draft *Mission Plan Amendment* (DOE, 1991a), a proposed revision to the repository schedule, and the report *Ensuring Public Trust and Confidence*, issued by the secretary of energy's Task Force on Civilian Radioactive Waste Management (DOE, 1991b), highlight the importance of regaining public trust. In view of the evidence of this volume, this is a long-awaited, salutary sign. It is, however, important to caution against optimism that a recognition of the problem will ensure a swift and certain solution. For another message of this volume is that the regaining of public trust will require not glossy education or public relations programs, as in the past, but fundamental change: change in the culture of responsible agencies, especially with respect to the role to be played by citizens. It will also require a thoughtful and persistent program of involvement accompanied by patience for demonstrable results. In the absence of such patience, the policy establishment may, frustrated by lack of results, choose, instead of greater citizen involvement, a draconian or authoritarian imposition on a chosen site, like Yucca Mountain. Should it pursue the latter strategy, there would be cause for deep concern on either of two grounds: on the one hand, the evidence analyzed in this volume warns against the success of this strategy and, attendantly, the waste in public resources of pursuing its dubious goal; on the other hand, the success of such a strategy would send an even graver warning—that the national state is willing to circumvent democratic process in order to achieve its own objectives; that democracy is indeed being eroded by technocracy.

References

Anderson, Christopher. 1993. "Clinton Asks for a Greener DOE." *Science.* 260:153.

Associated Press. 1991. "N-Dump Publicity to Target Nevada." *Spokane Spokesman-Review*, November 14.

Barber, Bernard. 1983. *The Logic and Limits of Trust.* New Brunswick, NJ: Rutgers University Press.

Baumgartner, Frank R., and Bryan D. Jones. 1991. "Agenda Dynamics and Policy Subsystems." *Journal of Politics.* 53:1044–74.

Boyer, Paul. 1985. *By the Bomb's Early Light.* New York: Pantheon.

Bupp, Irvin C., and Jean-Claud Derian. 1981. *The Failed Promise of Nuclear Power: The Story of Light Water.* New York: Basic.

Camus, Albert. 1955. "Helen's Exile." In *The Myth of Sisyphus and Other Essays.* New York: Vintage. 134–38.

Clarke, Lee. 1985. "The Origins of Nuclear Power: A Case of Institutional Conflict." *Social Problems.* 32:474–87.

Daily Executive Reporter. 1992. "Energy: Markup on House Energy Bill Could Occur Next Week, Staffer Says." March 3.

Dietz, Thomas M., and Robert W. Rycroft. 1987. *The Risk Professionals.* New York: Russell Sage Foundation.

Dietz, Thomas M., R. Scott Frey, and Eugene A. Rosa. 1993. "Risk, Technology, and Society. In Riley E. Dunlap and William Michelson, eds., *Handbook of Environmental Sociology.* Westport, CT: Greenwood.

Dunkle, Aerin. 1992. "Revitalization of the Nuclear Option: Political Attempts to Find a Role for Nuclear Power in the U.S." Unpublished manuscript, Department of Sociology, Washington State University.

Dunlap, Riley E., and Rik Scarce. 1991. "The Polls—Poll Trends: Environmental Problems and Protection." *Public Opinion Quarterly.* 55:651–72.

Flynn, James H. 1992. "How Not to Sell a Nuclear Waste Dump." *Wall Street Journal,* April 15.

Flynn, James H., C. K. Mertz, and Paul Slovic. 1991. *The Autumn 1991 Nevada State Telephone Survey.* Carson City, NV: Nevada Nuclear Waste Project Office.

Freudenburg, William R., and Eugene A. Rosa. 1984. "Are the Masses Critical?" In *Public Reactions to Nuclear Power: Are There Critical Masses?.* William R. Freudenburg and Eugene A. Rosa, eds. Boulder: Westview/American Association for the Advancement of Science. 331–48.

Gibson, Daniel. 1991. "Can Alchemy Solve the Nuclear Waste Problem?" *The Bulletin of the Atomic Scientists.* 47: 12–17.

Gould, Leroy C., Gerald T. Gardner, Donald R. DeLucca, Adrian R. Tiemann, Leonard W. Doob, and Jan A. J. Stolwijk. 1988. *Perceptions of Technological Risks and Benefits.* New York: Russell Sage Foundation.

Greenwald, John. 1991. "Time to Choose." *Time,* April 29, 54–61.

Hinman, George. 1991a. "New Methods for Radioactive Waste Processing to Reduce Repository Storage Times." Unpublished manuscript, Program in Environmental Science and Regional Planning, Washington State University.

———. 1991b. Personal communication.

Hughes, Thomas P. 1989. *American Genesis: A Century of Invention and Technological Enthusiasm.* New York: Penguin.

Idelson, Holly. 1992a. "Senate Energy Bill." *Congressional Quarterly* (March 7).

———. 1992b. "House Gives Energy Bill Big Win; Lengthy Conference Expected." *Congressional Quarterly,* May 30.

———. 1992c. "Nevada Senators Vow to Block Energy Bill Unless Appeased." *Congressional Quarterly,* June 13.

Ihde, Don. 1990. *Technology and the Lifeworld: From Garden to Earth.* Bloomington, IN: Indiana University Press.

Jacob, Gerald. 1990. *Site Unseen: The Politics of Siting a Nuclear Waste Repository.* Pittsburgh: University of Pittsburgh Press.

Kerr, Richard A. 1992. "Another Panel Rejects Nevada Disaster Theory." *Science*. 256:434–36.

Kraft, Michael E. 1993. "Environmental Gridlock, Searching for Consensus in Congress." In Norman Vig and Michael E. Kraft (eds.), *Environmental Policy in the 90s*, 2nd ed. Washington, D.C.: Congressional Quarterly Press.

LaPorte, Todd R. 1988. "The United States Air Traffic System: Increasing Reliability in the Midst of Rapid Growth." In *The Development of Large Technological Systems*. Edited by Renate Mayntz and Thomas P. Hughes. Boulder: Westview. 215–44.

Lawrence, George P. 1991. "High-Power Proton Linac for Transmuting the Long-Lived Fission Products in Nuclear Waste." Paper presented to the IEEE Particle Accelerator Conference, San Francisco, May 6–9.

Lipset, Seymour Martin, and William Schneider. 1987. *The Confidence Gap: Business, Labor, and Government in the Public Mind* (rev. ed.). Baltimore: Johns Hopkins University Press.

Mazur, Allan. 1990. "Nuclear Power, Chemical Hazards, and the Quantity-of-Reporting Theory of Media Effects." *Minerva*. 28:294–323.

———. 1981. *The Dynamics of Technological Controversy*. Washington, D.C.: Communication Press.

Metlay, Daniel S. 1978. "History and Interpretation of Waste Management in the United States." In W. I. Bishop Inihoor, W. Hilberry, and R. Watson, eds. *Essays on Issues Relevant to the Regulation of Waste*. Washington, D.C.: U.S. Nuclear Regulatory Commission. 1–19.

Mills, C. Wright. 1967. "The Structure of Power in American Society." In Irving Louis Horowitz, ed. *Power, Politics, and People: The Collected Essays of C. Wright Mills*. New York: Oxford University Press. 23–38.

Nealey, Stanley M., and John A. Hebert. 1983. "Public Attitudes Toward Radioactive Waste." In Charles A. Walker, Leroy C. Gould, and Edward J. Woodhouse, eds. *Too Hot To Handle? Social and Policy Issues in the Management of Radioactive Wastes*. New Haven: Yale University Press. 94–111.

Pigford, Thomas H. 1991. "Waste Transmutation and Public Acceptance." Unpublished paper, Berkeley: University of California, Department of Nuclear Energy.

Rankin, William L., and Stanley M. Nealey. 1978. "Attitudes of the Public About Nuclear Waste." *Nuclear News*. 21:112–17.

Rappaport, Roy A. 1988. "Toward Postmodern Risk Analysis." *Risk Analysis*. 8:189–91.

Rhodes, Richard. 1986. *The Making of the Atomic Bomb*. New York: Simon and Schuster.

Rogers, Kenneth C. 1991. "Safety is the Key to a Nuclear Future." *Public Utilities Fortnightly*. (April 1): 54–65.

Schneider, Keith. 1991. "Nuclear Industry Plans Ads to Counter Critics." *New York Times*, November 13.

Science. 1991. Briefing, "A Nuclear Cure for Nuclear Waste," 252:1613.

Scoville, Jerry. 1989. Testimony before the Washington State Nuclear Waste Advisory Council.

Short, James F., Jr. 1984. "The Social Fabric at Risk: Toward the Social Transformation of Risk Analysis." *American Sociological Review*. 49:711–25.

Shrader-Frechette, Kristin. 1993. *Uncertainty, Expert Error, and Radioactive Waste: The Case Against Geological Disposal*. Berkeley: University of California Press.

Stebben, Gregg. 1992. "Meet the Nobelists! This Month's Question: Can You Program Your VCR?" *Spy*, May.

Teich, Albert H., ed. 1990. *Technology and the Future.* 5th ed. New York: St. Martin's.

U.S. Department of Energy. 1991a. Office of Civilian Radioactive Waste Management. *Mission Plan Amendment.* Draft. Washington, D.C.: GPO, September.

———. 1991b. Task Force on Civilian Radioactive Waste Management. *Ensuring Public Trust and Confidence.* Washington, D.C.: GPO, November.

U.S. Senate. 1991a. "National Security Act of 1991: Titles XII and XIII." *Hearing Before the Committee on Energy and Natural Resources on S. 341.* March 5. 102nd Congress, 1st Session.

———. 1991b. "Summary of S. 341." *Congressional Digest* (May): 133–36.

Weart, Spencer R. 1988. *Nuclear Fear: A History of Images.* Cambridge, MA: Harvard University Press.

Weinberg, Alvin M. 1966. "Can Technology Replace Social Engineering?" *University of Chicago Magazine.* 59:6–10.

———. 1972. "Social Institutions and Nuclear Energy." *Science.* 177:27–34.

———. 1992a. "Social Institutions and Nuclear Energy II." Paper presented at the annual meetings of the American Association for the Advancement of Science, Chicago, February 6–11.

———. 1992b. Personal communication.

Winner, Langdon. 1986. *The Whale and the Reactor: A Search for Limits in an Age of High Technology.* Chicago: University of Chicago Press.

Index

American Nuclear Energy Council, 313

Antinuclear movement, 8, 44, 138; historical roots of, 32–40, 54–57

Argonne National Laboratory, 307

Atomic Energy Act: 1946, 33, 35; 1954, 34

Atomic Energy Commission (AEC), 33, 35, 36–37, 39, 46, 305, 312, 313

Battelle Human Affairs Research Centers, 53

Baxter, Rodney K., 24, 136

Brandeis, Louis (Justice), 292

Brody, Julia G., 23, 115, 294, 298

Brookhaven National Laboratory, 307

Brown's Ferry nuclear power plant, 38

Bryan, Richard, 209, 312

Bush, George: administration of, 12; nuclear waste and, 6

California, nuclear waste referendum in, 36

Cambridge Reports, 45, 47

Carter, Jimmy: interagency review group (IRG), 8–9

Chernobyl accident: effect on public attitudes, 16, 50

Citizen participation. See Public participation

Clary, Bruce B., 23, 89, 297

Cold War, 33

Confidence. See Public trust

Congress, U.S., 43; nuclear waste and, 6–10, 16, 89, 90, 311–312

Council on Environmental Quality, White House (CEQ), 8, 51

Court of Appeals, U.S. District of Columbia, 42, 310

Cramer, Lori A., 26, 263, 295, 302

Credibility. See Public trust and Department of Energy

Critical Mass, 46

Deaf Smith County. See Texas

Department of Energy, U.S., xiv, 4, 10, 11, 14–16, 19, 33, 38, 40–41, 64–66, 89, 292, 305, 309, 314, 317; Dose Reconstruction Project, 139. See also Public trust and Public hearings

Desvousges, William H., 24, 175, 294, 299–300

DOE. See Department of Energy

Downwinders, 139, 154

Dunlap, Riley E., 3, 24, 136, 291, 294–295, 298–299

Easterling, Douglas, xiii, 25, 209, 294, 300–310

Energy Policy Act of 1992 (Pub. L. 102–486), 6, 310–312

Energy Research and Development Administration, 33, 36

Farrington, Jerry, 54–55

Federal Communication Act of 1934, 34

Ford, Daniel, 36

Fleishman, Judy K., 23, 115, 294, 298

Flynn, James H., 22, 64, 294, 297
Freudenburg, William R., xiii, 21, 32

General Accounting Office, U.S., 36
Geologic Disposal, 8–9
Gertz, Carl, 210, 217
Goiania, Brazil, 161, 218
Ground Zero, 49

Hanford Education Action League, 139
Hanford Journal, 139
Hanford nuclear reservation, 23, 45, 307;
    Basalt Waste Isolation Program, 146. See
    also Washington State
High-level nuclear waste repository
    (HLNWR), 22; acceptance of, 291–320;
    images of, 70–80, 205–206; knowledge
    about, 145, 157–160, 163–167, 181–185.
    See also Department of Energy, Nevada,
    Nuclear waste
House of Representative, U.S., 311, 312

Idaho, Moscow, 139
Ihde, Don, 306
Impact Assessment, Inc., 142
Implementation, scenarios for, 18–20
Interagency Review Group (IRG), 8

Johnston, J. Bennett, 311
Justice; U.S. Department of, 43

Kraft, Michael E., 3, 23, 89, 291, 294, 297
Krannich, Richard S., 26, 263, 295, 302
Kunreuther, Howard, xiii, 24, 25, 175, 209,
    295, 300–301

Las Vegas convention industry, 25, 209–
    235; convention attendees, 214–216,
    227–228; Convention Planner Survey,
    219–221; effect of nuclear accident, 217–
    218; effect of nuclear repository, 231–
    232; forecasted actual losses, 223–227;
    impact of imagery, 213–215, 228–230.
    See also Nevada
Las Vegas Convention and Visitors Au-
    thority, 209
Las Vegas Review Journal, 210
Layman, Mark, 22, 64, 294, 297
League of Women Voters, 139
Little, Ronald L., 26, 263, 295, 302
Locally undesirable land uses (LULUS), 50
Los Alamos National Laboratory, 307, 308
Louis Harris and Associates, 45

Lyons, Kansas, 8

Mass media, coverage by, 35, 45, 46, 313,
    314–315
Merrill Lynch, 44
Metlay, Daniel, 320
Mitchell, Robert Cameron, 24, 136
Monitored Retrievable Storage (MRS), 16,
    83
Mushkatel, Alvin H., 25, 239, 295, 301–
    302

Nader, Ralph, 46
Nagasaki, Japan, 136
National Academy of Sciences, 4, 15–16,
    19, 127
National Energy Strategy, 6, 42, 311–312
National Environmental Policy Act (NEPA),
    9, 38
National Research Council, 45, 66, 206,
    260
National Science Foundation, 37
Nevada, 175–207; Clark County, 179, 240,
    265, 267, 268; Death Valley National
    Monument, 266; legislature, 65; Las
    Vegas, 179, 209–235, 239–260, 314,
    319; Lincoln County, 205, 265, 268; Nye
    County, 205, 265, 267; rural communi-
    ties, 263–284; Yucca Mountain, 15, 17,
    22, 64–84, 209–235, 239–260, 263–284
Nevada Resort Association, 210
Nevada Test Site, 217, 244–248, 265, 275,
    276; accident reporting and, 217–218,
    252
New Mexico: Alamogordo, 136; Carls-
    bad, 3
Nigg, Joanne M., 25, 239, 295, 301–302
Not in anyone's back yard (NIABY), 318
Not in my back yard (NIMBY), 21, 76, 92,
    96–100, 110, 142, 150–153, 318
Nuclear Information and Resource Ser-
    vice, 42
Nuclear Management and Resources
    Council, 42
Nuclear power; accidents as warning sig-
    nal, 57–58; commercial nuclear energy,
    5, 34; licensing process and problems,
    5–6, 37, 41–43
Nuclear proliferation, 309
Nuclear Regulatory Commission (NRC), 38,
    39, 305, 311; licensing procedures for
    power plants, 5–6, 41–43, 311–312

Nuclear subgovernments, 8, 33, 36, 312; knowledge fix and, 312–313, 316

Nuclear waste: agricultural impacts, 122–123; environmental impact, 121; knowledge about, 158–160; policy problem, 5–6; projected amounts of, 6, 64; saliency of issue, 7–8, 144–145, 140, 180–183; socioeconomic impacts, 116, 121–123, 124, 130, 178. See also Technological fix

Nuclear Waste Policy Act (NWPA) of 1982: history of, 7–11; implementation of, 11–17, 292–293, 310; provisions of, 9–10, 22, 239; public participation and, xiv, 10, 91–93, 110–12; state and local compensation and, 124, 133, 235

Nuclear Waste Policy Amendments Act of 1987, 292, 310; provisions of, 16–17, 64–65, 124, 133

Nuclear weapons: arms race, 48–50; secrecy, 33, 54, 317; testing, 244–248

Nuclear Weapons Freeze Campaign, 49

Oak Ridge National Laboratory, 118

Oregon, Portland, 138

Pijawka, K. David, 25, 239, 295, 301–302

PIMBY. See Not in my back yard; Put in my back yard

Plutonium. See Radioactive elements

Political ideology, 154, 156–157

Price Anderson Act of 1957, 34, 35

Public hearings; as citizen participation, 90; as data source, 90–91; implementation of NWPA and, 41, 89–114; environmental and social impacts discussed at, 105–106; emotionalism expressed at, 98–100; in nuclear waste siting process, 91–92; knowledge level revealed at, 96–98; political and social concerns expressed at, 103–105; risk acceptability expressed at, 106–108; technical criticisms made at, 102–103

Public interest groups, 33, 42, 44–45

Public opinion: and acceptance of repository, 291–302; characteristics of, 20–21; concept of 3–26; economic impact, 270–271, 274–275, 277, 280; and farmers, 127–133; and investors, 43–44; linkage to public policy, 89; nuclear power, 40, 47–58, 147–150; policy implementation and, 14; radioactive waste management, 45–58, 66–70; repository siting and, 21, 23–25, 93–96, 115, 119–121, 125–133, 166–167, 179, 183–185, 201–205, 269–270, 272–273, 278–280; response to accidents, 47–48, 50, 52–54; rural views, 263–284; safety and health, 147, 272, 275, 277, 296; sociodemographic characteristics, 154–157, 252–254, 271; stigmatization, 160–166; urban views, 239–260

Public participation: and Atomic Energy Act, 42; in decision making, 13–14, 38–40, 111; Nuclear Waste Policy Act and, 40, 91–92, 110–12; repository siting process and, 10–11, 316–17. See also Public hearings

Public perceptions; alienation, 304; attitudes, 20–21; dread, 56, 76, 177, 296, 318; imagery of nuclear hazards, 78–80; equity, 133, 318; mistrust of management, 81, 305, 318; nuclear fear, 66, 80; nuclear risks, 56, 257–259; stigmatization, 138–139, 160–166

Public relations, and nuclear waste, 36, 313–314

Public trust, xiv–xv, 64–84, 314–315, 316, 319; accident reporting, 252; confidence in institutions, 56–57, 304, 310; credibility of government, 100–101, 249–252, 276, 281, 304; Department of Energy and, 57, 68–70, 76–77, 80–84, 100–101, 109–110, 111–112, 157–160, 163–167, 296, 305; faith in science and technology, 156–157, 276; of industry and business, 271; nuclear waste management, 81–84

Put in my backyard (PIMBY), 142, 150–153

Radioactive elements: cesium 137, 308; iodine 129, 308; plutonium, 308; ruthenium 100, 308; stable xenon 130, 308; strontium 90, 308; technetium 99, 308

Radioactive waste. See Nuclear waste

Rasmussen, Norman, 46

Reactor Safety Study, 46

Reagan, Ronald: administration, 12, 15, 304; nuclear weapons, 49

Reid, Harry M., 312

Repository. See High-level nuclear waste repository

Resources for the Future, 51

Risk communication, 206

Risk perception, xv, 4–5, 24–25, 64–84, 175–207, 310; characteristics of, 187–190; cognitive maps and, 186–187;

Risk perception (*continued*)
  imagery of nuclear hazards, 22; models
  of, 124–125, 190–201; siting noxious
  facilities and, 127–129, 176; transportation accidents and, 26, 69–70, 121,
  147
Roper survey, 50
Rosa, Eugene A., 3, 21, 24, 32, 136, 175,
  291

Science (magazine), 310
Scoville, Jerry, 219
Senate, U.S., 311, 312
Seattle Times, 148
Slovic, Paul, 22, 24, 64, 175, 294, 297
Soviet Union, 49
Spokane Spokesman-Review, 139
State Utility Commissions, 43
Strauss, Lewis, 35
Supreme Court, U.S., 43, 311

Tacoma News Tribune, 142
Technological fix, 33, 306–310, 316; canisters, 307; reprocessing of radioactive
  wastes, 8; transmutation, 307, 308–309,
  310
Texas: Deaf Smith County, 15, 115–133
Three Mile Island (TMI), 8, 9, 35, 37, 39,
  43, 47–48, 52–53, 218

Transmutation. *See* Technological fix
Tri-Cities Nuclear Industrial Council, 142
Trinity bomb test, 136
Trust. *See* Public trust

Union of Concerned Scientists, 46
U.S. Department of Energy. *See* Department of Energy

Wallop, Malcolm, 311
Washington State; Cascade Mountains,
  138, 144; Columbia River, 23, 137, 138,
  161; Hanford, 15, 136–167; Kennewick,
  136, 137; Mount St. Helens, 139; Office
  of High-Level Nuclear Waste Management, 142; Olympia, 138; Pasco, 136,
  137; Pullman, 139; Referendum 40, 140,
  141, 142; Richland, 136, 137; Seattle-
  Tacoma area, 138; Spokane, 138, 139
Washington Public Power Supply System,
  44, 138, 150
Waste Isolation Pilot Plant (WIPP), 3
Waste management, 7–26
Watkins, James D., 3, 4
Weinberg, Alvin, 306
Wisconsin, 16, 92

Yucca Mountain. *See* Nevada

Contributors

Riley E. Dunlap is professor of sociology and rural sociology at Washington State University, where he has conducted research on the public's views of environmental and energy issues for the past two decades. A past chair of the American Sociological Association's Section on Environmental Sociology, he is co-editor of *American Environmentalism: The U.S. Environmental Movement, 1970–1990* (Taylor and Francis, 1992). He recently directed an international environmental opinion survey for the George H. Gallup International Institute, where he was appointed Gallup Fellow in Environment.

Michael E. Kraft is professor of political science and public affairs and Herbert Fisk Johnson Professor of Environmental Studies at the University of Wisconsin-Green Bay. Among other works, he is co-editor of *Environmental Policy in the 1990s: Toward a New Agenda* (Congressional Quarterly Press, 1990) and *Technology and Politics* (Duke University Press, 1988). His current research focuses on the politics of nuclear waste disposal in the United States as a case study in the democratic management of risky technologies.

Eugene A. Rosa is professor of sociology, member of the graduate faculty in environmental science and regional planning, and faculty associate in the Social and Economic Sciences Research Center at Washington State University. His research focuses on energy, technology, and risk, and he is co-editor of *Public Reactions to Nuclear Power: Are There Critical Masses?* (Westview/AAAS, 1984). He was twice appointed by Governor Booth Gardner to the Washington State Nuclear Waste Advisory Council.

Rodney K. Baxter received his Ph.D. in sociology from Washington State University, where he is now a research associate at the Social and Economic Sciences Research Center. His areas of interest are quantitative methodology and environmental sociology, and he has published articles on statistical methods and on public attitudes toward nuclear energy.

Julia G. Brody is associate scientist at Tellus Institute, an environmental research group in Boston. Her work focuses on public participation in decisionmaking about environmental risk. Brody served as deputy commissioner for policy development and communications at the Massachusetts Department of Environmental Management and as director of special programs for Texas Agriculture Commissioner Jim Hightower. She received a Ph.D. in psychology from the University of Texas at Austin.

Bruce B. Clary is a professor in the graduate program in public policy and management and a senior research associate at the Health Policy Center at the University of Southern Maine. In addition to his work on public testimony concerning high-level radioactive waste, Dr. Clary has written related articles on the NIMBY question, congressional decision making on Yucca Mountain, and the legislative history and implementation of the Nuclear Waste Policy Act of 1982. He has a general interest in technological and natural hazards.

Lori A. Cramer is Assistant Professor of Sociology (specializing in environment and natural resource sociology) at Oregon State University. Current research interests broadly concern the application of sociology to natural resource problems with a particular emphasis on the social impacts of siting hazardous facilities near rural communities.

William H. Desvousges is an economist who directs the risk communication program at Research Triangle Institute in North Carolina. He specializes in the use of focus groups and evaluation surveys in risk communication research. He coordinated the nuclear risk perception surveys for the national and Nevada studies reported in chapter 7 and has published in *Environment, Risk Analysis,* and other journals.

Douglas Easterling is a research associate with The Colorado Trust, a philanthropic foundation in Denver. He recently received a Ph.D. in public policy and management from the Wharton School at the University of Pennsylvania. His research has focused on psychological issues within environmental and health policy, especially risk perception and risk-avoidance behaviors, and has been published in journals such as *Journal of Applied Psychology, Journal of Policy Analysis and Management,* and *Risk Analysis.*

Judy K. Fleishman is currently a psychology fellow at Hennepin County Medical Center in Minneapolis, where she works with psychiatric and medical patients. She was previously an instructor at Southwest Texas State University and a research consultant for the Texas Department of Agriculture. Her research interests focus on psychological dimensions of public policy and program development. She received a Ph.D. in psychology from the University of Texas at Austin.

James H. Flynn is a research associate at Decision Research in Eugene, Oregon. He has conducted socioeconomic research on nuclear facilities for the Nuclear Regulatory Commission, Puget Sound Power and Light, the state of Mississippi, and as the project manager for the state of Nevada socioeconomic studies of Yucca Mountain. His recent publications on nuclear issues include articles in *Environment, Science, Risk Analysis, Issues in Science and Technology,* and *Forum for Applied Research and Public Policy.*

William R. Freudenburg is professor of rural sociology at the University of Wisconsin-Madison. He specializes in studying technological controversies and the social impacts of environmental and technological change, with a special emphasis on social-science aspects of risk assessment and risk management. He is co-editor of *Public Reactions to Nuclear Power* (with Eugene A. Rosa) and is the author of numerous articles in journals such as *Science, American Sociological Review, American Journal of Sociology, Annual Review of Sociology,* and *Social Forces.*

Richard S. Krannich is professor of sociology and forest resources at Utah State University. His research has focused on the social implications of natural resource use and resource development projects, and he has published several articles on these topics. More recently, he has been researching the social and cultural issues involved in siting hazardous facilities such as nuclear waste disposal facilities and toxic waste incinerators.

Howard Kunreuther is the Riklis Professor of Decision Sciences and co-director of the Wharton Risk Management and Decision Processes Center at the University of Pennsylvania. His current research is concerned with the roles of insurance, compensation, and regulation for dealing with technological and natural hazards. He is author and co-author of numerous articles and books concerned with low-probability/high-consequence events, including *The Dilemma of Siting a Nuclear Waste Repository*, with Douglas Easterling (Kluwer, 1993).

Mark Layman has a degree in chemistry and graduate work in the social sciences and has worked as a programmer and research analyst on numerous social science projects. His work with Decision Research on nuclear waste siting led to articles in *Science* and *Environment* (with James Flynn and Paul Slovic).

Ronald L. Little is professor of sociology at Utah State University where he specializes in environmental sociology. His research has focused on changes in community social structures, both anglo and native American, resulting from alterations in the use of natural resources, a topic on which he has published several articles and chapters. Specific research topics include studies of arid land utilization, energy exploration and production, and hazardous waste storage and disposal.

Robert Cameron Mitchell received a Ph.D. in sociology from Northwestern University. He was a senior fellow at Resources for the Future in Washington, D.C., before moving to his current position as professor of geography at Clark University. For the past two decades his research has focused on environmental perception, human response to environmental change, and the environmental movement. More recently he has collaborated with resource economists in using surveys to value environmental public goods, and is co-author of *Using Surveys to Value Public Goods: The Contingent Valuation Method* (Johns Hopkins University Press, 1988).

Alvin H. Mushkatel is currently the interim director of the Office of Hazards Studies at Arizona State University, where he is also professor in the School of Public Affairs. He received his Ph.D. in political science from the University of Oregon, and is the author or editor of four books and numerous articles examining various facets of hazards and environmental policy.

Joanne M. Nigg is professor of sociology and director of the Disaster Research Center at the University of Delaware. Her research interests include public perceptions of natural and technological hazards and the development of collective understandings of risk. Her current work focuses on public and expert judgments of responsibility for disaster consequences. She is co-author of *Waiting for Disaster* (University of California Press, 1986).

K. David Pijawka is research director for the Office of Hazards Studies at Arizona State University, where he is also assistant director of the Center for Environmental Studies and professor in the School of Public Affairs. He received a Ph.D. in geography from Clark University, and is the author or editor of several books on nuclear plant decommissioning and other topics dealing with environmental hazards.

Paul Slovic is the president of Decision Research in Eugene, Oregon, and is also professor of psychology at the University of Oregon. He is a past president of the Society for Risk Analysis and recipient of that society's distinguished contribution award, and is also a member of the board of directors of the National Council on Radiation Protection and Measurements. Slovic has written or co-written numerous articles and chapters on risk perception, and he is co-author of *Acceptable Risk* (Cambridge University Press, 1981).

Library of Congress Cataloging-in-Publication Data
Public reactions to nuclear waste : citizens' views of
repository siting / [edited by] Riley E. Dunlap,
Michael E. Kraft, Eugene A. Rosa.
Includes bibliographic references and index.
ISBN 0-8223-1355-3. — ISBN 0-8223-1373-1 (pbk.)
1. Radioactive waste sites—Location—Public opinion.
2. Radioactive wastes—Public opinion.   3. Public
opinion—United States.   I. Dunlap, Riley E.   II. Kraft,
Michael E.   III. Rosa, Eugene A.
TD898.15.P83   1993
363.72'89525—dc20   93-6980 CIP